Application of Big Data Mining, Machine Learning and Artificial Intelligence in Ore Deposits

Application of Big Data Mining, Machine Learning and Artificial Intelligence in Ore Deposits

Guest Editors

Yongzhang Zhou
Renguang Zuo

MDPI

Basel • Beijing • Wuhan • Barcelona • Belgrade • Novi Sad • Cluj • Manchester

Guest Editors
Yongzhang Zhou
School of Earth Science and Engineering
Sun Yat-Sen University
Guangzhou
China

Renguang Zuo
State Key Laboratory of Geological Processes and Mineral Resources
China University of Geosciences
Wuhan
China

Editorial Office
MDPI AG
Grosspeteranlage 5
4052 Basel, Switzerland

This is a reprint of the Special Issue, published open access by the journal *Minerals* (ISSN 2075-163X), freely accessible at: https://www.mdpi.com/journal/minerals/special_issues/ABDMMLAIOD.

For citation purposes, cite each article independently as indicated on the article page online and as indicated below:

Lastname, A.A.; Lastname, B.B. Article Title. *Journal Name* **Year**, *Volume Number*, Page Range.

ISBN 978-3-7258-5463-9 (Hbk)
ISBN 978-3-7258-5464-6 (PDF)
https://doi.org/10.3390/books978-3-7258-5464-6

Contents

About the Editors

Yongzhang Zhou

Yongzhang Zhou is a Professor of the School of Earth Sciences and Engineering, Chief Scientist of the Research Center for Earth Environment & Resources of Sun Yat-sent University, and Foreign Academician of both the Russian Academy of Engineering and the Russian Academy of Natural Sciences. He is the winner of the Excellent Teacher award of the National Education Ministry of China and the Felix Chayes Prize of the International Association for Mathematical Geosciences (2015). He has published more than 400 scientific papers and monographs, including the widely disseminated and studied *Big Data Mining & Machine Learning in Geoscience* (Guangzhou: Sun Yat-sen University Press, 2018) and *Mathematical Geosciences* (2012). Included in his research interests are big data mining, machine learning, mathematical geoscience, smart mineral exploration systems, and intelligent environment monitoring systems. He is currently chairing the Committee of Artificial Intelligence and Big Data in Earth Sciences at the Chinese Society of Mineralogy, Petrology, and Geochemistry.

Renguang Zuo

Renguang Zuo is a Professor of the China University of Geosciences (Wuhan). He received his Bachelor's and Ph. D. degrees from the China University of Geosciences (Wuhan) in 2004 and 2009, respectively. He served as a councilor of IAMG from 2020 to 2024 and was awarded the IAMG Distinguished Lecturer award. His contributions mainly focus on the big data artificial intelligence-driven spatial pattern recognition of geochemical elements and geochemical anomaly extraction, providing effective solutions for identifying geochemical anomalies in complex geological conditions and promoting a paradigm shift in big data artificial intelligence-driven exploration geochemistry.

minerals

MDPI

Article

Prediction of Prospecting Target Based on Selective Transfer Network

Yongjie Huang [1,†], Quan Feng [2,3,†], Wanting Zhang [1], Li Zhang [4] and Le Gao [1,*]

1 Faculty of Intelligent Manufacturing, Wuyi University, Jiangmen 529000, China
2 State Key Laboratory of Public Big Data, Guizhou University, Guiyang 550025, China
3 College of Computer Science and Technology, Nanjing University of Aeronautics and Astronautics, Nanjing 210016, China
4 College of Computer Science and Technology, Nanjing Forestry University, Nanjing 210095, China
* Correspondence: le.gao@nscc-gz.cn
† These authors contributed equally to this work.

Abstract: In recent years, with the integration and development of artificial intelligence technology and geology, traditional geological prospecting has begun to change to intelligent prospecting. Intelligent prospecting mainly uses machine learning technology to predict the prospecting target area by mining the correlation between geological variables and metallogenic characteristics, which usually requires a large amount of data for training. However, there are some problems in the actual research, such as fewer geological sample data and irregular mining features, which affect the accuracy and reliability of intelligent prospecting prediction. Taking the Pangxidong study area in Guangdong Province as an example, this paper proposes a deep learning framework (SKT) for prospecting target prediction based on selective knowledge transfer and carries out intelligent prospecting target prediction research based on geochemical data in Pangxidong. The irregular features of different scales in the mining area are captured by dilation convolution, and the weight parameters of the source network are selectively transferred to different target networks for training, so as to increase the generalization performance of the model. A large number of experimental results show that this method has obvious advantages over other state-of-the-art methods in the prediction of prospecting target areas, and the prediction effect in the samples with mines is greatly improved, which can effectively alleviate the problems of a small number of geological samples and irregular features of mining areas in prospecting prediction.

Keywords: transfer learning; dilated convolution; deep learning; prospecting target prediction

1. Introduction

Mineral resources are an essential strategic resource to ensure national stability and economic development. After years of geological exploration, surface minerals have gradually become scarce, resulting in a further increase in the difficulty of prospecting. The prospecting work started in deep surfaces and overburdened areas. In recent years, the development of machine learning technologies, especially deep learning, has achieved satisfactory success in many fields, e.g., in computer vision [1], natural language processing [2], synthetic aperture radar [3], and atmospheric prediction [4]. Meanwhile, it has also interested and been selected by geological researchers [5–8]. Therefore, the traditional geological prospecting method is gradually turning into an intelligent prospecting method based on machine learning.

An intelligent prospecting target refers to the use of machine learning technologies to mine the correlation between geological variables and metallogenic characteristics to predict the metallogenic target area [9,10]. Although quantitative prospecting target prediction based on machine learning technologies is in its infancy, some progress has been made.

These works can be divided into the following three categories according to the method of implementation:

(1) Methods based on ensemble learning combine multiple supervised learning algorithms for prospecting target prediction. For example, the authors in [11] determined the hyperparameters of the random forest by simulating the natural evolution process, which were used to improve the accuracy of the model in predicting the prospecting target area. The authors in [12] used the isolation forest algorithm to predict outliers to determine the prospecting target area. The authors in [13] proposed the use of metric learning in the random forest to project the sample features into the feature space, separating the background and mining targets, so as to improve the prediction accuracy of the model.
(2) Methods based on the support vector machine (SVM) divide mining samples and other samples through a hyperplane. For example, the authors in [14] separated the "mines" samples from "non-mines" samples through the optimal hyperplane and determined three prospecting target areas on both sides of the hyperplane. The authors in [15] used a genetic algorithm to optimize the hyperparameters of SVM to reduce its influence on prospecting target prediction.
(3) Methods based on depth neural networks project the geoscience data into the same depth neural network space and extract effective features through multiple nonlinear transformations for prospecting target prediction. For example, the authors in [16] used three-layer convolution to extract the features of a Zn-element concentration distribution map to predict the prospecting target area. The authors in [17] used AlexNet to extract the features of multiple ore-forming factor maps to determine four prospecting target areas.

Although the aforementioned methods have achieved success, problems such as the small number of geological samples and the irregular features of mining areas in prospecting target prediction have not been properly solved. Fortunately, in recent years, in-depth research has been carried out on these problems. These works can also be divided into the following two categories according to the method of implementation:

(1) Methods based on data augmentation increase the diversity of geological samples by cropping, changing the chromatic aberration and size, and distorting features. For example, the authors in [18] added random noises into geological data to predict a prospecting target by a deep convolutional network. The authors in [19] first oversampled the samples with mines and then used the random forest to determine the prospecting target areas. The authors in [20] proposed recombining pixel pairs of geological samples to assist in prospecting target prediction. These methods increase the number of geological samples but cannot cope with the irregular features of the mining areas.
(2) Methods based on multiscale feature transformation acquire mining area features of different scales through irregular sampling for training [21,22]. For example, the authors in [23] proposed to extract irregular features of different scales using multigroup convolution or a pooling operation and fused them for prospecting target prediction. The authors in [24] used four convolution operations with different sizes to extract and fuse irregular features of geological data to improve the prediction accuracy. However, these methods do not consider the small number of samples in the prospecting target area.

To summarize, the motivation behind our method comes from two aspects: (1) Selective transfer of label information learned from large tasks to small-shot tasks can effectively improve its generalization performance; therefore, we try to selectively transfer knowledge from a well-trained large prospecting target prediction task to the target task to assist training. (2) The other is to expand the receptive field of the convolution kernel without increasing the number of parameters. In this way, we can utilize it to deal with the irregular features of the mining areas. Therefore, we combine these two motives and propose a novel deep learning framework based on selective knowledge transfer (SKT) for prospecting target prediction to solve the problems existing in prospecting target prediction. As shown in Figure 1, we first introduce a large well-trained prospecting target prediction network as the source network and define multiple target networks with the same structure. Next, we use a soft mask to alleviate the differences in metallogenic effects that different elements may cause. Then, we use dilated convolution to capture features of different scales as inputs of the target networks. Based on this, we selectively transfer the weight parameters from the source network for the training of the different target networks. Finally, we perform top-down distillation on the large-scale and small-scale target network to mine the hidden knowledge between feature maps of different scales for small-scale target network learning. It is worth noting that the input size of $W \times H \times n$ is obtained by processing the geochemical data, where W denotes width, H denotes height, and n denotes the number of geochemical elements. Meanwhile, the output is generated by voting through all the target networks and contains two values representing the votes obtained for with and without mines, respectively. The whole SKT framework works in an end-to-end manner. Extensive experimental results show that our method is significantly competitive with state-of-the-art methods.

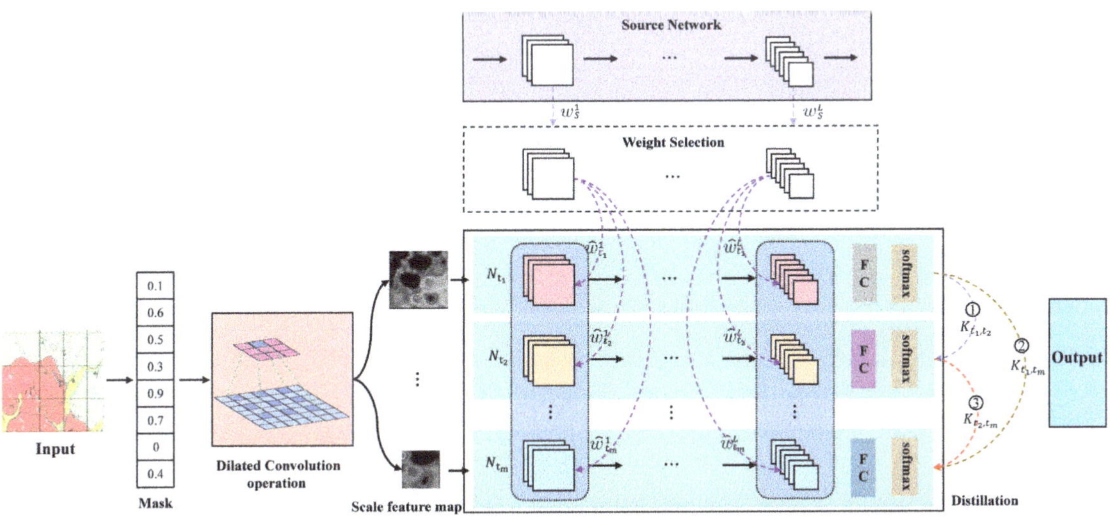

Figure 1. Schematic architecture of the SKT framework. w_S^l denotes the convolution kernels of the l-th convolution layer of the source network, $N_{t_v}, v = 1, 2, \ldots, m$ denotes the target networks, $\hat{w}_{t_v}^l$ denotes the convolution kernels transferred to N_{t_v} from the source network, and K_{t_i,t_j} denotes the direction of knowledge distillation, which means that N_{t_i} guides N_{t_j} to learn. Knowledge distillation proceeds from the top down, e.g., knowledge distillation performs in the order of ①, ②, ③. Weight selection makes the source network weights sparse through the Hadamard product and then transfers to the target network.

The main contributions of this paper include the following:

1. A deep learning framework for prospecting target prediction is proposed, which provides a new way for prospecting target prediction.
2. A novel selective knowledge transfer mechanism is designed, which selectively transfers knowledge from the source network to target networks, which increases the performance of the target networks in prospecting target prediction during testing without adding additional computational cost.
3. For the first time, a soft mask strategy is proposed to maintain the consistency of related mineral elements. Its purpose is to utilize the metallogenic indicative significance of the

main mineral elements and associated mineral elements to complete the prospecting target prediction task.

The rest of this paper is organized as follows. Section 2 introduces introduces the study area and data processing methods. Section 3 infers the structure of the SKT framework. Section 4 exhibits our experiment, including analysis of the experimental results and visualization. Section 5 summarizes the work of this paper.

2. Study Area and Data

2.1. Study Area

The study area is the Pangxidong study area in Guangdong Province, China. Figure 2 shows a simple geological map of the study area, where the red dots are the mining areas. It is a part of the Qinzhou–Hangzhou metallogenic system, and the metallogenic geological conditions are superior. Its tectonic background belongs to the eastern margin of the southern segment of the Qinzhou–Hangzhou metallogenic system. Since the late Paleozoic, it has been uplifted for a long time and is an important metallogenic area for precious metals and non-ferrous metals, with many kinds of minerals [25]. The minerals that have been explored or are being exploited include gold ore, silver ore, lead–zinc ore, tungsten and molybdenum ore, iron ore, etc. [26]. The types of deposits mainly include ductile shear zone silver–gold deposits, porphyry-type molybdenum–tungsten–copper polymetallic deposits, sedimentary reformation type, magmatic hydrothermal type, and contact metasomatic lead–zinc polymetallic deposits [27]. Minerals are mainly distributed in the northwest and southeast, along the Pangxitong fault zone and the Gucheng–Shachan fault zone, in a northeast direction.

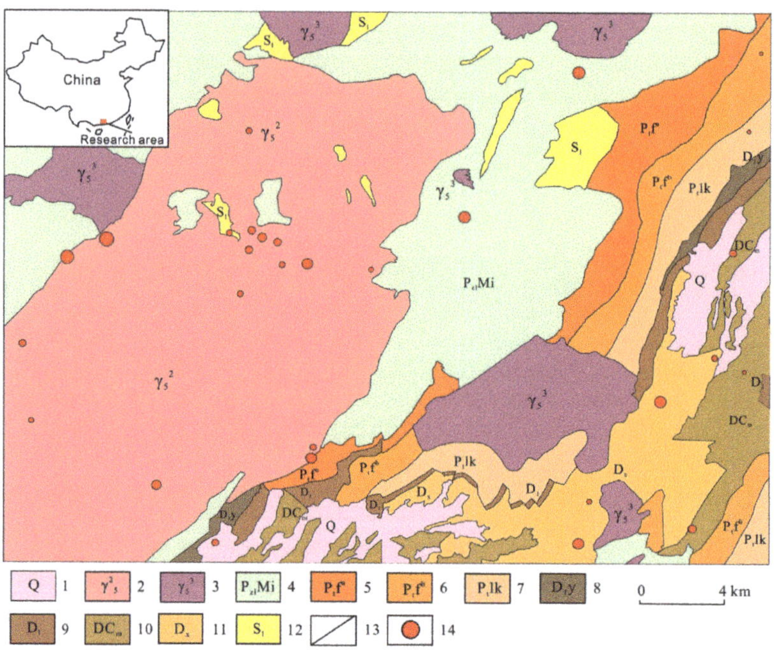

Figure 2. Simple geological map of the study area (1—Quaternary; 2—Early Yanshanian granite; 3—Late Yanshanian granite; 4—Upper Proterozoic migmatite; 5—Lower Member of Middle-Upper Proterozoic Fengdongkou Formation; 6—Upper Member of Middle-Upper Proterozoic Fengdongkou Formation; 7—Middle-Upper Proterozoic Lankeng Formation; 8—Devonian Yangxi Formation; 9—Devonian Laohutou Formation; 10—Devonian-Carboniferous Maozifeng Formation; 11—Devonian Xindu Formation; 12—Silurian Liantan Formation; 13—Faults; 14—Deposit).

2.2. Data Processing

The experimental data consist of geochemical data and were extracted from a 1:50,000 stream sediment survey. They were obtained from the Pangxidong study area in Guangdong Province. The sampling area of the stream sediment was 1694 km^2, and the average sampling density was 4.27 samples km^2. Sixteen chemical elements, including Au, B, Sn, Cu, Ag, Ba, Mn, Pb, Zn, As, Sb, Bi, Hg, Mo, W, and F, were analyzed from the stream sediment samples. Table 1 shows some of our original data, with X and Y as the coordinates of the sampling points and the rest as the values of the geochemical elemental content, totaling to 7237 points. The area under curve (AUC) is defined as the area under the receiver operating characteristic (ROC) curve enclosed by the coordinate axes [28]. Based on thesis [29], we leveraged SVM and statistical methods to screen out geochemical elements, which are indicative of mineralization. Specifically, we calculated the AUC of each element using the SVM algorithm, and then the standard deviation of the AUC was calculated by

$$S_{AUC} = \sqrt{\frac{AUC(1-AUC) + (C_p - 1)\left(Q_1 - AUC^2\right) + (C_n - 1)\left(Q_2 - AUC^2\right)}{C_p \times C_n}} \quad (1)$$

where S_{AUC} denotes the standard deviation of the AUC; C_p and C_n denote the numbers of samples with and without mines, respectively; and $Q_1 = \frac{AUC}{2-AUC}$ and $Q_2 = \frac{2AUC^2}{1+AUC}$ are temporary variables. To determine whether the AUC was significantly different from 0.5, we defined the following random variable:

$$Z_{AUC} = \frac{AUC - 0.5}{S_{AUC}} \quad (2)$$

Table 1. Geochemical dataof Pangxidong.

X	Y	Au	B	Sn	Cu	Ag	Ba	Mn	Pb	Zn	As	Sb	Bi	Hg	Mo	W	F
422.24	2418.80	0.9	3	8.7	4	0.025	33	147	27	26	1.17	0.31	0.23	0.04	2.67	0.79	212
421.37	2418.80	0.54	4	2.56	7	0.078	88	209	12	23	0.9	0.29	0.13	0.04	0.82	1.16	204
419.76	2418.25	0.81	3	1.52	5	0.043	1111	423	42	14	0.51	0.35	0.06	0.07	0.59	0.38	101
420.12	2418.40	0.37	2	1.65	6	0.046	941	498	38	17	0.53	0.31	0.1	0.02	0.57	0.33	111
420.55	2418.60	1.09	4	1.53	8	0.033	427	338	37	29	0.74	0.28	0.09	0.07	1.68	0.73	186
433.81	2397.92	2.31	121	2.2	4	0.075	365	239	16	18	4.31	0.96	0.43	0.066	0.77	3.01	186
424.17	2415.02	0.43	5	2.18	4	0.069	42	250	13	14	1.21	0.33	0.4	0.031	1.04	1.53	201
423.74	2415.31	0.51	5	4.85	7	0.004	30	242	47	31	0.5	0.26	0.32	0.016	1.02	3.07	210
425.14	2414.87	0.46	6	2.08	7	0.061	28	298	15	25	1.49	0.35	0.24	0.075	1.75	1.3	217
425.14	2415.15	0.47	6	1.95	7	0.055	54	420	9	18	1.07	0.37	0.14	0.042	1.08	0.85	108
424.86	2414.76	0.5	5	1.46	4	0.036	21	355	6	11	1.1	0.33	0.17	0.022	0.98	1.14	130
424.47	2414.47	0.59	6	2.6	4	0.038	29	170	6	21	1.01	0.33	0.2	0.039	1.32	2.02	192
424.82	2414.37	0.43	11	2.26	2	0.027	22	210	6	19	1.19	0.31	0.2	0.03	1.5	1.76	177
425.22	2414.46	1.05	45	4.2	3	0.065	39	125	6	25	1.94	0.39	0.54	0.046	2.17	3.04	396
424.41	2414.11	0.4	6	1.84	3	0.054	16	231	6	10	0.77	0.28	0.11	0.016	0.83	0.91	93
424.72	2413.83	0.9	7	3.87	9	0.094	135	231	48	45	2.15	0.41	0.73	0.069	1.82	2.54	327
424.35	2413.78	0.68	6	2.68	3	0.059	26	130	5	25	2.32	0.34	0.37	0.045	1.09	1.69	201
431.88	2411.24	0.81	4	3.58	15	0.053	140	143	83	34	2.76	0.36	2	0.052	2.05	8.94	241
432.90	2411.89	0.39	3	3.4	10	0.077	121	133	93	33	1.73	0.34	4.24	0.061	2.33	6.41	230
433.60	2410.63	0.42	5	2.62	1	0.042	81	152	12	21	1.55	0.31	0.3	0.054	0.77	0.94	135
433.91	2411.37	0.8	4	3.49	2	0.025	120	88	31	33	3.17	0.33	0.72	0.062	0.85	1.48	231
434.07	2410.91	0.42	6	2.99	4	0.057	109	117	12	22	2.38	0.32	0.27	0.048	0.63	2.19	210
434.73	2410.27	0.37	4	3.11	9	0.045	171	132	12	21	1.93	0.31	0.12	0.042	0.69	1.98	180
434.09	2409.69	0.36	5	2.92	6	0.044	135	123	5	21	1.76	0.29	0.09	0.046	0.71	0.6	156
432.45	2414.37	1.86	46	1.82	11	0.059	96	278	13	22	11.58	0.42	0.35	0.017	0.78	3.05	150
432.29	2414.59	3.76	65	6.04	4	0.036	41	164	67	20	9.37	0.79	0.67	0.047	2.42	7.75	486
432.60	2414.75	2.63	83	2.09	22	0.049	112	208	30	17	26.06	0.54	0.64	0.033	1.18	3.09	201

The random variable Z_{AUC} satisfies a standard normal distribution. The critical value is obtained by comparing Z_{AUC} with the standard normal distribution table. Based on Table 2, we chose a Z_{AUC} greater than the critical value of 2.58 when the significance level was 0.01. In other words, a total of eight elements, including Au, Sn, Cu, Ag, Ba, Sb, Hg, and Mo, were selected as indicative elements for prospecting target prediction.

Table 2. AUC and Z_{AUC} calculational results of 16 geochemical elements.

Element	AUC	Z_{AUC}	Element	AUC	Z_{AUC}
Au	0.6024	2.8395	B	0.5901	2.4839
Sn	0.6065	2.9595	Cu	0.6311	3.6977
Ag	0.6762	5.1563	Ba	0.6147	3.2020
Mn	0.5573	1.5617	Pb	0.5778	2.1341
Zn	0.5450	1.2232	As	0.5655	1.7893
Sb	0.5942	2.6017	Bi	0.5901	2.4839
Hg	0.6393	3.9516	Mo	0.5983	2.7203
W	0.5778	2.1341	F	0.5696	1.9037

We combined experimental results and the geological environment of the study area for analysis. The Pangxidong study area has faults, developed fold structures, and strong magmatic activity. The metallogenic geological environment is similar to many areas with rich mineral resources in the Qinzhou–Hangzhou metallogenic belt. The types of minerals involved are mainly Au, Ag, Pb, Zn, W, Mo, Fe, and Mn metal ores. Then, Pb, Zn, and Cu metal ores are visible in Devonian strata in the study area. In addition, the northeast Gucheng–Shachan and Pangxitong faults in the study area are the main ore-conducting and storage structures for Au, Ag, Sn, and other minerals [30]. In summary, most of the eight indicator elements we selected are consistent with the metallic ores in the study area. This shows the validity of our choice of indicator elements.

By referring to the processing method of the stream sediment data in [31], we used inverse distance weight interpolation to generate 3072 × 3072 grids (elemental content map) according to the values of the geochemical elemental content. Specifically, we calculated the distance between the nearby discrete points and the grid (x_0, y_0) by

$$D_i = \sqrt{(x_0 - x_i)^2 + (y_0 - y_i)^2} \quad (3)$$

where D_i denotes the distance from the i-th discrete point near (x_0, y_0), and (x_i, y_i) denotes the coordinates of the i-th discrete point.

Based on this, we estimated the value of the grid (x_0, y_0) as follows:

$$Z(x_0, y_0) = \sum_{i=1}^{N} \frac{1}{(D_i)^2} Z_i / \sum_{i=1}^{N} \frac{1}{(D_i)^2} \quad (4)$$

where $Z(x_0, y_0)$ denotes the estimated value at the grid (x_0, y_0), Z_i denotes the observed value at the i-th discrete point, and N denotes the number of discrete points involved in the calculation.

Figure 3 shows the elemental content maps of the eight elements. We normalized the values of the elemental content map, making its mean value 0 and its variance 1. For a value x in the elemental content map, the normalized value is as follows:

$$\hat{x} = \frac{x - \overline{x}}{\sigma} \quad (5)$$

where $\hat{x}$ denotes the normalized value, $\overline{x}$ denotes the mean of the values of the elemental content map, and σ denotes the variance of the values of the elemental content map.

We divided the elemental content map of size 3072 × 3072 into two parts, where the upper part (2560 × 3072) was used for making the training dataset, and the lower part

(512 × 3072) was used for making the test dataset. We defined two 256 × 256 windows to slide on the upper and lower elemental content maps with a step length of 128. There were 437 patches in the training dataset, where 78 patches were with mines and 359 patches were without mines. On the other hand, the test dataset contained 69 patches, including 17 patches with mines and 52 patches without mines. In addition, Gaussian noise with a mean value of 0 and a variance of 0.01 was added to augment the data [32,33]. In the end, the generated training dataset had 2169 patches, with 1092 patches with mines and 1077 patches without mines, and the generated test dataset had 275 patches, with 119 patches with mines and 156 patches without mines. We combined the elemental content maps of the eight elements, i.e., our experimental dataset consisted of 2444 patches with a size of 256 × 256 × 8. In addition, we would randomly clip the data into a 224 × 224 size before each iteration as the input of our model.

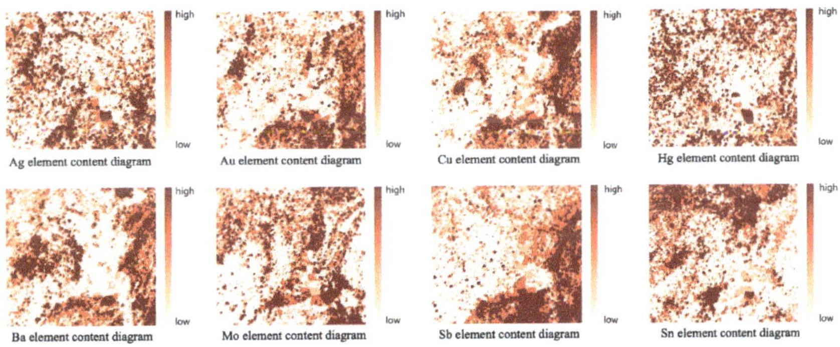

Figure 3. Elemental content diagrams.

3. Methodology

In this section, we propose a method of selective knowledge transfer for prospecting target prediction. Specifically, Section 3.1 formally addresses this issue. Section 3.2 explains how to use the soft mask to keep the weight between the associated mineral elements and the main mineral elements consistent and use dilated convolution to capture feature maps of different scales. Section 3.3 discusses selective knowledge transfer. Section 3.4 raises self-distillation to mine the hidden knowledge of target networks with different multiscale features. Section 3.5 exports the objective function of SKT used to deploy this mechanism.

3.1. Problem Formulation

In this section, we transform geochemical data into a geochemical elemental content map through inverse distance weight interpolation [31]. Then, we define a sliding window with a size of 256 × 256 by referring to [34]. At the same time, a geochemical dataset $\mathcal{D} = \{x_h^i, y_h^i\}_{h=1,i=1,2,\ldots,n}^{N}$ is constructed by cutting the elemental content map with a step size of 128. Here, N denotes the number of samples, i denotes the class of the geochemical element, and $x_h \in \mathbb{R}^d$ and $y_h \in \{0,1\}$ denote the feature vector and the corresponding label, where 0 and 1 mean "without mines" and "with mines", respectively, and d denotes the size of the sample feature space. We suppose that the SKT framework consists of a well-trained source task network $\mathcal{N}_S$ and multiple target task networks $\mathcal{N}_T = \{N_{t_1}, N_{t_2}, \ldots, N_{t_m}\}$. These networks have the same structure and L convolution layers, where the numbers of input and output channels in the l-th convolution layer are M_l and M_{l+1}, respectively. The convolution kernel at the l-th layer in the source network $\mathcal{N}_S$ is defined as $W_S^l = \left[w_{S1}^l, w_{S2}^l, \ldots, w_{SM_{l+1}}^l\right]$, where $W_S^l \in \mathbb{R}^{M_{l+1} \times M_l \times K \times K}$, $w_{Sj}^l \in \mathbb{R}^{M_l \times K \times K}$, and $K \times K$ is the size of the convolution kernel. For all convolution kernels at the l-th layer of $\mathcal{N}_T$, it is defined as $\mathcal{W}_T^l = \left[W_{t_1}^l, W_{t_2}^l, \ldots, W_{t_e}^l\right] \in \mathbb{R}^{e \times M_{l+1} \times M_l \times K \times K}$, where the convolution kernels of the task network t_v are $W^l_{t_v} = \left\{w^l_{t_v1}, w^l_{t_v2}, \ldots, w^l_{t_vM_{l+1}}\right\}$, and the setting of the convolution kernel is consistent with that in $\mathcal{N}_S$. We propose using selective transfer knowledge for prospecting target prediction.

3.2. Congruence of Related Mineral Elements

We consider that the concentration of geochemical elements has indicative significance for mineralization, and different elements have differences in predicting mineralization. Generally, mined minerals contain associated minerals. Their main mineral elements have an important indicative significance for mineralization, while the associated mineral elements also have a certain indicative significance [35]. Therefore, we try to introduce a soft mask $\mathbf{M} = [\mathcal{M}_1, \mathcal{M}_2, \ldots, \mathcal{M}_n]$ strategy based on the above consideration. Its purpose is to make the corresponding weight of the associated mineral elements as consistent as possible with that of the main mineral elements and increase the diversity of effective samples as follows:

$$\hat{x}^i_h = x^i_h \times \mathcal{M}_i \tag{6}$$

where x^i_h denotes the h-th sample in the i-th geochemical element, and $\mathcal{M}_i$ denotes the weight corresponding to the i-th geological element in $\mathbf{M}$, which is a scalar; $\hat{x}^i_h$ denotes the h-th sample in the i-th geochemical element by the mask operation. In this way, we obtain the feature maps of different geochemical elements.

To meet the challenge brought by the irregular features of the mining areas, we perform the dilated convolution operation on the features of different geochemical elements to generate feature maps of different scales. Specifically, we define a set of dilated coefficients $\rho = \{\rho_1, \rho_2, \ldots, \rho_m\}$, where each coefficient corresponds to a feature map and target network at different scales. Then, we perform a dilated convolution operation on the v-th scale feature map as the input of the target network N_{t_v}, where the value of the position $(p_{\text{row}}, p_{\text{col}})$ on the feature map is calculated by

$$\mathbf{f}^i_{t_v} = \sum_{u=-r}^{r} \sum_{g=-r}^{r} \hat{x}^i_{t_v}(\rho_{t_v}u + p_{\text{row}}, \rho_{t_v}g + p_{\text{col}}) \times o(u, g) \tag{7}$$

where $\mathbf{f}^i_{t_v}$ denotes the t_v-th sample scale feature map of the i-th geochemical element through dilated convolution, ρ_{t_v} denotes the corresponding dilated coefficient to N_{t_v}, $\hat{x}^i_h(\cdot,\cdot)$ denotes the weight of the h-th sample position of the i-th geochemical element, and $o(\cdot,\cdot)$ denotes the weight of the position on the dilated convolution kernel. u, g denotes the nearest neighbor units of the position $(p_{\text{row}}, p_{\text{col}})$, and $r = S - 1$, where S is the size of the dilated convolution kernel. For example, for a dilated convolution kernel of size 3, the nearest neighbor of u and g is 1.

3.3. Selective Knowledge Transfer

Due to the small number of samples, it easily leads to problems such as underfitting or nonconvergence in the training process of the prospecting target prediction model. Thus, we selectively transfer elements in the convolution kernels from the well-trained source task network $\mathcal{N}_S$ to assist target network learning, as shown in Figure 4. Specifically, we first define a matrix $P^l_{t_vj}$ with the same size as the convolution kernel in the target network N_{t_v}. Then, we calculate the Hadamard product of W^l_S as follows:

$$\hat{w}^l_{t_vj} = w^l_{Sj} \odot P^l_{t_vj} \tag{8}$$

where $\hat{w}^l_{Sj}$ denotes the selected convolution kernels, and $\odot$ denotes the Hadamard product.

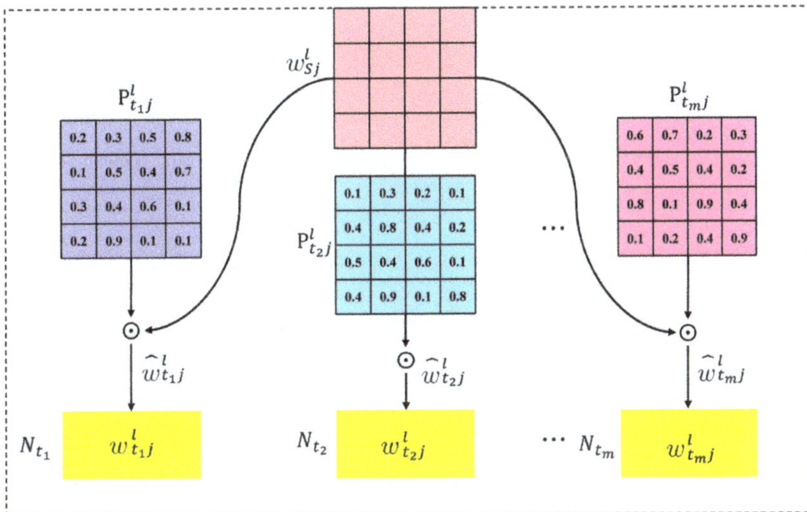

Figure 4. Description of the weight selection algorithm. w^l_{Sj} denotes the convolution kernels of the l-th convolution layer of the source network, $j = 1, 2, \ldots, M_{l+1}$ denotes the output channels, $P^l_{t_vj}$ denotes a matrix corresponding to the target network $N_{t_v}, v = 1, 2, \ldots, m$, $\odot$ denotes the Hadamard product, $\widehat{w}^l_{t_vj}$ denotes the selected convolution kernels, and $w^l_{t_vj}$ denotes the convolution kernels of N_{t_v}.

To further help target task network training, we transfer $\widehat{w}^l_{Sj}$ to $\mathcal{N}_T$ for training by

$$f_{t_v} = \left(\widehat{w}^l_{Sj} \times w^l_{t_vj}\right) * \mathbf{f}^i_{t_v} + b \tag{9}$$

where f_{t_v} denotes the convolution operation output, $*$ denotes the convolution operation, b denotes the convolution bias, and $\mathbf{f}^i_{t_v}$ denotes the t_v-th sample scale feature map of the i-th geochemical element.

3.4. Self-Distillation

To mine the hidden knowledge between the feature maps of different scales, we perform knowledge distillation from top to bottom according to the target network corresponding to the size of the feature maps. For example, there are three target networks, N_{t_1}, N_{t_2}, and N_{t_3}, and the size of their input feature maps decreases successively. We use N_{t_1} to guide N_{t_2} and N_{t_3}, and N_{t_2} to guide N_{t_3}. Specifically, we use Kullback–Leibler (KL) divergence to calculate the probability distribution of the softmax output between each pair of target networks by

$$\mathcal{L}_{t_vKD} = \sum_{j=1}^{v-1}\sum_{i=1}^{n} f\left(\mathbf{f}^i_{t_j}, \boldsymbol{\theta}^{N_{t_j}}\right) \log\left(\frac{f\left(\mathbf{f}^i_{t_j}, \boldsymbol{\theta}^{N_{t_j}}\right)}{f\left(\mathbf{f}^i_{t_v}, \boldsymbol{\theta}^{N_{t_v}}\right)}\right) \tag{10}$$

where $\mathcal{L}_{t_vKD}$ denotes the self-distillation loss of the target network N_{t_v}; the size of the input feature maps of N_{t_j} is larger than that of N_{t_v}; $\boldsymbol{\theta}^{N_{t_j}}$ and $\boldsymbol{\theta}^{N_{t_v}}$ denote the parameters of the target networks N_{t_j} and N_{t_v}, respectively; $f(\cdot)$ denotes the softmax operation; and $\mathbf{f}^i_{t_j}$ and $\mathbf{f}^i_{t_v}$ denote the t_j-th and t_v-th sample scale feature maps of the i-th geochemical element through the dilated convolution operation.

3.5. Objective Function

The target network maps features to the corresponding label space through the full connected layer. During the training, we designed the classification loss for each target network by

$$\mathcal{L}_{t_v} = \sum_{i=1}^{n} \mathrm{L}\left(\boldsymbol{\theta}^{N_{t_v}}, \mathbf{f}^i_{t_v}, \mathrm{P}^{1:L}_{t_v}, \mathrm{y}^i_{\mathrm{h}}\right) \tag{11}$$

where $\mathcal{L}_{t_v}$ denotes the classification loss of the target network N_{t_v}, $\mathrm{L}(\cdot)$ denotes cross entropy, $\boldsymbol{\theta}^{N_{t_v}}$ denotes the parameter of N_{t_v}, y^i_h is the label for input x^i_h, $\mathbf{f}^i_{t_v}$ denotes the t_v-th sample scale feature map of the i-th geological element obtained through the mask operation and dilated convolution operation, and $\mathrm{P}^{1:L}_{t_v}$ denotes the matrix associated with selective knowledge transfer.

In the end, the overall optimization objective of SKT is to minimize the classification and self-distillation losses by

$$\mathcal{L}_{total} = \sum_{v=1}^{m} \mathcal{L}_{t_v} + \beta \sum_{v=1}^{m} \mathcal{L}_{t_v KD} \tag{12}$$

where $\mathcal{L}_{total}$ denotes the objective function, and β denotes the coefficient of self-distillation loss.

4. Experiments

In this section, we present extensive experiments to verify our method. Specifically, Section 4.1 describes the experimental environment and settings. Section 4.2 presents the comparison of our method with state-of-the-art methods on the geochemical dataset and the analysis of the experimental results. Section 4.3 presents the experimental results based on the relevant modules and parameters of our method. Section 4.4 visualizes the prediction result of the Pangxidong study area in Guangdong Province.

4.1. Experimental Settings

The evaluation metrics used in the experiments include Accuracy, Recall, and F1-score. All experiments are programmed and implemented with the PyTorch framework and one GeForce RTX 3090 GPU.

The SKT framework is implemented based on the ResNet-18 architecture. During training, the model uses an SGD optimizer with a momentum of 0.01, a weight decay of 1×10^{-4}, and a mini-batch size of 128. The learning rate is initially 0.1 and decreases by half every 30 epochs. We set the dilated coefficients $\boldsymbol{\rho} = \{\rho_1, \rho_2, \rho_3, \rho_4, \rho_5, \} = \{1, 6, 12, 18, 24\}$, referring to DeepLab [36]. The model is trained for 150 epochs on the geochemical dataset. In addition, the coefficient β of the self-distillation loss is set to 0.7.

4.2. Experimental Results and Analysis

In this section, we compare machine learning algorithms and some state-of-the-art classification methods with SKT to demonstrate that it outperforms other models in prospecting target prediction tasks. Specifically, we compare the following methods, including the traditional methods SVM [37], KNN [38], RandomForest [39], and Decisiontree [40] and the deep learning methods ResNet-18, ShufflenetV2 [41], GoogLeNet [42], MobilenetV2 [43], Mnasnet [44], SCnet [45], Efficientnet-b0 [46], T2T-vit-14 [47], and SNL [48]. It is worth noting that for fairness of comparison, SCnet and SNL are implemented based on the ResNet-18 architecture. In a machine learning algorithm, we compress each data point in the geochemical dataset into a one-dimensional tensor as the input to the algorithm.

As shown in Table 3, our model obviously performs better than ResNet-18. Specifically, the Accuracy of our model increases by 12.30%, Recall increases by 15.83%, and F1-score improves by 11.79%. Furthermore, SKT outperforms the other methods in terms of Accuracy, Recall, and F1-score. This indicates that SKT has excellent performance in prospecting target prediction and has the highest improvement in the prediction of samples with mines. Meanwhile, this also proves that our method can effectively solve problems such as the small number of geological samples and the irregular features of mining areas in prospecting target prediction.

Table 3. Experimental results. The optimal performances are bolded.

Methods	Accuracy	Recall	F1-Score
SVM	49.51	17.64	43.73
KNN	51.45	35.29	50.09
RandomForest	59.70	25.49	54.27
Decisiontree	58.73	39.21	57.03
ResNet-18	56.79	24.50	53.21
ShufflenetV2	57.45	17.64	48.24
GoogLeNet	61.81	31.09	56.51
MobilenetV2	55.82	16.66	47.74
Mnasnet	59.22	17.64	50.61
SCnet	58.73	30.39	55.05
Efficientnet-b0	57.28	23.52	51.70
T2T-vit-14	57.76	39.21	56.19
SNL	59.70	35.29	57.07
Ours	69.09	40.33	65.00

Figure 5 is a confusion matrix, showing the number of True Negative (TN), False Positive (FP), False Negative (FN), and True Positive (TP) samples. This figure shows that (1) among these four values, TN is the highest, i.e., the number of correctly predicted the samples without mines is the largest. At the same time, FP is the lowest, i.e., the number of incorrectly predicted samples without mines is the smallest. We explain that before using Gaussian noise for data augmentation, the samples without mines are much more than the samples with mines, and the samples without mines have rich features, which are helpful to the STS framework for prediction. (2) TP denotes the number of correctly predicted samples with mines. On the contrary, FN denotes the number of incorrectly predicted samples with mines. TP is the third of the four values, lower than FN but much higher than FP. The reason for this result is the small number of samples with mines and their irregular features, which negatively affects the STS framework for predicting the samples with mines, but still has a certain performance. In conclusion, SKT can predict the samples without mines well, and it also has a certain ability to predict the samples with mines.

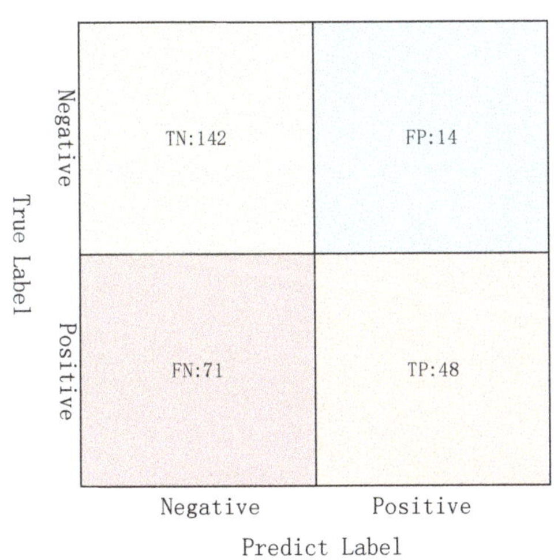

Figure 5. Confusion matrix.

In the SKT framework, the final prediction result is obtained by multiple target network voting. Table 4 shows the predicted results for the multiple target networks and voting. The following can be seen from this table: (1) As the dilated coefficient increases, the performance of the target networks first increases and then decreases. (2) When compared with the best target network, voting only has a small gap in Accuracy and F1-score, but there is a certain gap in Recall. The above result is caused by the performance of the model being affected with the increase in expansion coefficient, but at the same time, self-distillation can improve the performance of the target network, leading to first the increase and then the decrease. In addition, due to the contradiction between the dilated coefficient and self-distillation for the improvement of the model performance, we cannot determine the optimal target network. However, voting has a similar performance to the optimal target network, so the final prediction result obtained by voting is reasonable to a certain extent.

Table 4. Experimental results of the target networks and voting.

Target Network	Accuracy	Recall	F1-Score
$\rho = 1$	57.45	31.93	53.30
$\rho = 6$	65.45	41.17	62.08
$\rho = 12$	61.45	35.29	57.38
$\rho = 18$	70.18	47.05	67.34
$\rho = 24$	64.72	37.81	60.70
Voting	69.09	40.33	65.00

4.3. *Correlation Analysis Experiment*

We present the experimental results based on the relevant modules of SKT in Section 4.3.1. Then, in Section 4.3.2, we verify the effect of the coefficient of self-distillation loss on SKT.

4.3.1. Ablation Experiments

To evaluate the performance of the method proposed in this paper, we perform SKT ablation experiments and deploy various SKT variants. Specifically, we design the following ablation experiments: (1) remove soft mask (R-S-Mask), (2) remove dilated convolution (R-D-Convolution), (3) remove selective knowledge transfer (R-Sk-Transfer), and (4) remove self-distillation (R-S-Distillation). Based on SKT, we apply the control variable approach to soft mask, dilated convolution, and selective knowledge transfer one at a time. It is worth noting that we set the dilated coefficients $\boldsymbol{\rho} = \{\rho_1, \rho_2, \rho_3, \rho_4, \rho_5, \} = \{1, 1, 1, 1, 1\}$ in the second experiment. Finally, SKT is compared to the above methods.

Table 5 presents the experimental results. The following observations are made:

1. The soft mask makes the corresponding weight of the associated mineral elements as consistent as possible with that of the main mineral elements. Dilated convolution deals with the irregular features of the mining areas through different receptive fields. Selective knowledge transfer improves the model generalization performance to solve the problem of a small number of samples. Self-distillation mines the hidden knowledge between the feature maps of different scales. All of the aforementioned methods can improve the Accuracy, Recall, and F1-score of the prospecting prediction.
2. The contributions of these methods to SKT are different. According to the contribution from large to small, they are ranked as follows: dilated convolution, selective knowledge transfer, soft mask, and self-distillation.

Table 5. Ablation experiment results. The optimal performances are in bold.

Methods	Accuracy	Recall	F1-Score
R-S-Mask	64.72	29.41	55.43
R-D-Convolution	61.09	29.41	58.29
R-Sk-Transfer	62.54	30.25	56.83
R-S-Distillation	65.81	31.09	59.72
Ours	69.09	40.33	65.00

4.3.2. Parameter Analysis Experiments

The objective function of SKT includes self-distillation loss. To evaluate the effect of the coefficient of self-distillation loss β on SKT, we set $\beta = 0.1, 0.2, \ldots, 1$ to perform a total of 10 experiments. Figure 6 shows the experimental results. From this figure, we can find the following: (1) β has a certain effect on the SKT performance. (2) The best prospecting target prediction results can be obtained when β is between 0.6 and 0.8.

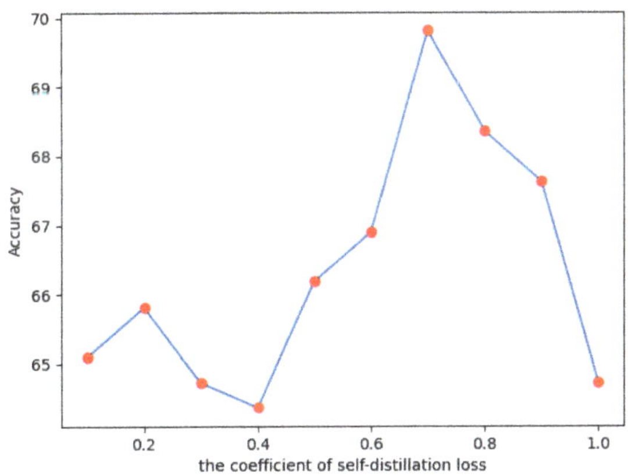

Figure 6. The coefficient of self-distillation loss experiment results.

4.4. Visualization

In this section, we use the SKT trained in Section 4.2 to predict the prospecting target area in the Pangxidong study area and visualize the prediction results. Specifically, we first cut the elemental content map of size 3072 × 3072 obtained in Section 2.2 into 12 × 12 patches, each with a size of 256 × 256. Then, we use the trained SKT to predict and visualize it. Figure 7 shows the visualization result. Based on the visualization, we come to the following conclusions: (1) The predicted prospecting target area is basically consistent with the actual mining area. (2) The prediction results of (row 7, column 4), (row 11, column 10), (row 12, column 4), and (row 12, column 11) are inconsistent with the actual mining area. In addition, we use the principal component analysis (PCA) method to reduce the dimensionality of geochemical data in the experiment to one dimension and visualize it [49]. As seen from Figure 8, we can find that (1) most of the mining areas correspond to high values, i.e., rich in these eight geochemical elements. This proves that geochemical elements have an important influence on mineralization, which is consistent with the conclusion of paper [50,51]. (2) Irregular features of geochemical-enriched regions in incorrectly predicted samples with mines (green boxes) make STS prediction difficult. In summary, we can conclude that SKT can fit the distribution of the prospecting target area in the Pangxidong study area. This again shows the effectiveness of SKT in prospecting target prediction.

Figure 7. Visualization of experimental results in Pangxidong study area.

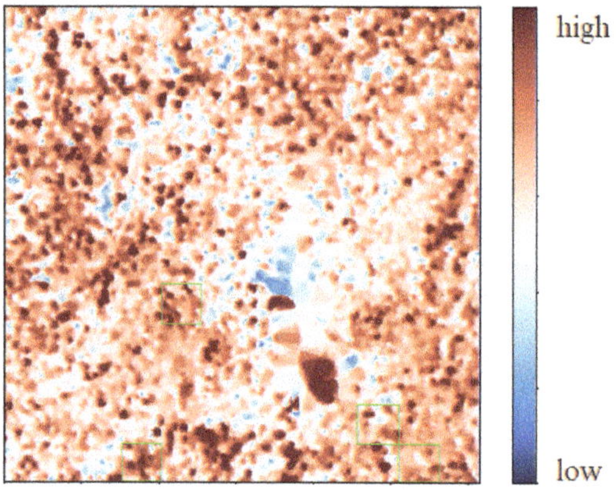

Figure 8. Visualization of geochemical data.

5. Conclusions

In this paper, a deep learning framework (SKT) for prospecting target prediction based on selective knowledge transfer is proposed to solve the problems of fewer geological sample data and irregular mining area features in the research of intelligent prospecting prediction. Taking the Pangxidong study area of Guangdong Province as an example, the prospecting target area is intelligently predicted by using the geochemical data of this area. Compared with other methods, the effectiveness of this method is proved. The main conclusions are:

(1) In view of problems such as the small number of geological samples and the irregular features of mining areas in the research of prospecting prediction, the deep learning framework (SKT) for prospecting target prediction based on selective knowledge transfer has greatly improved the prediction of the samples with mines, which is obviously superior to other methods.

(2) Soft mask makes the corresponding weight of associated mineral elements consistent with that of the main mineral elements as much as possible; dilation convolution enriches irregular features of the mining areas through capturing features at different scales; selective knowledge transfer improves the generalization performance of the model and solves the problem of a small number of samples; and self-distillation mines the hidden knowledge between different scale feature maps.

(3) Parameter analysis experiments show that dilation convolution, selective knowledge transfer, soft mask, and self-distillation can improve the accuracy of SKT prediction, but their contribution to SKT gradually weakens.

Author Contributions: Conceptualization and methodology, Y.H. and Q.F.; software and validation, Y.H.; resources and data curation, L.G.; writing—original draft preparation, Y.H. and Q.F.; writing—review and editing, Y.H. and L.Z.; visualization, W.Z.; funding acquisition, L.G. All authors have read and agreed to the published version of the manuscript.

Funding: This paper was produced by "the Natural Science Foundation of Guangdong Province (18zxxt52), the Wuyi University Youth Team Fund (2019td10), and the Wuyi University-Hong Kong-Macau Joint Fund (2019WGALH23)".

Data Availability Statement: The data presented in this study are available on request from the corresponding author.

Conflicts of Interest: The authors declare no conflict of interest.

References

1. Scharf, T.; Kirkland, C.; Daggitt, M.; Barham, M.; Puzyrev, V. AnalyZr: A Python application for zircon grain image segmentation and shape analysis. *Comput. Geosci.* **2022**, *162*, 105057. [CrossRef]
2. Middya, A.I.; Nag, B.; Roy, S. Deep learning based multimodal emotion recognition using model-level fusion of audio–visual modalities. *Knowl.-Based Syst.* **2022**, *244*, 108580. [CrossRef]
3. Cui, S.; Ma, A.; Zhang, L.; Xu, M.; Zhong, Y. MAP-Net: SAR and Optical Image Matching via Image-Based Convolutional Network With Attention Mechanism and Spatial Pyramid Aggregated Pooling. *IEEE Trans. Geosci. Remote. Sens.* **2022**, *60*, 1–13. [CrossRef]
4. Elashmawy, M.; Alatawi, I. Atmospheric water harvesting from low-humid regions of Hail City in Saudi Arabia. *Nat. Resour. Res.* **2020**, *29*, 3689–3700. [CrossRef]
5. Eppelbaum, L.; Eppelbaum, V.; Ben-Avraham, Z. Formalization and Estimation of Integrated Geological Investigations: An Informational Approach. *Geoinformatics* **2003**, *14*, 233–240. [CrossRef]
6. Siebels, K.; Goïta, K.; Germain, M. Estimation of Mineral Abundance From Hyperspectral Data Using a New Supervised Neighbor-Band Ratio Unmixing Approach. *IEEE Trans. Geosci. Remote. Sens.* **2020**, *58*, 6754–6766. [CrossRef]
7. Li, S.; Chen, J.; Liu, C. Overview on the Development of Intelligent Methods for Mineral Resource Prediction under the Background of Geological Big Data. *Minerals* **2022**, *12*, 616. [CrossRef]
8. Jooshaki, M.; Nad, A.; Michaux, S. A systematic review on the application of machine learning in exploiting mineralogical data in mining and mineral industry. *Minerals* **2021**, *11*, 816. [CrossRef]
9. Duda, R.O.; Hart, P.E. *Pattern Classification and Scene Analysis*; Wiley: New York, NY, USA, 1973; Volume 3.
10. Zekri, H.; Cohen, D.R.; Mokhtari, A.R.; Esmaeili, A. Geochemical prospectivity mapping through a feature extraction–selection classification scheme. *Nat. Resour. Res.* **2019**, *28*, 849–865. [CrossRef]
11. Daviran, M.; Maghsoudi, A.; Ghezelbash, R.; Pradhan, B. A new strategy for spatial predictive mapping of mineral prospectivity: Automated hyperparameter tuning of random forest approach. *Comput. Geosci.* **2021**, *148*, 104688. [CrossRef]
12. Zhang, S.; Carranza, E.J.M.; Xiao, K.; Wei, H.; Yang, F.; Chen, Z.; Li, N.; Xiang, J. Mineral Prospectivity Mapping based on Isolation Forest and Random Forest: Implication for the Existence of Spatial Signature of Mineralization in Outliers. *Nat. Resour. Res.* **2021**, 1–19. [CrossRef]
13. Wang, Z.; Zuo, R.; Dong, Y. Mapping of Himalaya Leucogranites Based on ASTER and Sentinel-2A Datasets Using a Hybrid Method of Metric Learning and Random Forest. *IEEE J. Sel. Top. Appl. Earth Obs. Remote. Sens.* **2020**, *13*, 1925–1936. [CrossRef]
14. Wang, Y.; Zhou, Y.; Xiao, F.; Wang, J.; Wang, K.; Yu, X. Numerical Metallogenic Modelling and Support Vector Machine Methods Applied to Predict Deep Mineralization:A Case Study from the Fankou Pb-An ore Deposit in Northem Guangdong. *Geotecton. Metallog.* **2020**, *44*, 9. [CrossRef]
15. Mandana, T.; Behnam, B.; Saeed, D. Intelligent geochemical exploration modeling using multiclass support vector machine and integration it with continuous genetic algorithm in Gonabad region, Khorasan Razavi, Iran. *Arab. J. Geosci.* **2021**, *14*, 1–15. [CrossRef]
16. Liu, Y.; Zhu, L.; Zhou, Y. Experimental Research on Big Data Mining and Intelligent Prediction of Prospecting Target Area Application of Convolutional Neural Network Model. *Geotecton. Metallog.* **2020**, *44*, 1–11. [CrossRef]
17. Li, S.; Chen, J.; Xiang, J. Applications of deep convolutional neural networks in prospecting prediction based on two-dimensional geological big data. *Neural Comput. Appl.* **2020**, *32*, 2037–2053. [CrossRef]
18. Li, T.; Zuo, R.; Xiong, Y.; Peng, Y. Random-Drop Data Augmentation of Deep Convolutional Neural Network for Mineral Prospectivity Mapping. *Nat. Resour. Res.* **2020**, *30*, 27–38. [CrossRef]

19. Li, T.; Xia, Q.; Zhao, M.; Gui, Z.; Leng, S. Prospectivity Mapping for Tungsten Polymetallic Mineral Resources, Nanling Metallogenic Belt, South China: Use of Random Forest Algorithm from a Perspective of Data Imbalance. *Nat. Resour. Res.* **2020**, *29*, 203–227. [CrossRef]
20. Zhang, C.; Zuo, R.; Xiong, Y. Detection of the multivariate geochemical anomalies associated with mineralization using a deep convolutional neural network and a pixel-pair feature method. *Appl. Geochem.* **2021**, *130*, 104994. [CrossRef]
21. Li, D.; Yao, A.; Chen, Q. Learning to learn parameterized classification networks for scalable input images. In Proceedings of the European Conference on Computer Vision, Glasgow, UK, 23–28 August 2020; Springer: Berlin/Heidelberg, Germany, 2020; pp. 19–35.
22. Yang, T.; Zhu, S.; Chen, C.; Yan, S.; Zhang, M.; Willis, A. Mutualnet: Adaptive convnet via mutual learning from network width and resolution. In Proceedings of the European Conference on Computer Vision, Glasgow, UK, 23–28 August 2020; Springer: Berlin/Heidelberg, Germany, 2020; pp. 299–315.
23. Yang, N.; Zhang, Z.; Yang, J.; Hong, Z.; Shi, J. A Convolutional Neural Network of GoogLeNet Applied in Mineral Prospectivity Prediction Based on Multi-source Geoinformation. *Nat. Resour. Res.* **2021**, *30*, 3905–3923. [CrossRef]
24. Wang, X. Metallogenic Pattern and Mineral Prospectivity Modeling of the Dashui Gold Concentration District. Ph.D. Thesis, China University of Geosciences, Beijing, China, 2020.
25. Xiao, F.; Wang, K.; Hou, W.; Erten, O. Identifying geochemical anomaly through spatially anisotropic singularity mapping: A case study from silver-gold deposit in Pangxidong district, SE China. *J. Geochem. Explor.* **2020**, *210*, 106453. [CrossRef]
26. Zhou, Y.; Li, X.; Zhen, Y.; Shen, W.; He, J.; Yu, P.; Niu, J.; Zeng, C. Geological settings and metallogenesis of Qinzhou Bay-Hangzhou Bay orogenic juncture belt, South China. *Acta Petrol. Sin.* **2017**, *33*, 667–681.
27. Lin, Z.; Zhou, Y.; Qin, Y.; Zheng, Y.; Liang, Z.; Zou, H.; Niu, J. Ore-controlling structure analysis of Panxidong-Jinshan silver-gold orefield, southern Qin-Hang belt: Implications for furthern exploration. *Miner. Depos.* **2017**, *36*, 866–878.
28. Lobo, J.M.; Jiménez-Valverde, A.; Real, R. AUC: A misleading measure of the performance of predictive distribution models. *Glob. Ecol. Biogeogr.* **2008**, *17*, 145–151. [CrossRef]
29. Zhang, S. Deep Learning for Mineral Prospecitivity Mapping of Lala-Type Copper Deposit in the Huili Region, Sichuan. Ph.D. Thesis, China University of Geosciences, Beijing, China, 2020.
30. Zhang, Y.; Zhou, Y.Z.; Wang, L.F.; Wang, Z.H.; He, J.G.; An, Y.F.; Li, H.Z.; Zeng, C.Y.; Liang, J.; Lü, W.C.; et al. Mineralization-related geochemical anomalies derived from stream sediment geochemical data using multifractal analysis in Pangxidong area of Qinzhou-Hangzhou tectonic joint belt, Guangdong Province, China. *J. Cent. South Univ.* **2013**, *20*, 184–192. [CrossRef]
31. Zeyu, Z.; Qingying, Z.; Shixian, L. Comparison of two machine learning algorithms for geochemical anomaly detection. *Glob. Geol.* **2018**, *37*, 1288–1294.
32. Kumar, V.; Gupta, P. Importance of statistical measures in digital image processing. *Int. J. Emerg. Technol. Adv. Eng.* **2012**, *2*, 56–62.
33. Jia, M.; Dong, M. Analysis and comparison of Gaussian noise denoising algorithms. *J. Phys. Conf. Ser.* **2021**, *1846*, 012069. [CrossRef]
34. Zuo, R.; Peng, Y.; Li, T.; Xiong, Y. Challenges of geological prospecting big data mining and integration using deep learning algorithms. *Earth Sci.* **2021**, *46*, 350–358.
35. Zuo, R.; Wang, J.; Xiong, Y.; WANG, Z. Progresses of researches on geochemical exploration data processing during 2011–2020. *Bull. Mineral. Petrol. Geochem.* **2021**, *40*, 81–93.
36. Chen, L.C.; Papandreou, G.; Kokkinos, I.; Murphy, K.; Yuille, A.L. DeepLab: Semantic Image Segmentation with Deep Convolutional Nets, Atrous Convolution, and Fully Connected CRFs. *IEEE Trans. Pattern Anal. Mach. Intell.* **2018**, *40*, 834–848. [CrossRef] [PubMed]
37. Lyu, P.; He, L.; He, Z.; Liu, Y.; Deng, H.; Qu, R.; Wang, J.; Zhao, Y.; Wei, Y. Research on remote sensing prospecting technology based on multi-source data fusion in deep-cutting areas. *Ore Geol. Rev.* **2021**, *138*, 104359. [CrossRef]
38. Chen, Y.; Zhao, Q.; Lu, L. Combining the outputs of various k-nearest neighbor anomaly detectors to form a robust ensemble model for high-dimensional geochemical anomaly detection. *J. Geochem. Explor.* **2021**, *231*, 106875. [CrossRef]
39. Wang, C.; Pan, Y.; Chen, J.; Ouyang, Y.; Rao, J.; Jiang, Q. Indicator element selection and geochemical anomaly mapping using recursive feature elimination and random forest methods in the Jingdezhen region of Jiangxi Province, South China. *Appl. Geochem.* **2020**, *122*, 104760. [CrossRef]
40. Dai, L.M.; Chen, Y.L.; Zhou, Y.G.; Liu, B.; Lou, D. A decision tree model for mineral potential mapping. *Prog. Geophys.* **2009**, *24*, 1081–1087.
41. Ma, N.; Zhang, X.; Zheng, H.T.; Sun, J. Shufflenet v2: Practical guidelines for efficient cnn architecture design. In Proceedings of the European Conference on Computer Vision (ECCV), Munich, Germany, 8–14 September 2018; pp. 116–131.
42. Szegedy, C.; Liu, W.; Jia, Y.; Sermanet, P.; Reed, S.; Anguelov, D.; Erhan, D.; Vanhoucke, V.; Rabinovich, A. Going deeper with convolutions. In Proceedings of the IEEE Conference on Computer Vision and Pattern Recognition, Boston, MA, USA, 7–12 June 2015; pp. 1–9.
43. Sandler, M.; Howard, A.; Zhu, M.; Zhmoginov, A.; Chen, L.C. Mobilenetv2: Inverted residuals and linear bottlenecks. In Proceedings of the IEEE Conference on Computer Vision and Pattern Recognition, Salt Lake City, UT, USA, 18–23 June 2018; pp. 4510–4520.

44. Tan, M.; Chen, B.; Pang, R.; Vasudevan, V.; Sandler, M.; Howard, A.; Le, Q.V. Mnasnet: Platform-aware neural architecture search for mobile. In Proceedings of the IEEE/CVF Conference on Computer Vision and Pattern Recognition, Long Beach, CA, USA, 15–20 June 2019; pp. 2820–2828.
45. Liu, J.J.; Hou, Q.; Cheng, M.M.; Wang, C.; Feng, J. Improving convolutional networks with self-calibrated convolutions. In Proceedings of the IEEE/CVF Conference on Computer Vision and Pattern Recognition, Seattle, WA, USA, 13–19 June 2020; pp. 10096–10105.
46. Tan, M.; Le, Q. Efficientnet: Rethinking model scaling for convolutional neural networks. In Proceedings of the International Conference on Machine Learning, Long Beach, CA, USA, 9–15 June 2019; pp. 6105–6114.
47. Yuan, L.; Chen, Y.; Wang, T.; Yu, W.; Shi, Y.; Jiang, Z.H.; Tay, F.E.; Feng, J.; Yan, S. Tokens-to-token vit: Training vision transformers from scratch on imagenet. In Proceedings of the IEEE/CVF International Conference on Computer Vision, Montreal, QC, Canada, 11–17 October 2021; pp. 558–567.
48. Zhu, L.; She, Q.; Li, D.; Lu, Y.; Kang, X.; Hu, J.; Wang, C. Unifying Nonlocal Blocks for Neural Networks. In Proceedings of the IEEE/CVF International Conference on Computer Vision, Montreal, QC, Canada, 10–17 October 2021; pp. 12292–12301.
49. Amaral, T.G.; Pires, V.F.; Pires, A.J. Fault detection in PV tracking systems using an image processing algorithm based on PCA. *Energies* **2021**, *14*, 7278. [CrossRef]
50. Zhou, S.G.; Zhou, K.F.; Wang, J.L. Geochemical metallogenic potential based on cluster analysis: A new method to extract valuable information for mineral exploration from geochemical data. *Appl. Geochem.* **2020**, *122*, 104748. [CrossRef]
51. Ayari, J.; Barbieri, M.; Barhoumi, A.; Belkhiria, W.; Braham, A.; Dhaha, F.; Charef, A. A regional-scale geochemical survey of stream sediment samples in Nappe zone, northern Tunisia: Implications for mineral exploration. *J. Geochem. Explor.* **2022**, *235*, 106956. [CrossRef]

minerals

MDPI

Article

Construction and Application of a Knowledge Graph for Gold Deposits in the Jiapigou Gold Metallogenic Belt, Jilin Province, China

Yao Pei [1,2,*], Sheli Chai [1], Xiaolong Li [3], Jofrisse Cremilda Samuel [1], Chengyou Ma [1], Haonan Chen [1], Renxing Lou [4] and Yu Gao [1]

1 College of Geoexploration Science and Technology, Jilin University, Changchun 130026, China
2 Geoscience Big Data Analysis and Application Technology Innovation Center, Ministry of Natural Resources, Changchun 130026, China
3 Jilin Branch of China National Geological Exploration Center of Building Materials Industry, Changchun 130033, China
4 College of Earth Sciences, Jilin University, Changchun 130061, China
* Correspondence: peiyao@jlu.edu.cn

Abstract: Over the years, many geological exploration reports and considerable geological data have been accumulated during the prospecting and exploration of the Jiapigou gold metallogenic belt (JGMB). It is very important to fully utilize these geological and mineralogical big data to guide future gold exploration. This work collects the original textual data of different gold deposits in JGMB and constructs a knowledge graph (KG) for deposits based on deep learning (DL) and natural language processing (NLP). Based on the metallogenic geological characteristics of deposits, a visual construction method of a KG for deposits and a calculation of the similarity between deposits are proposed. In this paper, 20 geological entities and 24 relationship categories are considered. By condensing the key KG information, the metallogenic geological conditions and factors controlling the ore in 14 typical deposits in the JGMB are systematically analyzed, and the metallogenic regularity is summarized. By calculating the deposits' cosine similarities based on the KG, the mineralization types of deposits can be divided into two categories according to the industrial types of ore bodies. The results also show that the KG is a cutting-edge technology that can extract the rich information of ore-forming regularity and prospecting criteria contained in the textual data to help researchers quickly analyze the mineralization information.

Keywords: the Jiapigou gold metallogenic belt; knowledge graph; natural language processing; metallogenic geological characteristics; the similarity between deposits

1. Introduction

The Jiapigou gold metallogenic belt (JGMB), located in southeastern Huadian, southern Jilin Province, Northeast China, is not only an important gold cluster area in the eastern segment of the northern margin of the North China Craton (NCC), but also the main gold-producing area with proven gold reserves of more than 180 tons [1] in China, in which a dozen gold deposits (Figure 1) and more than 160 gold occurrences have been discovered since 1820. The JGMB has great potential for gold prospecting with the development of gold exploration in the depth of the existing gold deposits, and the new gold deposits have been funded along its southeast extension in recent years. Many studies have been performed on the regional ore-forming geological background, petrology, mineralogy, geochemistry, and genesis of some typical gold deposits and occurrences in the JGMB [1–5]. It can be concluded from previous studies that (1) the gold deposits are of three mineralized types: quartz vein (Banmiaozi deposit, Bajiazi deposit, Jiapigou deposit, and Haigou deposit), altered rock (Songjianghe deposit and Liupiye deposit), and breccia

(Toudaoliuhe deposit and Binhugou deposit); (2) the origins of gold deposits belong to the orogenic type [2,6–9] and the mesothermal one [1,4]; (3) the gold deposits exhibit a multistage gold mineralization processes, with a time span ranging from approximately 230 to 68 Ma; (4) the gold deposits are mainly controlled by the NW-striking Jiapigou fault belt and its related secondary faults; and (5) the gold deposits are closely related to Yanshanian tectono-magmatism [10,11].

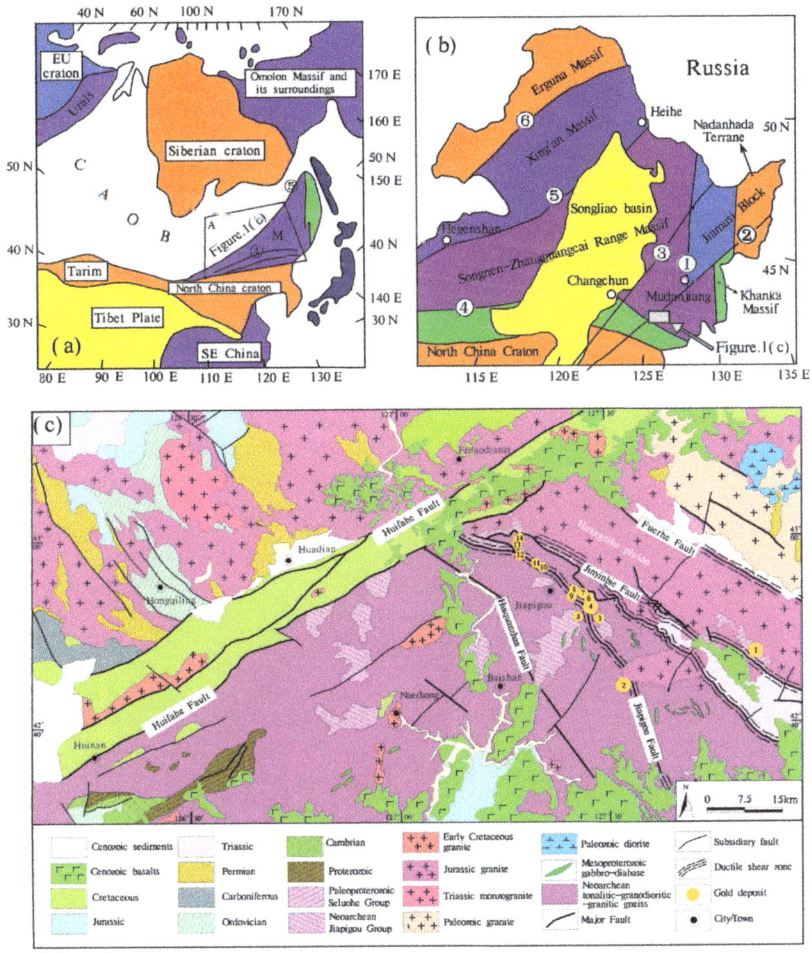

Figure 1. (**a**) Location of NE China with respect to the main tectonic units of China and Russia [9,12,13], 'A' and 'M' represent the 'Altaids' and 'Manchurides', respectively. (**b**) Tectonic units of NE China [9,12,13]. (**c**) Regional geological map of the JGMB and distribution of major gold deposits [14,15]. Gold deposits: 1. Songjianghe; 2. Liupiye; 3. Bajiazi; 4. Jiapigou; 5. Erdaogou; 6. Sidaocha; 7. Sandaocha; 8. Xiaobeigou; 9. Daxiangou; 10. Laoniugou; 11. Caiqiangzi; 12. Banmiaozi; 13. Damiaozi; 14. Yuanchaogou. Main faults in (**a**,**b**): ① Mudanjiang Fault; ② Dunhua–Mishan Fault; ③ Yilan–Yitong Fault; ④ Xar Moron–Changchun Fault; ⑤ Hegenshan–Heihe Fault; ⑥ Tayuan–Xiguitu Fault. Reproduced with permission from Elsevier, Journal of Asian Earth Sciences; published by Elsevier, 2016. Reproduced with permission from Elsevier, Gondwana Research; published by Elsevier, 2013. Reproduced with permission from Elsevier, Journal of Asian Earth Sciences; published by Elsevier, 2007.

These studies further deepen our understanding of the gold mineralization characteristics and gold metallogeny of the JGMB. However, the ages, origin, metallogenic processes, etc. of gold deposits in the JGMB are still debated partly because of the lack of regional comparisons between gold deposits in different spatial locations, of various styles of gold ore mineralization hosted in different country rocks, and with different ore-forming ages. A large number of research papers published in Chinese and English journals, scientific reports on geology and specific reports on ore exploration have been accumulated since the 1960s in JGMB. However, it is difficult to quickly extract some valuable information about ore-forming and ore-prospecting from numerous papers and reports of a dozen or a hundred gold deposits and occurrences with the manual curation method, which is a time-consuming process that requires some experts to read geological papers and reports, and annotate professional terms or sentences that assert some relationship between geological factors. Therefore, how to intelligently read and process from the papers and reports as text files from a dozen or a hundred gold deposits and occurrences and how to quickly extract the necessary information used for investigating the regional gold metallogeny and exploring gold deposits in a gold cluster area are great challenges for geologists and experts in artificial intelligence (AI).

Natural language processing (NLP) and deep learning (DL) are subsets of AI. Automated approaches rely on DL or NLP to rapidly detect terms and sentences of interest containing geological papers and reports. NLP makes it possible for humans to talk with machines, and its goal is to construct systems that can make sense of the written text and automatically perform tasks such as translation, keyword extraction, and topic classification [16]. NLP includes two parts: natural language understanding and natural language generation, which can recognize structured and unstructured textual data.

The concept of a knowledge graph (KG) was proposed by Google in 2012 [17]. NLP and KGs often need to be applied together to give full play to their maximum efficiency. A KG contains entities, concepts, attributes, relations, and other information and can use language understood by both humans and machines to describe the real world in the form of graphs, making the knowledge structure clearer. However, it is single-minded to understand the KG simply as a graph because in the face of massive data, it is usually impossible to display all knowledge structures in the form of a graph on one screen. Therefore, the KG is a semantic network formed by the semantic computing of many texts illustrating knowledge. Its main purpose is to construct a graph according to its relationship with knowledge and search by using the knowledge structure. As AI has developed, KGs have gradually penetrated various fields [18–21].

Previous scholars have explored the application of NLP and KGs in geological work [22–24]. Wang et al. [25] used NLP technology to analyze mineral resources and introduced the workflow of Chinese documents, proving the effectiveness of the designed workflow, and showed the potential of NLP technology in geoscience. Li et al. [26] used the Lala copper mine as an example and studied the method of mining prospecting information in text based on a convolutional neural network (CNN). Holden et al. [27] developed GeoDocA, a geological document analysis system, that can use text mining technology to visually analyze the geological exploration reports. Enkhsaikhan et al. [28] studied the method of understanding ore-forming conditions using the machine reading of text.

The research papers on ore exploration in the JGMB contain a large number of gold deposit-related geological factors and their relations. Identifying these geological factors and relations is of positive significance to rapidly improve the analysis of geological texts. There is a strict corresponding relationship between a deposit and its ore-controlling factors that has the advantage of constructing a KG. However, KG research in the geoscience field is only beginning, and there is a lack of corpora for annotated entities and relations. In this study, the original textual data such as geological exploration reports, published journals and dissertations related to typical deposits in JGMB, are collected. This work carries out research on KGs based on DL and NLP. A visual KG construction method for deposits and a method for calculating the similarities between deposits are proposed. Based on the analysis of texts by AI, rich information about metallogenic regularity, prospecting criteria and deposit genesis can be obtained. Through KG visualization, the relation between geological entities can be displayed. The research shows that this method has important theoretical significance and application value and can rapidly analyze texts and mine potential knowledge.

2. Geological Settings

The JGMB lies between the northern margin of the NCC and the eastern section of the Central Asian Orogenic Belt (CAOB), as shown in Figure 1a,b [12,29,30]. In the study area, the NCC, also known as the Longgang block, is separated from the eastern extension of the CAOB (locally named the Xing'an–Mongolian orogenic belt, a continental margin accretionary belt) by the Jiapigou and Jinyinbie faults with NW strikes, as shown in Figure 1b,c. The strata exposed in the study area are the Neoarchean Jiapigou Group, the Paleoproterozoic Seluohe Group, the early Paleozoic, and the later Paleozoic, Mesozoic, and Cenozoic (Figure 1c). The Neoarchean Jiapigou Group is composed of the Neoarchean Sandaogou Formation, which is exposed in the Longgang block and occurred as inclusions or lenticular bodies in the Neoarchean TTG (trondhjemite, tonalite, and granodiorite) metamorphic complex. The Sandaogou Formation consists of chlorite-hornblende schist, granitic gneiss, chlorite-actinolite, etc., which underwent the metamorphism of low amphibolite to greenschist facies. The Sandaogou Formation hosts most gold deposits in the JGMB such as the Jiapigou deposit and Banmiaozi deposit. The Seluohe Group is distributed in the northern margin of the Longgang block along or around the NW-striking Jinyinbie fault. The Seluohe Group underwent metamorphism of greenschist facies, which is composed mainly of a suite of metavolcanic and metaclastic sedimentary rocks, with rock types of plagioclase amphibolite, amphibole schist, mica chlorite schist, sericite quartz schist, metasandstone, meta-andesite, and tremolite marble.

The LA-ICP-MS zircon U-PB dating data in plagioclase amphibolite range from 2543 to 2527 Ma [31], belonging to the Neoarchean rather than the formerly considered Paleoproterozoic. The Seluohe Group is the ore-hosting strata of the Songjianghe deposit. The early Paleozoic is distributed only in the southwest of the Longgang block and is comprised mainly of detrital sedimentary rocks. The later Paleozoic metavolcanic and metasedimentary rocks are exposed in the Xing'an–Mongolian orogenic belt to the north of the Huifahe fault and Fuerhe fault. Mesozoic volcanic and sedimentary rocks are exposed in the Xing'an–Mongolian orogenic belt to the north of the Huifahe fault or along the Huifahe fault, and the rock assemblage of Mesozoic volcanic rocks is mainly andesite and dacite of the calc-alkaline series. Cenozoic basalts are distributed in the Changbai Mountains and along the faults of Huifahe and Fuerhe. Quaternary deposits are composed mainly of Holocene alluvial and diluvial sediments.

The intrusive activities in the study area can be grouped into five stages: (1) Neoarchean TTG complex, occupying the majority of the Longgang block; (2) Paleozoic diorite and granite intrusions, exposed mainly as batholiths or stocks in the Xing'an–Mongolian orogenic belt to the north of the Fuerhe fault; (3) Triassic monzogranite intrusions, which occurred as stocks in the Xing'an–Mongolian orogenic belt; (4) Jurassic granite intrusions, extensively exposed as batholiths or stocks in the Xing'an–Mongolian orogenic belt, showing a close temporal–spatial relationship with Au mineralization; and (5) Early Cretaceous granite, distributed as only small stocks in the central part of the Longgang block and along the Huifahe fault. Gold deposits in the JMGB are controlled mainly by the NW-striking Jiapigou shear zone and its related faults such as the Jiapigou fault and Jinyinbie fault (Figure 1c) [4,7,32]. All the gold deposits are confined in a zone 40 km long and 4–10 km wide. The gold deposits from the south to the north of the JMGB are Songjianghe, Liupiye, Bajiazi, Jiapigou, Erdaogou, Sidaocha, Sandaocha, Xiaobeigou, Daxiangou, Laoniugou, Caiqiangzi, Banmiaozi, Damiaozi, and Yuanchaogou. The gold deposits in the JMGB can be divided into two types: gold-bearing quartz veins (abovementioned deposits except for the Songjianghe deposit and Liupiye deposit) and disseminated ores (Songjianghe and

Liupiye); between them, the former predominates, accounting for approximately 85% of the gold reserve and production [33]. The gold deposits in the JMGB are derived from mesothermal magmatic–hydrothermal processes, with ages ranging from approximately 240 to 150 Ma [31,34].

3. Related Work

The KG can visually show the relationship between geological characteristics. However, at present, research on geoscience KGs is in its infancy. Both the combination of knowledge systems and the construction of large-scale KGs need in-depth study, especially for the following core issues.

3.1. Named Entity Recognition (NER)

NER refers to the entity with specific significance in the text including the name of the deposit, metallogenic geological bodies, metallogenic structure, and metallogenic characteristics. The main task of NER is to detect named entities from the text and classify them into predefined categories. Traditional entity labeling based on statistics considers the frequency or probability of co-occurrences between words and predicts the optimal segmentation sequence through statistical modeling methods such as support vector machine (SVM) [35], hidden Markov model (MHH) [36], and conditional random field (CRF) [37]. With the integration of DL, many new methods have been developed in recent years such as CNNs [38,39], long short-term memory (LSTM) [40], bidirectional long short-term memory (Bi-LSTM) [41,42], recurrent neural networks (RNNs) [43], autoencoders [44,45], and generative adversarial networks (GANs) [46]. Qiu et al. [47] proposed a neural network approach, namely, attention-based bidirectional long short-term memory (Att-Bi-LSTM) with a CRF layer, and used it to extract informational entities describing geoscience information in reports. This method obtained a 91.47% average F1 score in the NER task. Moreover, TensorFlow, PyTorch, and other development environments provide tools for the application of DL.

Labeled data are the basis of supervised learning with neural networks. The quantity and quality of labeled data directly determine the prediction result quality. Geoscience vocabulary has strong professionalism and is greatly different from daily language. At present, there are no systematic labeled geoscience data. However, traditional entity extraction methods must annotate data manually, which results in massive time and labor consumption.

This study used JGMB text data and adopted the NER method based on Bi-LSTM and CRF [28,48]. The Bi-LSTM-CRF model includes three components: the input layer (character embedding), Bi-LSTM layer, and CRF layer, as shown in Figure 2. LSTM is an RNN with a long short-term memory unit. Bi-LSTM is composed of a forward LSTM and a backward LSTM. Both are often used to model contextual information in NLP tasks. Among them, character embeddings are given to a Bi-LSTM. l_i represents the word i and its left context, and r_i represents the word i and its right context. Concatenating these two vectors yields a representation of the word i in its context, c_i [49]. To improve the recognition effect of named entities, each character in sentence X is transformed into a vector composed of character embeddings, which are randomly initialized. These character embeddings are the input of the Bi-LSTM-CRF model, and the output is the corresponding prediction label of each character. This work uses the torch.nn.Embedding statement in PyTorch to train the character vectors. The prediction results of the Bi-LSTM layer are input into the CRF layer, which can improve the legitimacy of these predicted labels by adding some constraints. Using this method, a more robust geoscience system based on knowledge extraction can be obtained with low-resource text [50].

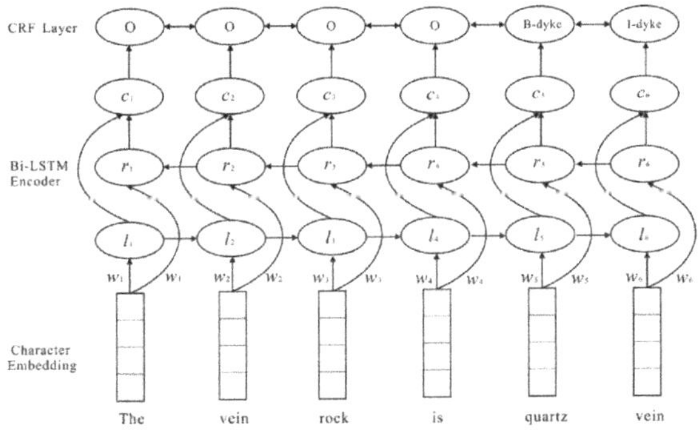

Figure 2. The neural network architectures of Bi-LSTM and CRF [48].

3.2. Relation Extraction

As an important task of information extraction, entity relation extraction refers to extracting predefined entity relations from unstructured text based on entity recognition [51]. The relation of entities is made up of triples ⟨e1, R, e2⟩, where e1 and e2 are entities and R is a relation.

Classic relation extraction methods are divided into four categories: supervised, semi-supervised, weakly supervised, and unsupervised. Supervised relation extraction is divided into feature-based methods and kernel function-based methods. Alokaili et al. [35] and Choi et al. [52] used SVM as a classifier to study the influence of lexical, syntactic, and semantic features on entity semantic relation extraction. In the face of unlabeled data, supervised methods need to manually annotate the training data, which wastes considerable time. Therefore, Huang et al. [53] and Quan et al. [54] then proposed relation extraction methods based on semi-supervised, weakly supervised, and unsupervised methods to solve this problem. Additionally, Brin [55] used the bootstrapping method to extract the relation between named entities; Wang et al. (2018) [24] studied the method of using weakly supervised learning to improve the accuracy of target detection; Hasegawa et al. [56] proposed an unsupervised relation extraction method between named entities. Mintz et al. [57] applied distant supervision in relation extraction.

In traditional relation extraction, the use of NLP leads to error propagation layer-by-layer and affects the effect of relation extraction. Therefore, distant supervised entity-relation extraction methods based on DL including CNN, RNN, LSTM, and other network structures [58–60] have become a research focus due to their ability to alleviate the propagation of labels and feature extraction errors. In recent years, scholars have proposed a variety of improvements based on the above basic methods such as the fusion of piecewise convolutional neural networks (PCNNs) and multi-instance learning [61], the COTYPE model [62], and the residual network proposed by Huang and Wang [63], which all enhance the effect of relation extraction.

In 2014, based on the mechanism of human visual attention, Bahdanau et al. applied the attention model to NLP [64,65]. Moreover, attention models have been applied to machine translation based on neural networks and have achieved good results [66]. With the development of DL, attention mechanisms have been widely used due to their excellent performance. Attention-based bidirectional long short-term memory networks (Att-Bi-LSTM) [67] add an attention layer after the Bi-LSTM structure, which can capture the most important semantic information in a sentence.

In this work, Att-Bi-LSTM was used to recognize the entity relation of KGs for gold deposits. Figure 3 shows the neural network architecture of the Att-Bi-LSTM model. This model contains five components: (1) the input layer inputs the sentence to this model, x_i represents each word in the sentence, and T represents the number of words in the sentence; (2) the embedding layer maps each word x_i in the sentence to a low dimensional vector e_i; (3) the LSTM layer, the Bi-LSTM layer inputs the word vectors of the training set obtained from the embedding layer into the forward and backward LSTM, respectively, to obtain high-level features; (4) the attention layer produces a weight vector, and weights the output of each LSTM step to obtain the sentence-level feature vector; and (5) in the output layer, the softmax activation function is used to obtain the final probability of each classification.

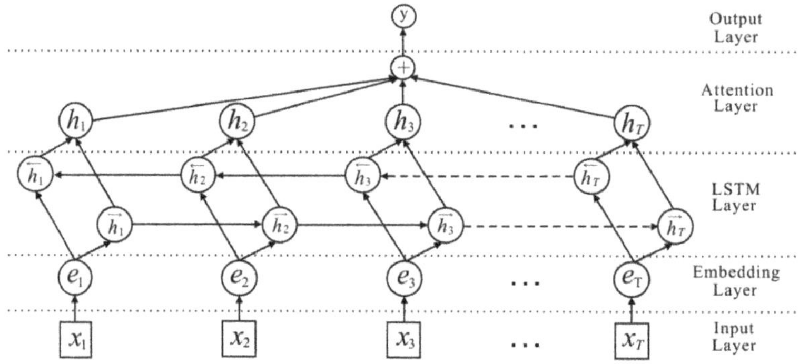

Figure 3. The bidirectional LSTM model with attention [67].

3.3. Entity Alignment

After relation extraction, the objective of obtaining the entity and relation information from unstructured data is achieved. However, in geological KGs, it is common for entities to have multiple words and one meaning. For example, the Jiapigou fault can also be named the Dalazi–Jiapigou fault. Therefore, this knowledge needs to be integrated through entity alignment to improve its quality.

Entity alignment can also be called coreference resolution. The goal of entity alignment is to distinguish whether these entities are the same entity by comparing and calculating the similarity between different entities, in order to solve the problem that one entity corresponds to multiple names. The similarity calculation expresses the similarity degree between entities through mathematical calculation. Traditional similarity calculation methods include the Levenshtein distance [68], Jaccard similarity coefficient [69], cosine similarity [70], term frequency-inverse document frequency (*TF-IDF*) [71,72], and Jaro-Winkler [73]. Volz et al. (2009) [74] provided a number of similarity metrics to calculate the similarity values of strings, URLs, numeric, and dates. In 2014, Vrandecic and Denny et al. [75] defined some feature templates for alignment and achieved good results.

The computation process of the entity alignment algorithm based on the similarity of string and manually defined features is relatively simple. However, because only character-level features are used to measure the similarity of entity categories, the implicit semantic information cannot be captured, resulting in low alignment accuracy. Subsequently, many graph-based entity alignment methods have been proposed. In 2009, Niu X and Rong S et al. [76] studied the measurement index of word sense disambiguation based on the graph method. The similarity of entities based on KGs uses the representation learning algorithm related to the graph structure, which is divided into network representation learning and KG representation learning. Network representation learning includes LINE, node2vec, and DeepWalk [77–79]. KG representation learning includes TransE, TranSparse, TransR, and TransH [80–83]. Using representation learning, entities in KGs can be vectorized. Compared with the similarity calculation based on the textual level, the entity vector obtained by this method can significantly improve the effect.

In recent years, word embedding, as a method of mining the deep-seated related semantics of words, has attracted great attention [84–86]. In particular, when using NLP to calculate the text similarity, the text needs to be vectorized first, so the finer-grained text is represented as word embedding. Different word embedding construction methods have different influences on sentence similarity calculation results [87,88]. In the actual application process, it is necessary to choose the appropriate algorithm according to the text situation. Word embedding is based on the core idea [89] that "words with similar context also have similar semantics". Characters or words are mapped into a vector space, so that words with similar semantics have similar directions in the vector space. Therefore, the alignment method based on word embedding can learn the deep semantic information of words from the corpus to effectively improve the alignment accuracy. Santos et al. [90] calculated the similarity based on the word vector of place names. The results show that this method is superior to the traditional alignment method based on the edit distance.

In this context, this work takes the geological entities of the JMGB as the research object and carries out joint entity alignment method research. This work considers the similarity between entities from two aspects. One is the edit distance similarity of entity names: the entities used in this paper are all geological domain entities, and the entity names are all string types. If the similarity of entity names is high, the probability that these two entities are the same entity is also high. The second is vector similarity: if two entities with similar semantics have a higher similarity of word vectors, then they are also more likely to be the same entity. Based on the above two cases, the weighted sum of the edit distance similarity and the word vector similarity is taken as the final similarity of the two entities. This method can consider the similarity of entities at both the string and semantic levels. The formula is as follows:

$$sim(a_i, b_i) = \alpha sim_{lev}(a_i, b_i) + (1 - \alpha) sim_{wor}(a_i, b_i) \tag{1}$$

where a_i and b_i are two entities to be aligned; α is the weight coefficient; $sim(a_i,b_i)$ is the similarity of entities to be aligned; $sim_{lev}(a_i,b_i)$ is the edit distance similarity; and $sim_{wor}(a_i,b_i)$ is the word vector similarity. We set the value of α and selected p as the threshold for whether the entity pair was aligned, that is $sim(a_i,b_i) \geq p$, and two entities were considered the same entity.

3.3.1. Edit Distance Similarity

The Levenshtein distance is a kind of edit distance. This algorithm was proposed by Levenshtein [91]. Entity similarity is calculated according to the Levenshtein distance algorithm, which is called the Levenstein ratio, and the formula is as follows:

$$sim_{lev}(a_i, b_i) = 1 - \frac{Idist}{length(a_i) + length(b_i)} \tag{2}$$

where *Idist* is the class edit distance, which refers to the minimum number of edits needed to change string a_i into string b_i. In this case, the insertion or deletion of a string is edited once, and the replacement of *a* string is edited twice; $length(a_i)$ is the length of string a_i; and $length(b_i)$ is the length of string b_i. We used the *Levenshtein* toolkit in Python and used the levenshtein.ratio statement to compute the similarity of two entities.

3.3.2. Vector Similarity

The word2vec proposed by Mikolov et al. [92] is widely used in comparing the similarity between words. In this paper, word2vec was used to transform the entities to be aligned into word vectors, and then cosine similarity was used to calculate their similarity, which represents the semantic similarity of entities. The cosine similarity formula is as follows:

$$sim_{wor}(a_i, b_i) = \frac{\sum_{i=1}^{n} a_i \times b_i}{\sqrt{\sum_{i=1}^{n} a_i} \times \sqrt{\sum_{i=1}^{n} b_i}} \tag{3}$$

Formula (3) shows that the range of cosine similarity values is [0, 1], and the greater the value, the higher the similarity. a_i and b_i represent the word vector set of the geological entities. In this paper, the *gensim* toolkit in Python was used to train word vectors by Gensim.models. The Word2Vec statement and the cosine similarity between entities were calculated by the cosine_similarity statement. The alignment entities were obtained by setting the threshold.

3.4. Visualization of KGs

Zhou et al. [93] considered that the geoscience KG is a clear display of all knowledge nodes and their relations in the field of earth science. Therefore, it is an important challenge to construct a geoscience KG by integrating the complex characteristics, computational attributes, and knowledge relations and rules from the nature of the graph structure. At present, typical KGs include DBpedia, YAGO, Wikidata, OpenCyc, and Freebase [94,95]. These resources cover knowledge in different fields, and the content is constantly enriched as human knowledge grows. KGs are usually stored in triples: <concepts, relationships, attributes>. Common graph databases mainly include OrientDB, Infinite Graph, Neo4j, and Titan [96]. In addition to being a visualization tool, Protégé [97] is used for constructing ontologies, relationships, attributes, and instances in the semantic web. As a science mapping tool, CiteSpace [98] is designed to facilitate the detection of emerging trends and abrupt changes in the scientific literature.

Among them, Neo4j is very suitable as the storage database of KG because of its good support for graph data, large amount of data storage, and support of multiple retrieval methods. By using Cypher, developers can query relevant data and display them in the control window, and return the results in the JSON, XML, and table formats. KG is widely used in NLP such as for auxiliary decision-making, semantic search, and intelligent question answering [99–101]. Neo4j creates the contents of the database through Cypher statements and supports external data import from CSV. In this work, entities, attributes and their relations to each deposit are classified and sorted into CSV files, and they are imported in batches through the LOAD CSV statement, new nodes are added to the database through the MERGE statement, and the relations between nodes are created through the MATCH statement.

Due to the deposit formation complexity, it is difficult to effectively integrate and analyze massive geological and mineralogical big data. There is still a lack of research on the intelligent mining of origin information and the discovery of metallogenic geological regulation from large geological and mineralogical big data. In this study, visualization technology based on KGs was used to explain and analyze geological textual data, and the node-edge structural model was used to associate the origin and evolution of gold deposits. The research was used to establish KGs for deposits based on Neo4j and then discover the overall formation mode of mineral resources. Visualization based on KGs can facilitate researchers in analyzing the spatial distribution of data and enhancing the interpretation ability of large amounts of geological data.

4. Construction of KG for Ore Deposits

4.1. Basic Ideas and Algorithm Flow

Figure 4 illustrates the framework for KG construction. Through the knowledge acquisition, annotation, and extraction of entities, relations and attributes in the text describing typical gold deposits in the JGMB, a KG for gold deposits with application functions was constructed. The construction process was divided into the following steps. (1) Text on relevant deposits in the JGMB was input, the geological entities of the corresponding deposits were extracted through DL, and an NER dataset of deposits was formed. (2) The relationship network between entities was constructed through relation extraction. (3) The entities were aligned. (4) The entities and relations were stored in Neo4j. The *TF-IDF* values of these geological entities were extracted based on establishing the KG, and the similarity between different deposits was calculated by cosine similarity.

The core idea of deposit visualization and interpretability based on a KG is to show the relations between deposits through graphs and distinguish the differences between deposits through the similarity of geological characteristics. Based on these design ideas, the proposed method includes three cores: (1) extracting text-based metallogenic information; (2) establishing a KG for the deposit model, and (3) constructing interpretable deposit characteristics. Among them, "constructing interpretable deposit characteristics" aims at the problem that the existing geoscience KG model lacks quantitative interpretation. Based on the geological information extraction of typical deposits in the JGMB, the relations between deposits were constructed, and the similarity between deposits was calculated. Then, quantitative interpretation of the mapping relations between deposits in the metallogenic belt was realized.

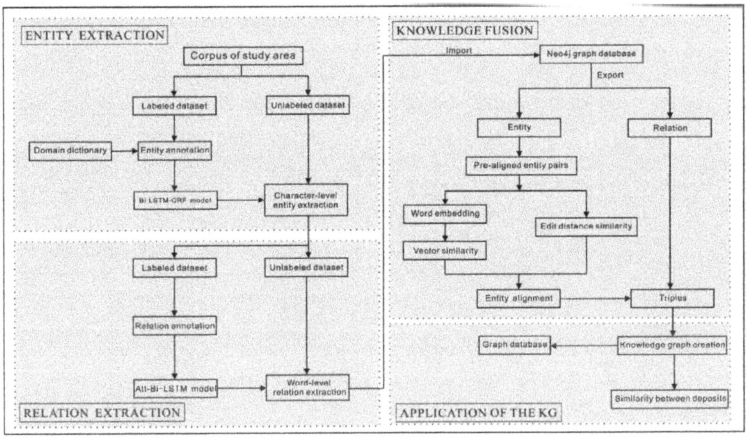

Figure 4. The KG construction framework.

4.2. Entity Extraction

Entity extraction is the key input of KG visualization based on DL. At present, there is no mature entity recognition method and no public dataset available for the construction of geoscience KGs. In this work, by collecting the textual data of gold deposits in the JGMB, the corresponding entity extraction was carried out under the guidance of domain experts, and the NER dataset of these deposits was constructed.

Corpus construction. A total of 266 papers and dissertations related to typical gold deposits in JMGB were retrieved and downloaded from the China National Knowledge Infrastructure (CNKI). Among them, 120 were randomly selected for model training and the other 146 were used for model prediction. Since these collected documents were stored in PDF format, the document format needed to be converted to text format data first. The contents of these documents were cleaned, and after removing the drawings and tables, 7014 sentences and 527,449 characters were obtained in the labeled dataset. The unlabeled dataset contained 8802 sentences and 671,489 characters.

Entity category. An entity in the KG refers to the general name of various geological characteristics related to ore deposits. Based on the metallogenic geological body, metallogenic structure, and mineralization characteristics, a total of 20 categories of entities were extracted, as follows in Figure 5.

Basic deposit information was extracted: (1) location such as Dunhua City or Huadian City; and (2) the name of the deposit such as the Songjianghe gold deposit and Liupiye gold deposit.

Metallogenic geological bodies refer to geological bodies that are closely related to the deposits in time, space, and origin and include the following seven entities: (1) group such as the Jiapigou Group and Longgang Group; (2) formation such as the Laoniugou Formation and Xinkaihe Formation; (3) metamorphic rocks such as amphibolite and marble; (4) plutons such as the Huangniling pluton and Wudaoliuhe pluton; (5) intrusive rocks such as granodiorite and adamellite; (6) extrusive rocks such as basalt and rhyolite; and (7) sedimentary rocks such as dolomite and siltstone.

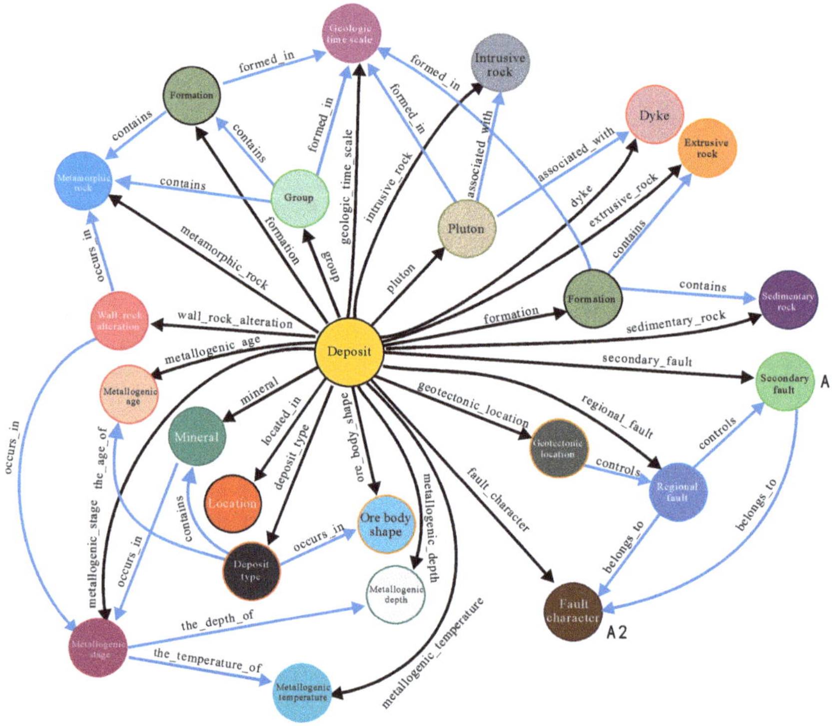

Figure 5. The construction structure of the KG for gold deposits. A1 is the node representing the secondary fault. A2 is the node representing the fault character.

Metallogenic structure refers to the structure that controls the spatial position, shape, scale, occurrence, and internal structure of geological bodies. The metallogenic structure includes the following four entities: (1) geotectonic locations such as the northeastern margin of the NCC and the Xingmeng orogenic belt; (2) regional faults such as the Huifahe fault and Ji'an–Songjiang fault; (3) secondary faults such as the Jiapigou fault and Jinyinbie fault; and (4) fault characteristics such as brittle fault, brittle–ductile fault, and ductile fault.

Mineralization characteristics refer to those that can directly indicate the locations of ore bodies and have special significance for prospecting prediction. The mineralization characteristics include the following seven entities: (1) metallogenic stages such as the milky-quartz stage and quartz-pyrite stage; (2) deposit types such as altered rock-type and quartz vein-type; (3) geological time scales such as Cretaceous and Jurassic; (4) minerals such as pyrite and hematite; (5) dykes such as diorite porphyrite and syenite porphyry; (6) wall rock alterations such as chloritization and pyritization; and (7) ore body shapes such as veins, lenticular veins, and thin veins.

Data annotation. A total of 527,449 characters in 7014 sentences were annotated as a corpus, and the corresponding domain dictionary was generated. After manual annotation and verification, an NER dataset of ore deposits in the JGMB was formed. The annotation rules were set as follows: (1) entity boundary detection, for example, correctly identifying the Seluohe Group rather than Seluohe; and (2) determining the category of the entity, for example, pyrite is a mineral entity rather than a rock entity. In this paper, the BIO annotation method was used. B represents the beginning of an entity, I represents the inside of an entity, and O represents a nonentity. These data are divided into the training dataset, verification dataset, and test dataset at a ratio of 8:1:1, as shown in Table 1.

Table 1. Labeling with the domain dictionary with the resolution rules applied.

Entity Category	Beginning of an Entity	Inside of an Entity	Number of Entities in Training Dataset	Number of Entities in Validation Dataset	Number of Entities in Test Dataset
Deposit type	B-deposit_type	I-deposit_type	278	6	19
Dyke	B-dyke	I-dyke	1326	139	176
Extrusive rock	B_extrusive_rock	I-extrusive_rock	129	18	6
Fault character	B-fault_character	I-fault_character	672	50	83
Formation	B-formation	I-formation	196	33	24
Geological time scale	B-geological_time_scale	I-geological_time_scale	1600	302	165
Geotectonic location	B-geotectonic_location	I-geotectonic_location	620	163	66
Group	B-group	I-group	340	36	33
Intrusive rock	B-intrusive_rock	I-intrusive_rock	1050	239	71
Location	B-location	I-location	1014	91	113
Metallogenic stage	B-metallogenic_stage	I-metallogenic_stage	120	2	29
Metamorphic rock	B-metamorphic_rock	I-metamorphic_rock	1451	200	90
Mineral	B-mineral	I-mineral	2705	191	482
Name of the deposit	B-deposit	I-deposit	2296	185	340
Orebody shape	B-orebody_shape	I-orebody_shape	1040	44	228
Pluton	B-pluton	I-pluton	112	8	4
Regional fault	B-regional_fault	I-regional_fault	110	27	22
Secondary fault	B-secondary_fault	I-secondary_fault	264	45	17
Sedimentary rock	B-sedimentary	I-sedimentary	105	18	13
Wall rock alteration	B-wall_rock_alteration	I-wall_rock_alteration	797	48	136

Experiment. The essence of NER is a sequence labeling problem, that is, classifying each word in the sequence. In this paper, accuracy (ACC), precision (P), recall (R), and F1 [102] were used for the NER evaluation metrics. The configuration of the experimental environment is shown in Table 2. The hyperparametric settings of model training are shown in Table 3.

Table 2. The configuration of the experimental environment.

Operating System	Windows 10
CPU	Intel Core i9-10900F @ 2.80 GHz
GPU	Nvidia GeForce RTX 3080 (10 GB)
Python	3.6
Pytorch	1.7.0

Table 3. The model parameters.

Hyperparameters	Value
Batch size	128
Learning rate	0.001
Epochs	250
Character embedding dimension	100
The number of hidden units	128
Dropout rate	0.5
Optimizer	Adam

Evaluation. We selected the Bi-LSTM-CRF model and compared it with the HMM, CRF, and Bi-LSTM models to compare the accuracy of different methods on the geological NER task. As shown in Table 4, the accuracy of HMM was low. CRF was better than Bi-LSTM, but slightly lower than Bi-LSTM-CRF. The experimental results showed that the Bi-LSTM-CRF model had a better entity recognition effect than other models, which can meet the requirements of the construction of the deposit KG.

Table 4. The test results of the entity extraction models.

Model	ACC (%)	P (%)	R (%)	F1 (%)
MHH	95.32	66.26	80.17	72.55
CRF	99.16	94.62	93.25	93.93
Bi-LSTM	99.09	91.70	94.65	93.15
Bi-LSTM-CRF	**99.27**	**97.01**	**96.83**	**96.92**

Note: The best results are highlighted in bold.

4.3. Relation Extraction

Discrete nodes are obtained after entity extraction, so it is necessary to determine the relationship between entities to form a network knowledge system. The Att-Bi-LSTM model was used to ensure the accuracy of relation extraction. By default, this method knows the relationship categories contained in all texts. In this case, relation extraction is a problem of text classification. A sentence and two entities contained in the sentence are the inputs, and the category of relation is the output.

Dataset. Currently, there is a lack of public datasets in the field of geology. In this paper, 7014 sentences in the dataset for constructing the NER task in Section 4.2 were used for the relation extraction experiments. The approach taken in this paper was to identify the entities contained in each sentence according to the geological entity dictionary. Through the permutation and combination of entities in sentences, the possible relations are found, and then the relation dataset is constructed. The label categories in the dataset include sentences, entities, and relations. A sentence may contain more than two geological entities. To solve this situation, sentences containing multiple entities are exhaustive until all possible situations are traversed. For example, the sentence "the types of wall rock alteration of the Songjiang gold deposit include pyritization, chloritization, etc." includes two cases, namely, <Songjiang gold deposit, wall rock alteration, pyritization> and <Songjiang gold deposit, wall rock alteration, chloritization>, which need to be repeated twice. The dataset constructed in this paper contained a total of 80,011 triples, of which 60,000, 10,000, and 10,011 were randomly selected as the training dataset, validation dataset, and test dataset, respectively. Based on the output, relations were defined under the guidance of domain experts. In terms of the definition of the relation type, this paper divided the entity-relation based on the results of DL into two categories: (1) the relationship between ore deposit entities and their geological characteristic entities indicates that geological characteristics play a role in the ore deposits, and the relationship points from the deposit name to the geological characteristic entities such as wall_rock_alteration, metallogenic_stage and secondary_fault; (2) the relationship between the geological characteristic entities such as contains, associated_with and occurs_in. Figure 5 shows the KG for the deposit model based on the algorithm in this paper. Finally, 24 predefined relations and 1 other relation are constructed, as shown in Table 5.

Table 5. The relation types for the geological KG.

Relation Category	Number of Triples		
	Training Dataset	Validation Dataset	Test Dataset
associated_with	23	4	3
belongs_to	86	8	17
contains	565	86	97
controls	93	15	10
deposit_type	60	8	9
dyke	557	81	90
extrusive_rock	3	1	3
fault_character	276	48	56
formation	86	10	11
formed_in	244	35	42
geologic_time_scale	422	64	53
geotectonic_location	120	16	15
Group	131	22	22
intrusive_rock	177	37	39
located_in	46	11	12
metallogenic_stage	45	11	9
metamorphic_rock	302	53	55
Mineral	584	101	97
occurs_in	543	94	85
orebody_shape	118	22	17
Other	55,057	9206	9195
pluton	24	4	5
regional_fault	26	6	2
secondary_fault	153	24	21
wall_rock_alteration	259	44	35

Word embedding. According to the characteristics of the Chinese language, the Jieba toolkit in Python was used to segment sentences to improve the performance of the model. Due to the large number of proper nouns in the geological field, to improve the accuracy of word segmentation, this paper used the trained Bi-LSTM-CRF model to identify the entities in the prediction dataset in the corpus, added these entities to the domain dictionary, and built a dictionary of stopping words to reduce the word segmentation errors.

Word vectors were trained using the continuous bag of words (CBOW) model in word2vec. The CBOW model parameters were set as follows: the sliding window size was 5, the number of training iterations was 5, and the dimensions of the word vector were 50, 100, and 200. Under the supervision of the domain dictionary, all 15,816 sentences in the corpus were segmented and trained to obtain 13,645 word vectors.

Model test. In this paper, the Att-Bi-LSTM model proposed by Zhou et al. (2016) [67], the CNN model proposed by Nguyen et al. (2015) [103], and the RNN model proposed by Zhang et al. (2015) [104] were used for relation extraction. The effect of different word vector dimensions on the training accuracy was verified. The three models used the same hyperparameters, as shown in Table 6. Among them, the heights of the filters used in the CNN were 3, 4, and 5.

Table 6. The model parameters.

Hyperparameters	Value
Batch size	32
Learning rate	0.001
Epochs	100
Word embedding dimension	50, 100, 200
Size of hidden state	256
Size of position embedding	50
Dropout_rate	0.5
Optimizer	adadelta

Experimental. As seen in Table 7, the ACC, P, R, and F1 scores of the Att-Bi-LSTM model were slightly improved compared with those of the other two models. In contrast, this method could better extract the relationship information between geological entities. Experimental results demonstrated that for the dataset used in this paper, the larger the word vector dimension, the better the training effect. When the word vector dimension was 200, the F1 score of the Att-Bi-LSTM model could reach 91.01%.

Table 7. The test results of the relation extraction models.

Word Embedding Dimension	Model	ACC (%)	P (%)	R (%)	F1 (%)
50	CNN	97.95	88.96	82.21	85.46
100		98.19	87.18	88.81	87.99
200		98.27	87.15	90.30	88.70
50	Bi-LSTM + Pooling	98.25	86.82	90.92	88.82
100		98.40	89.84	89.05	89.44
200		98.36	87.16	92.04	89.53
50	Att-Bi-LSTM	98.59	90.06	90.17	90.12
100		98.48	88.21	**92.16**	90.15
200		**98.67**	**90.73**	91.29	**91.01**

Note: The best results are highlighted in bold.

4.4. Knowledge Fusion

Data cleaning. AI can maximize the exploration of entities and relations in texts to save labor costs. In this paper, the trained Bi-LSTM-CRF and Att-Bi-LSTM models were used to extract entities and relations from the predicted dataset, respectively. After integrating with the training dataset, the triples for constructing the KG were formed. As shown in Table 8, there are two main problems. First, relatively few relations are effectively annotated. More than 90% of the relations in the dataset were labeled Other, meaning that the relations between entities in these sentences were not predefined. Second, the relation types were not balanced. Common relations have more labeled sentences, while some uncommon relations only have a few labeled sentences.

Table 8. The data statistics of the geological KG in the JGMB.

Name	Labeled Dataset	Unlabeled Dataset	Total
Number of entities	20,187	25,514	45,701
Number of relations	80,011	101,049	181,060
Number of predefined relations	6553	8803	15,356
Number of sentences	7014	8802	15,816
Number of characters	527,449	671,489	1,198,938

Therefore, the triples predicted by the model had redundant or even wrong information, and the relations should be further cleaned. At this stage, the following processes were carried out: (1) the same triplets were combined, 15,356 were integrated into 7295 non-repeated triples, and the occurrence times of each triplet were recorded; and (2) according to predefined entity categories and relation categories, we used Neo4j to establish nodes and relations between nodes, and exported them using the Export CSV statement. The version of the Neo4j Community Edition used in this work was 4.2.5. This step can remove these incorrect triples. For example, the semantic relation of triples <Jiapigou gold deposit, dyke, rhyolite> was not in the predefined semantic relationship. After this step, 5168 valid triples remain after cleaning.

Entity alignment. After data cleaning, entities from different sources need to be aligned to ensure the reliability of the KG. The entity alignment method used in this paper included five components. (1) In the face of a large number of entities in the corpus, if all entities are considered as candidate entities and the similarity is compared one by one, the efficiency of the entity alignment will be affected. To improve efficiency, this paper classified entities according to their categories and aligned entities of the same category to establish a candidate set. (2) Word vectors that can express the semantic information of geological entities are generated according to the corpus. The experiments in Section 4.3 have proven that when the dimension of the word vector is set to 200, it can express more semantic information. Therefore, the dimension of the word vector used for entity alignment was also set to 200. (3) According to the generated word vector, the semantic similarity of different entities was calculated. (4) According to the edit distance of the entity name, the similarity of different entities is calculated. (5) The joint entity alignment results were calculated and evaluated according to vector similarity and edit distance similarity.

Comparison of model architectures. Since there were no labeled data, to verify the effectiveness of the proposed entity alignment method, we annotated the entity alignment candidate set. If two entities pointed to one entity, we labelled it as 1. Otherwise, it was labeled 0, and a total of 28,824 annotated data were obtained. The entity alignment methods based on edit distance and word vector proposed in this paper were all unsupervised algorithms. We conducted experiments and compared the effects of these methods on the same set.

The experimental results of the ACC, P, R, and F1 scores of each entity alignment method are shown in Table 9. Table 9 shows that the joint entity alignment method used in this paper achieved the best effect by setting the threshold as 0.67. The experimental results showed that the algorithm worked best when α was set to 0.1, as shown in Figure 6. The method can also be ranked according to the similarity score, which is convenient for experts to further screen, save time, and improve accuracy. The principle of entity alignment used in this work is that if A is aligned with B and B is aligned with C, then the A and C entities can be considered as aligned regardless of whether the similarity score between A and C exceeds the threshold. Finally, the knowledge base of the KG was verified manually, and then the KG was constructed by Neo4j. There were 350 nodes and 2181 edges in the KG.

Table 9. The performance of the models.

Model	Threshold	ACC (%)	P (%)	R (%)	F1 (%)
Edit distance	0.670	93.14	**95.15**	97.53	96.32
Word vector	0.965	85.94	92.18	92.53	92.36
The proposed method	0.670	**93.21**	95.03	**97.72**	**96.36**

Note: The best results are highlighted in bold, and the threshold is the optimization threshold of each method.

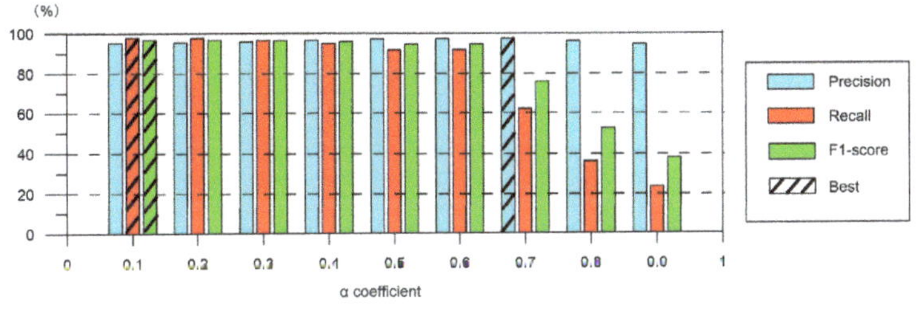

Figure 6. The performance of the α coefficient on the proposed method.

4.5. The Application of the KG

Due to the spatiotemporal nature of geoscience data, almost all entities can be visually displayed on the KG. Therefore, constructing the expression form of the KG for deposits based on the "entity-relation" can be used to indicate the complex coupling relationship between the geological characteristics related to gold deposits in the metallogenic belt. As shown in Figure 5, A1 and A2 are two nodes representing the fault structure, and the edge between them represents the comparative relationship between a secondary fault and the fault characteristics in a specific state. The fault characteristics of the secondary fault are constantly changing over time; for example, the Jiapigou fault may have three different fracture properties: brittle fault, ductile fault, and brittle–ductile fault. Therefore, there is a complex correlation between seemingly independent geological entities.

Extracting key geological entities. In the JGMB, two deposits with the same origin usually have highly similar geological characteristics. Two deposits of different origins may have different geological characteristics. These differences between deposits can be distinguished by KGs. In this paper, the idea of *TF-IDF* was analogized, and a key geological entity extraction method based on KGs for deposits was proposed.

TF-IDF is a statistical method [105]. *TF-IDF* is often used to assess how important a word is to a document. The more important the word, the more likely it is to be the keyword of the document. *TF-IDF* is often used in keyword extraction. In this work, the concept of *TF-IDF* was introduced into the importance identification of geological entities, where *TF* refers to the frequency of a geological entity appearing in all deposits, and *IDF* indicates the general importance of a geological entity to all deposits in the dataset.

By multiplying *TF* and *IDF*, the *TF-IDF* of all geological entities of each deposit can be obtained. The higher the importance that a geological entity is to the deposit, the greater its *TF-IDF* value, and the better discriminant ability of the entity for a given deposit. The formula is as follows:

$$TF \times IDF(i,j) = TF_{ij} \times IDF_i = \frac{n_{ij}}{\sum_k n_{kj}} \times \log(\frac{|D|}{1+|D_i|}) \quad (4)$$

where n_{ij} is the number of occurrences of geological entity i in deposit j, and $\sum\limits_{k} n_{kj}$ is the total number of occurrences of all geological entities in deposit j. $|D|$ is the total number of deposits, and $|D_i|$ is the number of deposits containing geological entity i. This method identifies the importance of each geological entity to all deposits and specific deposits by calculating the frequency of the occurrence of each geological entity in the KG. The entity of geological characteristics is the text type. Textual entities can represent the characteristics as a collection of characteristic words and then count their *TF-IDF* values.

Similarity and distance between deposits based on cosine similarity. An important problem in the KG for deposits is, given two deposits, how to assess whether they are similar and their degree of similarity. This paper introduces the idea of cosine similarity and proposes a calculation method of cosine similarity between deposits based on the *TF-IDF* value of geological entities. This similarity representation is close to the method of human understanding, and it is easy to add other external human knowledge with strong scalability. The core idea is to use the same geological entities and their *TF-IDF* values of different deposits in the KG to represent the similarity between deposits. Theoretically, the greater the same geological entities between the two deposits and the greater the *TF-IDF* value of the geological entities, the higher the similarity between the two deposits.

The steps to calculate the cosine similarity of ore deposits are as follows: (1) the geological entities of each ore deposit are counted; (2) geological entities are mapped into *TF-IDF* vectors; and (3) the similarity of ore deposits is compared pairwise by a nested loop. In this work, the sklearn toolkit in Python was used to calculate *TF-IDF* and cosine similarity, which were completed by the TfidfVectorizer statement and cosine_similarity statement, respectively.

Figure 7 illustrates the method by taking some geological entities of deposit A and deposit B as examples. The nodes in Figure 7 represent the two deposits and their geological entities, and the values on the edge represent the *TF-IDF* values of deposits with different geological entities. The similarity between deposit A and deposit B depends on the number of common geological entities and their *TF-IDF* values. After calculation, the cosine similarity between these two deposits was approximately 0.47. Gephi software [106] was used to visualize the distance of the similarity matrix. The greater the similarity, the closer the distance between deposits. The visualization of clustering results among typical gold deposits in the JGMB can be obtained.

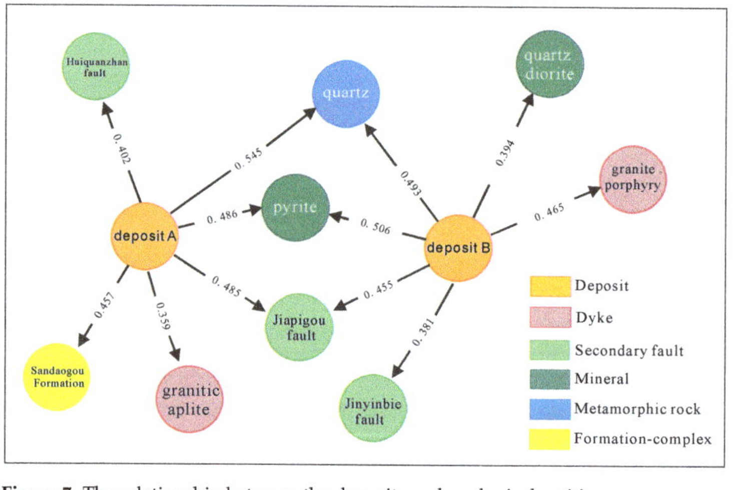

Figure 7. The relationship between the deposits and geological entities.

5. Application of the KG in the JGMB

5.1. Visualization of the KG

Regional magmatic rocks and dykes. There is a strong magmatic activity in the JGMB, and magmatic rocks of different ages and types are exposed including Archean TTG and phanerozoic granites and dykes. The exposed rocks in the study area include mainly granodiorite, granite, syenite, monzonite, granite porphyry, diabase, lamprophyre, and many Au-bearing quartz veins. Figure 8 clearly shows the relationship between the deposits and magmatic rocks in the study area. In terms of relationships, dykes are closely related to gold deposits and are an important magmatic rock condition for gold mineralization, as shown in Figure 9. These results show that magmatic evolution is closely related to regional tectonism and mineralization [2,3,107].

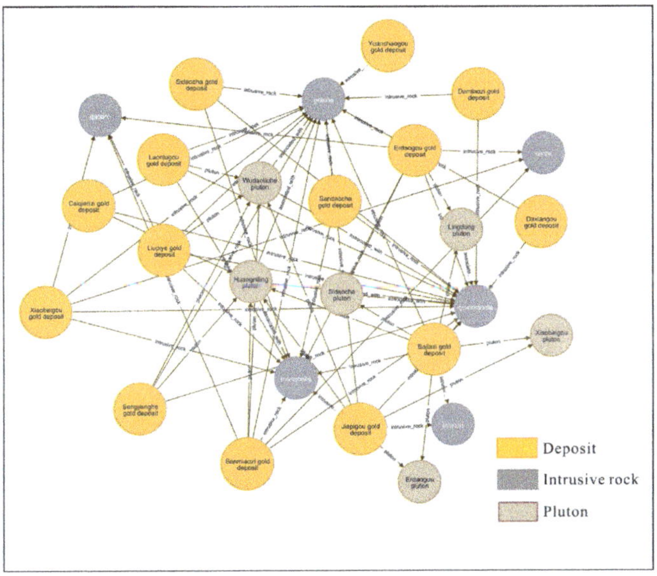

Figure 8. A subgraph for the plutons, intrusive rocks, and ore deposits.

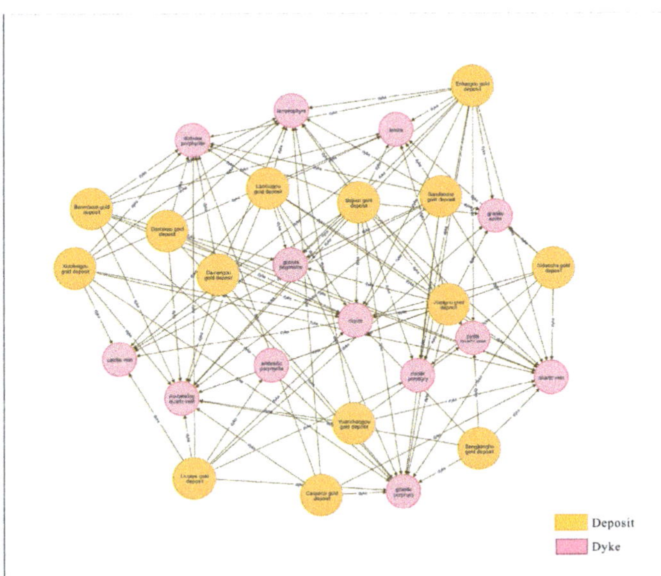

Figure 9. A subgraph for the dykes and ore deposits.

Regional metamorphic rocks, extrusive rocks, sedimentary rocks. The metamorphic rocks exposed in the study area mainly include mylonite, gneiss, schist, slate, and granulite; the extrusive rocks mainly include andesite, basalt, and rhyolite; and the sedimentary rocks mainly include limestone, mudstone, and glutenite. The relationship between gold deposits and the outcropping rocks of each stratum is shown in Figure 10.

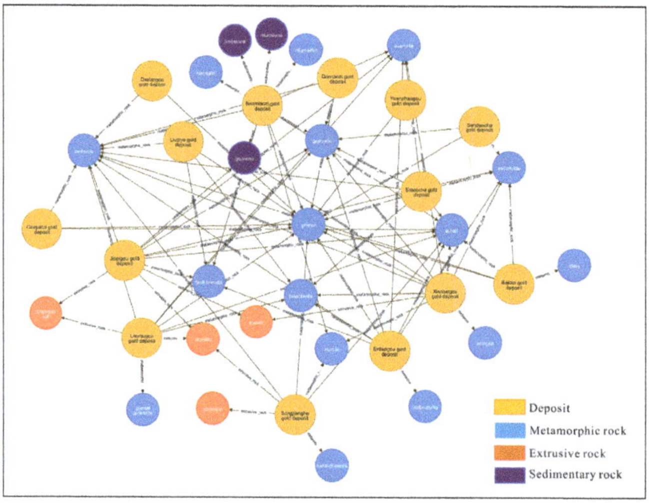

Figure 10. A subgraph for metamorphic rocks, extrusive rocks, sedimentary rocks, and ore deposits.

Regional structures. The regional structure of the JGMB includes folds and faults, as shown in Figure 11. Among them, the fold structures are the Banmiaozi anticline, Hongqiling syncline, Bajiazi anticline, etc. The regional faults include the Huifahe and Ji'an-Songjiang faults. Among them, the Huifahe fault activity is the most intense. The distribution direction of Mesozoic and Cenozoic faulted basins, sedimentary formations, and late Yanshanian granites in the area are controlled by this fault [2].

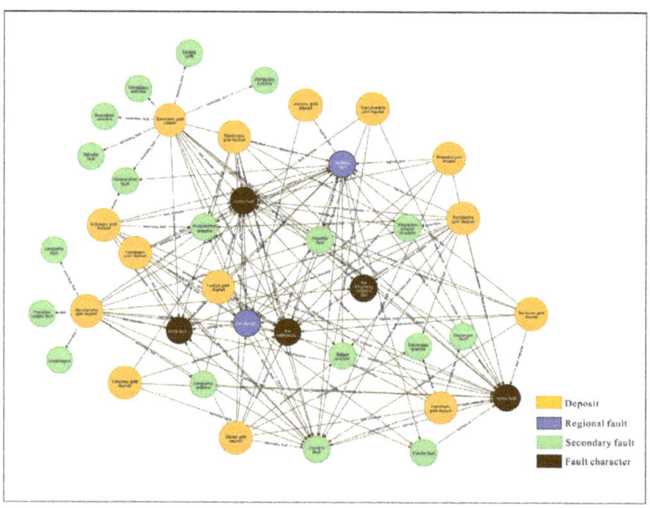

Figure 11. A subgraph for the regional fault, secondary fault, fault character, and ore deposits.

Three parallel faults are successively distributed from NE to SW: the Fuerhe, Jinyinbie, and Jiapigou faults; these faults are part of the secondary Huifahe fault. Among them, the Jiapigou fault controls the spatial distribution of deposits in this metallogenic belt.

The analysis of metallogenic-tectonic settings is an important way to distinguish the types of deposits. As shown in Figures 1c and 11, from the ore-controlling structural conditions, each deposit in the western section of the JGMB is adjacent to the Huifahe fault. The Songjiang gold deposit, located in the eastern section of the JGMB, is controlled by the Jinyinbie fault and has the characteristics of ductile activity. The gold deposits in the metallogenic belt are generally superimposed on brittle–ductile deformation and brittle deformation based on early ductile deformation, and the ore bodies are located mainly in the late superimposed brittle fault system.

5.2. Extraction of Key Geological Characteristic Entities

For two similar deposits, the coincidence rate of geological entities may be very high. Therefore, this paper explored whether the characteristics of different deposits could be accurately distinguished by geological entities and their relations. This section uses all of the entities in the KG to calculate the *TF-IDF* values of the geological entities of each deposit. Table 10 lists the geological entities of the top 10 *TF-IDF* values of each deposit and arranges them from high to low according to the *TF-IDF* values.

Table 10. The *TF-IDF* values of the top 10 geological characteristic entities.

Number	*TF-IDF*	Deposit	Geological Characteristics Entities
1	0.632	Songjianghe	Seluohe Group, Jurassic, Permian, ductile fault, Jinyinbie fault, Proterozoic, Dunhua City, schist, Triassic, brittle–ductile fault
2	0.446	Banmiaozi	Diorite, Zhenzhumen Formation, Diaoyutai Formation, fault breccia, Jiapigou block, diabase, marble, granite, Jiapigou fault, Huadian City
3	0.432	Laoniugou	Jiapigou block, gneiss, Laoniugou Formation, Archean, granulite, amphibolite, Sandaogou Formation, granite, diorite, quartzite
4	0.431	Liupiye	Granite, brittle–ductile fault, Archean, diorite, diabase, Jiapigou block, gabbro, Jurassic, Mesozoic, Biotite
5	0.403	Yuanchaogou	Pyrite, quartz vein, galena, Au-bearing quartz vein, sphalerite, Huadian City, Jiapigou block, Laoniugou Formation, lamprophyre, chalcopyrite
6	0.399	Damiaozi	Granite, Jiapigou fault, Huadian City, Jiapigou block, ductile fault, Sandaogou Formation, granodiorite, Archean, lamprophyre, breccia
7	0.393	Daxiangou	Archean, Jiapigou granite-greenstone belt, lenticular, ductile fault, diorite, granite porphyry, brittle fault, Jiapigou block, Daxiangou syncline, vein
8	0.390	Xiaobeigou	Gneiss, Jiapigou block, granite, quartz vein-type, quartz vein, Archean, ductile fault, quartz, Jiapigou fault, Jiapigou Group
9	0.385	Erdaogou	Diorite, quartz vein, gneiss, quartz, Archean, Jiapigou block, zircon, quartz vein-type, granodiorite, granite
10	0.370	Jiapigou	Archean, Mesozoic, granite, ductile fault, Jiapigou block, Jiapigou fault, quartz vein, Huadian City, quartz, gneiss
11	0.353	Caiqiangzi	Schist, mylonite, Jiapigou fault, compressional structure, Huadian City, diabase, Archean, granite, silicification, granodiorite
12	0.351	Bajiazi	Granite porphyry, pyrite, quartz, diorite, zircon, Triassic, quartz vein, granite aplite, quartz vein-type, Biotite
13	0.344	Sidaocha	Quartz vein, Archean, Au-bearing quartz vein, quartz vein-type, granite porphyry, gneiss, Jiapigou fault, diorite, amphibolite, Sandaogou Formation
14	0.313	Sandaocha	Brittle fault, granite porphyry, quartz vein, diorite, quartz vein-type, quartz, ductile fault, Archean, pyrite, Jiapigou fault

Although the valuable information contained in the text is often related to high-frequency content words [108], some low-frequency words are also important [109]. *TF-IDF* is a method to extract low-frequency words that contain important information, and to evaluate the importance of a word to a document. In other words, when a geological entity occurs many times in one deposit, but rarely in others, it has some discriminative power. These geological entities can be called key geological entities.

The *TF-IDF* values of geological entities will change with the size of the dataset. To quantify this difference, this work took the complete dataset of 14 deposits and all of their geological entities as a benchmark, and changed the size of the dataset by reducing a certain number of typical deposits and their geological entities. The mean square error (*MSE*) index was used to measure this bias.

$$MSE = \frac{1}{m}\sum_{i=1}^{m}(x_i - y_i)^2 \tag{5}$$

where m is the total number of geological entities; x_i is the *TF-IDF* value of each geological entity in the complete dataset; and y_i is the *TF-IDF* value of the corresponding geological entity after changing the dataset. The blue line in Figure 12 is the result of the quantification test using the MSE index. The data in the figure show that when the dataset contained 13 deposits, the MSE index was 6.668×10^{-7}, and the MSE deviation from the complete dataset was relatively minimal. In contrast, when the dataset contained five deposits, the MSE index was 432×10^{-7}. The overall trend was that the relative change in the MSE index was greater with the decrease in ore deposits. To ensure the accuracy of the experiment, we obtained the above experimental results by randomly reducing the range of the deposits and repeating the experiment several times to take the average.

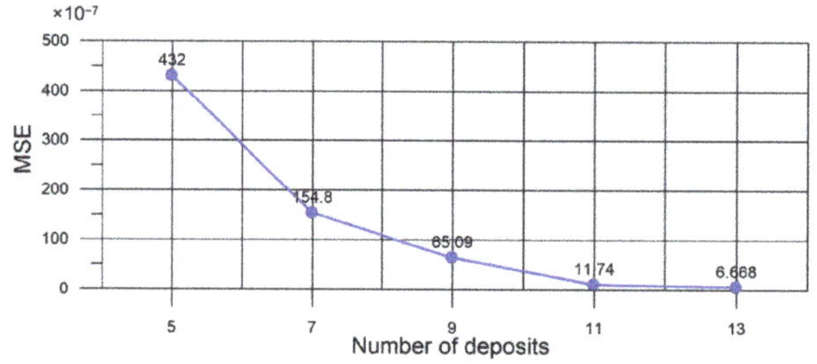

Figure 12. The MSE deviation of the *TF-IDF* values among different datasets.

5.3. Similarity and Distance between Deposits

The principles of ore deposit type identification are as follows: (1) If the cosine similarity between two deposits is greater than the threshold value, they are of similar type; (2) if the cosine similarity between deposit A and deposit B is greater than the threshold, and the cosine similarity between deposit B and deposit C is also greater than the threshold, it is considered that the types of deposit A, deposit B, and deposit C are similar; and (3) if the cosine similarity between deposit A and any deposit in the dataset is less than the threshold, it is considered that this deposit is not similar to other deposits.

The Jaccard coefficient and cosine similarity were used to compare the similarity recognition accuracy of different methods for deposits. The experimental results are shown in Figure 13. The cosine similarity calculation results are as follows. When all geological entities were used, 13 deposits were correctly identified when the threshold value was set as 0.6. Only the Liupiye gold deposit, as an altered rock-type gold deposit, was connected to other quartz vein-type gold deposits with cosine similarities greater than 0.6, which is not consistent with the known results, so the overall accuracy was 92.86%. When the ten geological entities with the highest *TF-IDF* value were used, the recognition accuracy was also 92.86%. After removing the geological entities with the highest ten *TF-IDF* values of each deposit, the accuracy of cosine similarity between deposits was reduced to 42.86%.

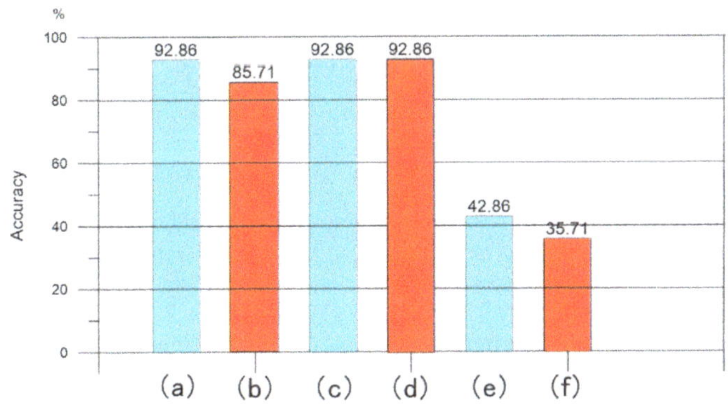

Figure 13. The accuracy of different similarity calculation methods: (**a**) the recognition accuracy of the cosine similarity using all geological entities with their *TF-IDF* values; (**b**) the recognition accuracy of the Jaccard coefficient using all geological entities; (**c**) the recognition accuracy of the cosine similarity using the geological entities with the top ten *TF-IDF* values; (**d**) the recognition accuracy of the Jaccard coefficient using the geological entities with the top ten *TF-IDF* values; (**e**) the recognition accuracy of the cosine similarity after removing the geological entities with the top ten *TF-IDF* values; (**f**) the recognition accuracy of the cosine similarity after removing the geological entities with the top ten *TF-IDF* values.

When all geological entities of each deposit were used, the recognition accuracy of the Jaccard coefficient was 85.71%. The Damiaozi gold deposit and Sidaocha gold deposit, as quartz vein-type gold deposits, have no connection with other quartz vein-type gold deposits with a similarity greater than 0.6. When using the ten geological entities with the highest *TF-IDF* value of each deposit, the recognition accuracy increased to 92.86%. After removing the geological entities with the highest ten *TF-IDF* values of each deposit, the accuracy of the Jaccard similarity between deposits decreased to 35.71%.

In conclusion, the cosine similarity is more accurate than the Jaccard coefficient in measuring the similarity between ore deposits. When the ten key geological entities with the highest *TF-IDF* value of each deposit were used, the accuracy of the similarity recognition results between deposits was higher. This is because the Jaccard coefficient does not account for the number of occurrences of geological entities in different deposits, nor does it consider the problem that key geological entities may bring more information. The vector space model accounts for the two points, so the cosine similarity has a better effect. The geological entity with a higher *TF-IDF* value is more inclined to be the main identification feature of the deposit, which has a greater impact on the similarity of the deposit. However, the geological entity with a low *TF-IDF* value has weak pertinence for the deposit and is difficult to use as the main deposit identification feature.

The *TF-IDF* values listed in Table 10 are the highest *TF-IDF* values for each deposit. Taking the Yuanchaogou gold deposit as an example, pyrite, galena, quartz veins, and gold-bearing quartz veins describe the important metallogenic characteristics of this deposit. The known data show that natural gold is distributed mainly in the microcracks of pyrite. The quartz pyrite stage is the main mineralization stage of the Yuanchaogou gold deposit, and the Au-bearing symbiotic association is natural gold, pyrite, chalcopyrite, and galena. Therefore, the analysis based on the *TF-IDF* values can better evaluate the most significant geological characteristics of each deposit. The number of geological entities in each deposit is shown in Figure 14.

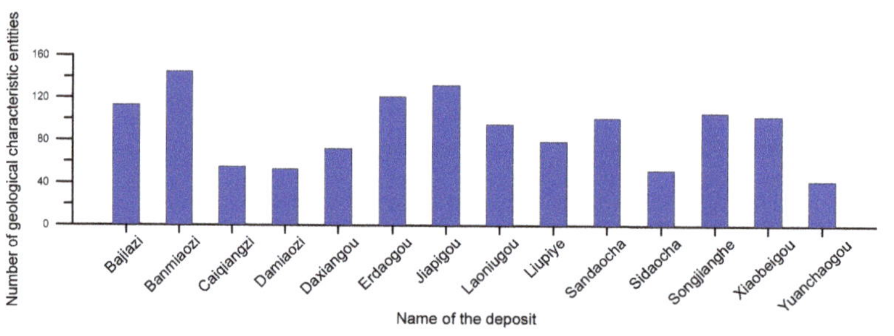

Figure 14. The distribution of the geological characteristic entities in the KG for each deposit.

To verify the effectiveness of the similarity calculation method between deposits based on the KG, the similarity of the typical deposits in the JGMB was calculated, and the types of these deposits were analyzed. The cosine similarity between deposits was calculated according to the *TF-IDF* value of the geological entities. Figure 15 shows the calculated clustering effect of typical deposits. In this figure, the nodes represent different deposits, while the edges represent the relationships between deposits. The figure shows that the similarity between the Songjianghe gold deposit and other deposits was low, and the distance was large. The Jiapigou gold deposit and 12 other gold deposits were closely adjacent, and there were many connections with similarities greater than 0.6 between them. In the cluster diagram showing the deposit distance, the similarity between the Jiapigou, Erdaogou, Xiaobeigou, Bajiazi, Laoniugou, Sidaocha and Sandaocha gold deposits was greater than 0.7. By calculating the similarity and distance, the deposits could be divided into two categories: the Songjianghe gold deposit was classified into one category, and the other 13 deposits were classified into another category.

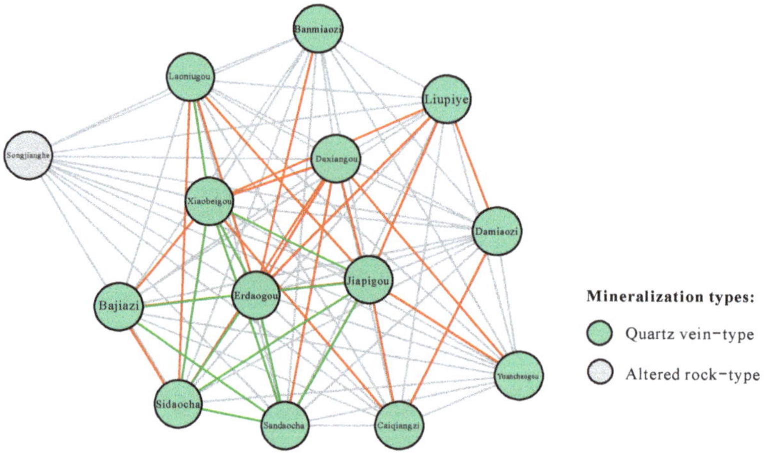

Figure 15. The clustering effect between typical deposits calculated by similarity. Colors of the connecting lines: green indicates similarity values ≥ 0.7, red indicates similarity values from 0.6–0.69, and gray indicates similarity values < 0.6.

To verify the accuracy of the proposed method for ore deposit type identification, the cosine similarity, Jaccard similarity coefficient, TransE, and the proposed method were compared. In the experiment, we used all of the geological entities of each deposit in the KG, and the similarity recognition accuracy of each method is shown in Figure 16. For the cosine similarity, we first converted the deposit names into 200-dimensional vectors according to the Word2Vec results in Section 4.3, and then calculated the similarity between them. The Jaccard similarity coefficient determines the similarity by calculating the ratio of the intersection and union of geological entities between two ore deposits. TransE, on the other hand, vectorizes geological entities and relationships in the KG into 200-dimensional vectors, and then calculates the cosine similarity between vectors. The experimental results showed that the proposed method based on the KG could accurately classify the deposits by distinguishing the differences in geological characteristics and then infer the origin of the exploration object in the research area by using the typical deposits of known types.

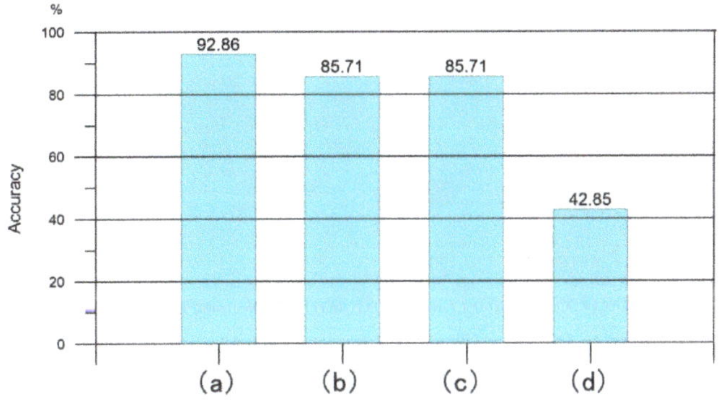

Figure 16. The accuracy comparison between the proposed method and the already studied methods: (**a**) the proposed method; (**b**) cosine similarity; (**c**) Jaccard coefficient; (**d**) TransE.

5.4. Visualization of the Regional Metallogenic Model Based on the KG

In prospecting prediction, due to different geological environments, there are also significant differences in metallogenic ore bodies, metallogenic structures, and metallogenic characteristics between deposits. The geological model for prospecting prediction is based on the existing mineral exploration, which fully expresses all of the known and inferred geological characteristics of the deposits and ore bodies in the exploration area and can effectively guide the deployment of prospecting engineering [110]. It is generally summarized in the form of drawings, words, and tables. Based on previous work, this paper explored the construction method of a metallogenic model based on KGs for gold deposits.

Taking the Songjianghe gold deposit as an example, 106 geological entities related to this deposit can be visualized through Neo4j. To facilitate a clear display, the geological entities of the top 69 *TF-IDF* values were selected to construct a visual metallogenic model of the Songjiang gold deposit based on the KG (Figure 17). The figure shows that the hierarchical information between geological features can be clearly expressed.

The ore-searching clues of the Songjianghe gold deposit based on KG are as follows: this deposit is located between the northeastern margin of the NCC and the Xingmeng orogenic belt and is controlled by the Ji'an-Songjiang, Huiquanzhan, Jiapigou, Jinyinbie and Fuerhe faults. Among them, the Jinyinbie fault has the highest *TF-IDF* value, making it the main rock- and ore-controlling structure in the mining area. The structural traces retained in this deposit are mainly ductile deformation and brittle–ductile deformation, and the fold structure in the area is mainly the Langchaihe anticline. The industrial type of ore bodies is altered rock-type. The host rock is mainly the Seluohe Group, and the lithology includes mainly gneiss and mylonite. Sulfide assemblages in ores include mainly pyrite, galena, chalcopyrite, and sphalerite. Wall rock alterations include silicification, sericitization, mylonitization, carbonation, chloritization, and potash feldspathization. The metallogenic stages are divided into the milky-quartz stage, quartz-pyrite stage, polymetallic sulfide stage, and quartz-calcite stage. Among them, the quartz-pyrite stage and polymetallic sulfide stage are the main enrichment and mineralization stages of gold.

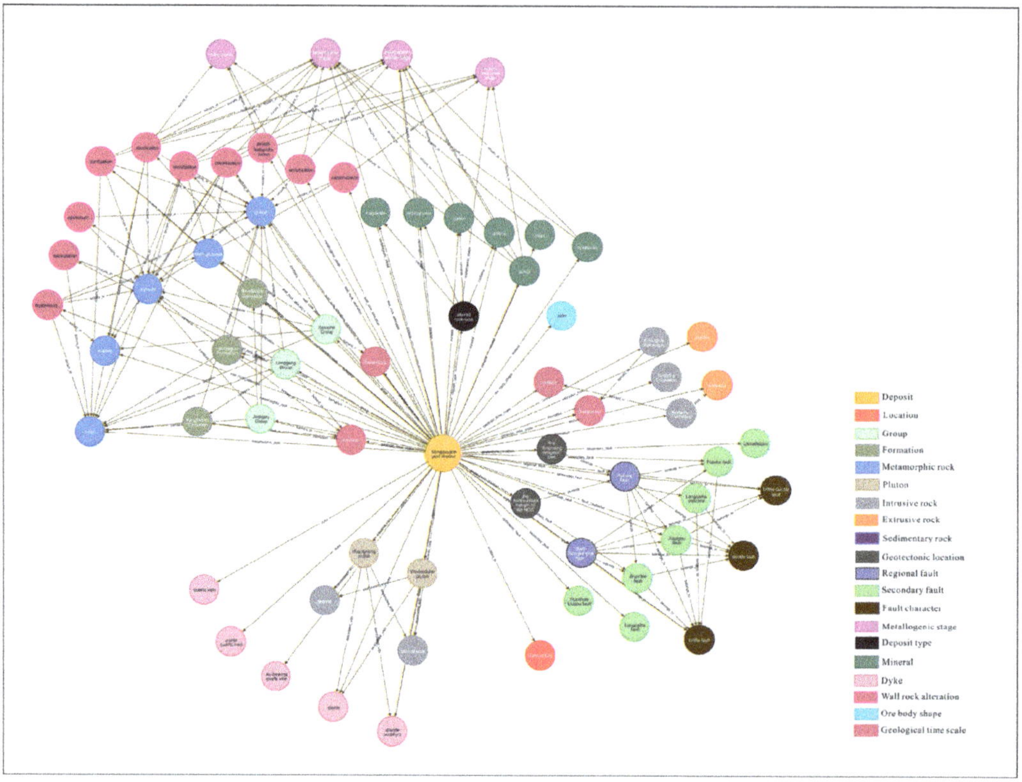

Figure 17. Visualization of the Songjiang gold deposit metallogenic model based on the KG.

Comparison of similar geological entities between different deposits. The Songjianghe, Jiapigou, and Liupiye gold deposits were selected for comparison, and the main similarities and differences between these three deposits were analyzed (Figure 18). According to the comparison of geological entities based on the KG, the mineralization characteristics of these three deposits are similar in general, but show some differences in the details. These were arranged based on the *TF-IDF* values, from large to small. The comparison of the main ore-controlling factors and metallogenic geological conditions of these three deposits is shown in Table 11.

The origin of ore deposits and prospecting criteria. The study of the KG for the deposits showed that the metallogenic geological conditions of the main gold deposits in the JGMB were similar, which revealed that these gold deposits were the products of the same gold mineralization and were related to the ductile shear zone. The difference in the spatial position and occurrence of each ore body was mainly caused by the secondary faults tracking different trends in the process of the rising and migration of the ore-forming fluid. There is a great possibility of altered rock-type ore bodies in the deep part of Jiapigou and other quartz vein-type gold deposits. The geological model for the prospecting prediction of the typical deposits in the JGMB based on KG is shown in Figure 19. Based on the analysis of the mineralization characteristics and ore-controlling conditions of each deposit, the geological characteristics of most typical deposits can be used as regional prospecting criteria. As shown in Table 12.

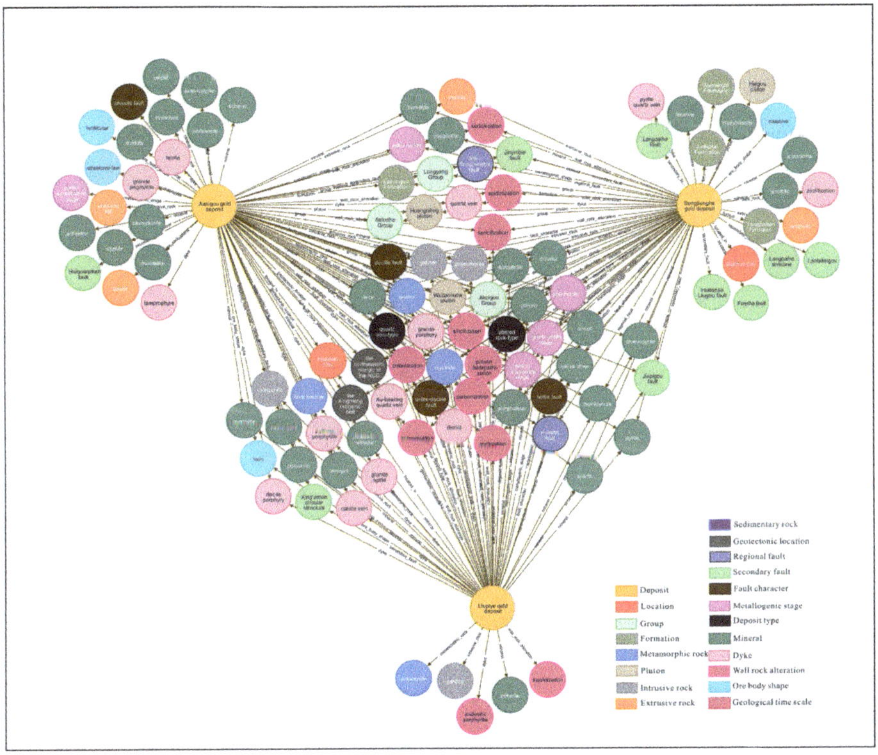

Figure 18. A comparison of the geological entities of the Songjianghe, Liupiye, and Jiapigou gold deposits based on the KG.

Table 11. A comparison of the main geological characteristics of the Jiapigou, Liupiye, and Songjianghe gold deposits.

Deposit	Jiapigou	Liupiye	Songjianghe
Country rocks	Gneiss, hornblende, mylonite, TTGs	Gneiss, mylonite, TTGs	Mylonite, gneiss, hornblende, TTGs
Intrusive rocks and dykes	Granite, quartz vein, diorite, granite porphyry, granite pegmatite, granodiorite, Au-bearing quartz vein	Granite, diorite, diabase dyke, gabbro, granodiorite, granodiorite, granite porphyry, Au-bearing quartz vein	Granite porphyry, diorite, granite, granodiorite
Structures	Jiapigou fault, Huiquanzhan fault, Jinyinbie fault, Xing'antun circular structure	Jiapigou fault, Xing'antun circular structure	Jinyinbie fault, Jiapigou fault, Langcaihe anticline
Mineralization types	Quartz vein-type	Altered rock-type	Altered rock-type
Metal minerals	Pyrite, sphalerite, galena, chalcopyrite, magnetite	Pyrite, chalcopyrite, galena	Pyrite, molybdenite, chalcopyrite, sphalerite
Ore body shapes	Vein, stratiform-like, lenticular	Vein	Vein
Wall rock alterations	Silicification, sericitization, chloritization, potassium, pyritization	Pyritization, clayization, carbonatization, sericitization, potassium, mylonitization	Silicification, sericitization, mylonitization, carbonatization, chloritization, epidotization, potassium, pyritization, boilerization
Main metallogenic stages	Quartz-pyrite stage, polymetallic sulfide stage	Quartz-pyrite stage, polymetallic sulfide stage	Quartz-pyrite stage, polymetallic sulfide stage

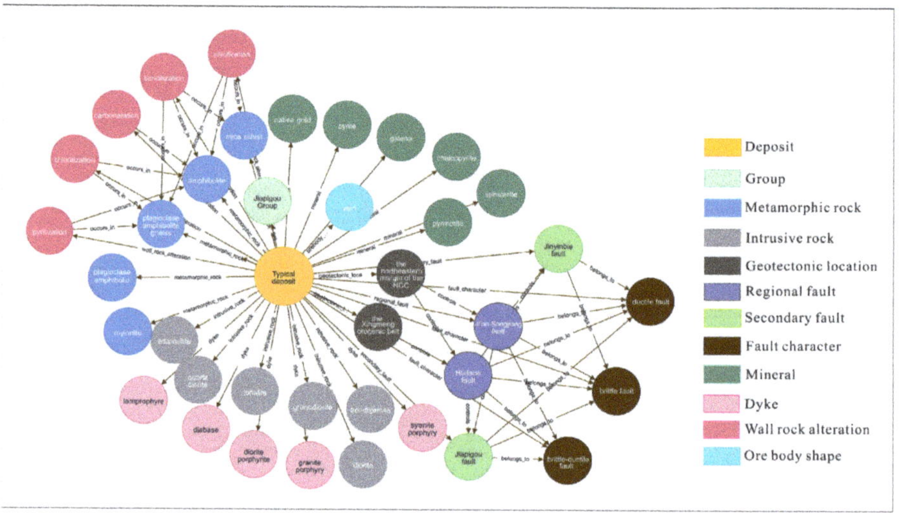

Figure 19. The geological model for the prospecting prediction of typical deposits in the JGMB based on the KG.

Table 12. Regional prospecting criteria of the JGMB.

	Regional Prospecting Criteria
Country rocks	TTGs, amphibolite, gneiss, Mesozoic granite.
Wall rock alterations	Silicification, carbonation, sericitization, chloritization, pyritization
Dykes	Syenite porphyry, lamprophyre, diabase, diorite, diorite porphyrite
Minerals	Natural gold, pyrrhotite, pyrite, chalcopyrite, galena, sphalerite
Strata	Jiapigou Group, Seluohe Group
Structures	Jiapigou fault, Jinyinbie fault
Ore body shape	Vein

6. Discussion

6.1. Benefits

The KG of deposits based on DL and NLP can reveal the relationship between the earth system and the origin of ore deposits, automatically extract geological characteristic entities, discover metallogenic regularity, and help researchers quickly analyze mineralization information. Based on the lack of quantitative interpretation and analysis based on the KG in the existing research, this paper proposed a visualization and interpretation method based on the KG, and the excellent effect of this method was shown through experiments. The work of this paper had two main advantages. First, the similarity between KG models for each deposit was quantified. Second, a more comprehensive visual interpretation system was constructed for the metallogenic KG model for deposits. In this work, the geological entities of strata, structures, magmatic rocks, metamorphic rocks, and wall rock alterations in the study area were collected. According to these basic data and KGs, the origin of the exploration object can be determined by a comparison with those deposits whose types are known.

6.2. Limitations

The entities and relationship types extracted in this work were relatively limited. To use the KG for deposits in a wider range, we need to strengthen the construction of the geological dictionary and model database. If we want to study the relationship between regional magmatic evolution and mineralization, we also need to increase the construction of the KG model based on geochemical data.

6.3. Compared with the Previous Work

At present, the construction and application of deposit KGs are only starting, and the construction method of KGs is not mature. This work was based on the classification method of the KG for the deposits, considered the relationship between geological entities, and calculated the similarities of the deposits based on geological facts. The construction of the KG for typical deposits in the JGMB was systematically studied. This paper discusses the feasibility and importance of the KG in the study of gold mineralization prediction, solves the problem of screening prospecting criteria, and expands the methods of prospecting prediction.

6.4. Future Work

In the future, it is also necessary to increase the ability of cross-language geological characteristic entities and the relationship extraction of KGs. The prospecting criteria of geophysics, geochemistry, and remote sensing geology have been established to express the relationship more accurately between deposits. The comprehensive and systematic geoscience KG has wide application prospects. It can not only deepen the existing big data geoscience analysis, but also expand the prospecting space, help to find different types of gold deposits, and expand the target minerals from gold to silver, copper, tungsten, and other minerals. By studying ore-controlling factors, prospecting criteria, and metallogenic information, various metallogenic models can be summarized to further clarify the prospecting direction and guide the prediction of the target area. Then, this information can be extended to the construction of the KG of the Laoling metallogenic belt and Yanbian Mesozoic tectonomagmatic rock belt around the JGMB.

7. Conclusions

In this work, through the knowledge acquisition, annotation, and extraction of entities and the relationships and attributes in the texts of typical gold deposits in the JGMB, a KG for gold deposits was constructed. Based on the calculation of the *TF-IDF* index of the geological characteristic entities of each deposit in the KG, the similarities and distances between different deposits were calculated. In this work, the accuracy of NER model was 91%, the accuracy of relationship extraction model was 92%, the accuracy of entity alignment model was 92%, and the accuracy of deposit type recognition based on the KG was 92%. The results show that the method proposed in this work can distinguish the differences in the geological characteristics of different gold deposits and then accurately classify the types of deposits through cosine similarity. Based on the analysis of the mineralization characteristics and metallogenic geological conditions of the deposits in the JGMB, the geological characteristics of most typical deposits can be used as prospecting criteria. Brittle and brittle–ductile faults are widely developed in the gold deposit in the western section of the JGMB. The main industrial type of deposits in this area is the quartz vein type. There may be altered rock-type gold deposits in the deep part of the known deposits.

Author Contributions: Conceptualization, Y.P. and S.C.; Methodology, Y.P.; Software, Y.P.; Validation, J.C.S., C.M. and H.C.; Formal analysis, Y.G.; Investigation, X.L.; Data curation, R.L.; Writing—original draft preparation, Y.P.; Writing—review and editing, S.C.; Visualization, Y.P.; Supervision, S.C. All authors have read and agreed to the published version of the manuscript.

Funding: This research was supported by the Science Advances Innovation Project of Jilin University (Grant No. 419080600070), the Jilin Province Higher Education Research Project (Grant No. JGJX2022B5), the Strategic Research Project of Science and Technology Commission of the Ministry of Education (Grant No. 20210602), and the College Students' Innovative Entrepreneurial Training Plan Program (Grant No. S202210183421).

Data Availability Statement: Not applicable.

Acknowledgments: We are very thankful to all of the editors and reviewers who have helped us improve and publish this paper.

Conflicts of Interest: The authors declare no conflict of interest.

References

1. Han, J.L.; Sun, J.G.; Liu, Y.; Zhang, X.T.; He, Y.P.; Yang, F.; Chu, X.L.; Wang, L.L.; Wang, S.; Zhang, X.W.; et al. Genesis and age of the Toudaoliuhe breccia-type gold deposit in the Jiapigou mining district of Jilin Province, China: Constraints from fluid inclusions, H-O-S-Pb isotopes, and sulfide Rb-Sr dating. *Ore Geol. Rev.* **2020**, *118*, 103356. [CrossRef]
2. Miao, L.C.; Qiu, Y.M.; Fan, W.M.; Zhang, F.Q.; Zhai, M.G. Geology, geochronology, and tectonic setting of the Jiapigou gold deposits, southern Jilin Province, China. *Ore Geol. Rev.* **2005**, *26*, 137–165. [CrossRef]
3. Deng, J.; Yang, L.Q.; Gao, B.F.; Sun, Z.S.; Guo, C.Y.; Wang, Q.F.; Wang, J.P. Fluid Evolution and Metallogenic Dynamics during Tectonic Regime Transition: Example from the Jiapigou Gold Belt in Northeast China. *Resour. Geol.* **2009**, *59*, 140–152. [CrossRef]
4. Zeng, Q.D.; Wang, Z.C.; He, H.Y.; Wang, Y.B.; Zhang, S.; Liu, J.M. Multiple isotope composition (S, Pb, H, O, He, and Ar) and genetic implications for gold deposits in the Jiapigou gold belt, Northeast China. *Miner. Deposita* **2014**, *49*, 145–164. [CrossRef]
5. Zhang, X.T. Research on Geology, Geochemistry and Metallogenesis of the Gold Deposits of the Jiapigou Ore Field in the Continental Margin of Northeast China. Ph.D. Thesis, Jilin University, Changchun, China, 2018.
6. Yang, L.Y.; Yang, L.Q.; Yuan, W.M.; Zhang, C.; Zhao, K.; Yu, H.J. Origin and evolution of ore fluid for orogenic gold traced by D-O isotopes: A case from the Jiapigou gold belt, China. *Acta Petrol. Sin.* **2013**, *29*, 4025–4035, (In Chinese with English Abstract).
7. Deng, J.; Yuan, W.M.; Carranza, E.J.M.; Yang, L.Q.; Wang, C.M.; Yang, L.Y.; Hao, N.N. Geochronology and thermochronometry of the Jiapigou gold belt, northeastern China: New evidence for multiple episodes of mineralization. *J. Asian Earth Sci.* **2014**, *89*, 10–27. [CrossRef]
8. Goldfarb, R.J.; Groves, D.I. Orogenic gold: Common or evolving fluid and metal sources through time. *Lithos* **2015**, *233*, 2–26. [CrossRef]
9. Li, L.; Sun, J.G.; Men, L.J.; Chai, P. Origin and evolution of the ore-forming fluids of the Erdaogou and Xiaobeigou gold deposits, Jiapigou gold province, NE China. *J. Asian Earth Sci.* **2016**, *129*, 170–190. [CrossRef]
10. Mao, J.W.; Zhang, Z.H.; Wang, Y.T.; Jia, Y.F.; Kerrich, R. Nitrogen isotope and content record of Mesozoic orogenic gold deposits surrounding the North China craton. *Sci. China Ser. D* **2003**, *46*, 231–245. [CrossRef]
11. Dai, J.Z.; Wang, K.Y.; Cheng, X.M. Geochemical features of ore-forming fluids in the Jiapigou gold belt, Jilin province. *Acta Petrol. Sin.* **2007**, *23*, 2198–2206.
12. Yu, J.J.; Wang, F.; Xu, W.L.; Gao, F.H.; Tang, J. Late Permian tectonic evolution at the southeastern margin of the Songnen-Zhangguangcai Range Massif, NE China: Constraints from geochronology and geochemistry of granitoids. *Gondwana Res.* **2013**, *24*, 635–647. [CrossRef]
13. Wu, F.Y.; Zhao, G.C.; Sun, D.Y.; Wilde, S.A.; Yang, J.H. The Hulan Group: Its role in the evolution of the Central Asian Orogenic Belt of NE China. *J. Asian Earth Sci.* **2007**, *30*, 542–556. [CrossRef]
14. Li, C.D.; Zhang, F.Q.; Miao, L.C.; Jie, H.Q.; Hua, Y.Q.; Xu, Y.W. Reconsideration of the Seluohe Group in Seluohe area, Jilin Province. *J. Jilin Univ. Earth Sci. Ed.* **2007**, *37*, 841–847, (In Chinese with English Abstract).
15. Zeng, Q.D.; Wang, Z.C.; Zhang, S.; Wang, Y.B.; Yang, Y.H.; Liu, J.M. The ore-forming epoch of the Jiapigou gold belt, NE China: Evidences from the zircon LA–ICP–MS U–Pb dating of the intrusive rocks. *Acta Geol. Sin.* **2014**, *88* (Suppl. S2), 1029–1030, (In Chinese with English Edition). [CrossRef]
16. Sarker, I.H.; Hoque, M.M.; Uddin, M.K.; Alsanoosy, T. Mobile Data Science and Intelligent Apps: Concepts, AI-Based Modeling and Research Directions. *Mobile Netw. Appl.* **2021**, *26*, 285–303. [CrossRef]
17. Singhai, A. Introducing the Knowledge Graph: Things, Not Strings. Available online: https://blog.google/products/search/introducing-knowledge-graph-things-not/ (accessed on 16 May 2012).
18. Ma, X.G.; Ma, C.; Wang, C.B. A new structure for representing and tracking version information in a deep time knowledge graph. *Comput. Geosci.* **2020**, *145*, 104620. [CrossRef]
19. Wang, C.S.; Hazen, R.M.; Cheng, Q.M.; Stephenson, M.H.; Zhou, C.H.; Fox, P.; Shen, S.Z.; Oberhansli, R.; Hou, Z.Q.; Ma, X.G.; et al. The Deep-Time Digital Earth program: Data-driven discovery in geosciences. *Natl. Sci. Rev.* **2021**, *8*, nwab027. [CrossRef]
20. Cheng, B.J.; Zhang, J.; Liu, H.; Cai, M.L.; Wang, Y. Research on Medical Knowledge Graph for Stroke. *J. Healthc. Eng.* **2021**, *2021*, 5531327. [CrossRef]
21. Yu, C.M.; Wang, F.; Liu, Y.H.; An, L. Research on knowledge graph alignment model based on deep learning. *Expert Syst. Appl.* **2021**, *186*, 115768. [CrossRef]
22. Zhu, Y.Q.; Zhou, W.W.; Xu, Y.; Liu, J.; Tan, Y.J. Intelligent Learning for Knowledge Graph towards Geological Data. *Sci. Program.* **2017**, *2017*, 5072427. [CrossRef]
23. Qiu, Q.J.; Xie, Z.; Wu, L.; Tao, L.F. Automatic spatiotemporal and semantic information extraction from unstructured geoscience reports using text mining techniques. *Earth Sci. Inform.* **2020**, *13*, 1393–1410. [CrossRef]
24. Wang, C.B.; Ma, X.G.; Chen, J.G.; Chen, J.W. Information extraction and knowledge graph construction from geoscience literature. *Comput. Geosci.* **2018**, *112*, 112–120. [CrossRef]

25. Wang, C.B.; Ma, X.G.; Chen, J.G. Ontology-driven data integration and visualization for exploring regional geologic time and paleontological information. *Comput. Geosci.* **2018**, *115*, 12–19. [CrossRef]
26. Li, S.; Chen, J.P.; Xiang, J. Prospecting Information Extraction by Text Mining Based on Convolutional Neural Networks—A case study of the Lala Copper Deposit, China. *IEEE Access* **2018**, *6*, 52286–52297.
27. Holden, E.J.; Liu, W.; Horrocks, T.; Wang, R.; Wedge, D.; Duuring, P.; Beardsmore, T. GeoDocA–Fast analysis of geological content in mineral exploration reports: A text mining approach. *Ore Geol. Rev.* **2019**, *111*, 102919. [CrossRef]
28. Enkhsaikhan, M.; Holden, E.J.; Duuring, P.; Liu, W. Understanding ore-forming conditions using machine reading of text. *Ore Geol. Rev.* **2021**, *135*, 104200. [CrossRef]
29. Wang, Z.W.; Pei, F.P.; Xu, W.L.; Cao, H.H.; Wang, Z.J.; Zhang, Y. Tectonic evolution of the eastern Central Asian Orogenic Belt: Evidence from zircon U-Pb-Hf isotopes and geochemistry of early Paleozoic rocks in Yanbian region, NE China. *Gondwana Res.* **2016**, *38*, 334–350. [CrossRef]
30. Wu, F.Y.; Sun, D.Y.; Ge, W.C.; Zhang, Y.B.; Grant, M.L.; Wilde, S.A.; Jahn, B.M. Geochronology of the Phanerozoic granitoids in northeastern China. *J. Asian Earth Sci.* **2011**, *41*, 1–30. [CrossRef]
31. Zhang, X.T.; Sun, J.G.; Yu, Z.T.; Song, Q.H. LA-ICP-MS zircon U-Pb and sericite Ar-40/Ar-39 ages of the Songjianghe gold deposit in southeastern Jilin Province, Northeast China, and their geological significance. *Can. J. Earth Sci.* **2019**, *56*, 607–628. [CrossRef]
32. Han, J.L.; Deng, J.; Zhang, Y.; Sun, J.G.; Wang, Q.F.; Zhang, Y.M.; Zhang, X.T.; Liu, Y.; Zhao, C.T.; Yang, F.; et al. Au mineralization-related magmatism in the giant Jiapigou mining district of Northeast China. *Ore Geol. Rev.* **2022**, *141*, 104638. [CrossRef]
33. Huang, Z.X.; Yuan, W.M.; Wang, C.M.; Liu, X.W.; Xu, X.T.; Yang, L.Y. Metallogenic epoch of the Jiapigou gold belt, Jilin Province, China: Constrains from rare earth element, fluid inclusion geochemistry and geochronology. *J. Earth Syst. Sci.* **2012**, *121*, 1401–1420. [CrossRef]
34. Zhang, X.T.; Sun, J.G.; Han, J.L.; Feng, Y.Y. Genesis and ore-forming process of the Benqu mesothermal gold deposit in the Jiapigou ore cluster, NE China: Constraints from geology, geochronology, fluid inclusions, and whole-rock and isotope geochemistry. *Ore Geol. Rev.* **2021**, *130*, 103956. [CrossRef]
35. Alokaili, A.; Menai, M.E. SVM ensembles for named entity disambiguation. *Computing* **2020**, *102*, 1051–1076. [CrossRef]
36. Bikel, D.M.; Schwartz, R.; Weischedel, R.M. An algorithm that learns what's in a name. *Mach. Learn* **1999**, *34*, 211–231. [CrossRef]
37. Lu, J.L.; Kato, M.P.; Yamamoto, T.; Tanaka, K. Entity Identification on Microblogs by CRF Model with Adaptive Dependency. *IEICE Trans. Inf. Syst.* **2016**, *E99d*, 2295–2305. [CrossRef]
38. He, C.H.; Zhang, C.; Hu, S.Z.; Tan, Z.; Zhu, H.M.; Ge, B. Chinese News Text Classification Algorithm Based on Online Knowledge Extension and Convolutional Neural Network. In Proceedings of the 2019 16th International Computer Conference on Wavelet Active Media Technology and Information Processing, Chengdu, China, 14–15 December 2019; pp. 204–211.
39. Qin, Y.; Shen, G.W.; Zhao, W.B.; Chen, Y.P.; Yu, M.; Jin, X. A network security entity recognition method based on feature template and CNN-BiLSTM-CRF. *Front. Inf. Technol. Electron. Eng.* **2019**, *20*, 872–884. [CrossRef]
40. Hochreiter, S.; Schmidhuber, J. Long short-term memory. *Neural. Comput.* **1997**, *9*, 1735–1780. [CrossRef]
41. Yang, G.; Xu, H.Z. A Residual BiLSTM Model for Named Entity Recognition. *IEEE Access* **2020**, *8*, 227710–227718. [CrossRef]
42. Tian, G.; Wang, Q.B.; Zhao, Y.; Guo, L.T.; Sun, Z.L.; Lv, L.Y. Smart Contract Classification With a Bi-LSTM Based Approach. *IEEE Access* **2020**, *8*, 43806–43816. [CrossRef]
43. Levine, S.; Pastor, P.; Krizhevsky, A.; Quillen, D. Learning Hand-Eye Coordination for Robotic Grasping with Large-Scale Data Collection. *Int. J. Robot. Res.* **2017**, *1*, 173–184.
44. Majumdar, A. Graph structured autoencoder. *Neural Netw.* **2018**, *106*, 271–280. [CrossRef] [PubMed]
45. Yu, T.Z.; Guo, C.X.; Wang, L.F.; Xiang, S.M.; Pan, C.H. Self-Paced AutoEncoder. *IEEE Signal Process. Lett.* **2018**, *25*, 1054–1058. [CrossRef]
46. Lei, J.P.; Ouyang, D.T.; Liu, Y. Adversarial Knowledge Representation Learning Without External Model. *IEEE Access* **2019**, *7*, 3512–3524. [CrossRef]
47. Qiu, Q.J.; Xie, Z.; Wu, L.; Tao, L.F.; Li, W.J. BiLSTM-CRF for geological named entity recognition from the geoscience literature. *Earth Sci. Inform.* **2019**, *12*, 565–579. [CrossRef]
48. Chen, Y.; Zhou, C.J.; Li, T.X.; Wu, H.; Zhao, X.; Ye, K.; Liao, J. Named entity recognition from Chinese adverse drug event reports with lexical feature based BiLSTM-CRF and tri-training. *J. Biomed. Inform.* **2019**, *96*, 103252. [CrossRef] [PubMed]
49. Lample, G.; Ballesteros, M.; Subramanian, S.; Kawakami, K.; Dyer, C. Neural Architectures for Named Entity Recognition. In Proceedings of the 2016 Conference of the North American Chapter of the Association for Computational Linguistics: Human Language Technologies, San Diego, CA, USA, 12–14 June 2016; pp. 260–270.
50. Enkhsaikhan, M.; Liu, W.; Holden, E.J.; Duuring, P. Auto-labelling entities in low-resource text: A geological case study. *Knowl. Inf. Syst.* **2021**, *63*, 695–715. [CrossRef]
51. Gao, X.; Tan, R.; Li, G.H. Research on text mining of material science based on natural language processing. In *IOP Conference Series: Materials Science and Engineering*; IOP Publishing: Bristol, UK, 2020; p. 768072094.
52. Choi, Y.; Ryu, P.M.; Kim, H.; Lee, C. Extracting Events from Web Documents for Social Media Monitoring Using Structured SVM. *IEICE Trans. Inf. Syst.* **2013**, *E96d*, 1410–1414. [CrossRef]
53. Huang, W.; Shao, Z.F.; Luo, M.Y.; Zhang, P.; Zha, Y.F. A novel multi-loss-based deep adversarial network for handling challenging cases in semi-supervised image semantic segmentation. *Pattern Recogn. Lett.* **2021**, *146*, 208–214. [CrossRef]

54. Quan, C.Q.; Wang, M.; Ren, F.J. An Unsupervised Text Mining Method for Relation Extraction from Biomedical Literature. *PLoS ONE* **2014**, *9*, e102039. [CrossRef]
55. Brin, S. *Extracting Patterns and Relations from the World Wide Web*; Springer: Berlin/Heidelberg, Germany, 1999; Volume 1590, pp. 172–183.
56. Hasegawa, T.; Sekine, S.; Grishman, R. Discovering relations among named entities from large corpora. In Proceedings of the 42nd Annual Meeting on Association for Computational Linguistics, Barcelona, Spain, 21–26 July 2004; pp. 415–422.
57. Mintz, M.; Bills, S.; Snow, R.; Jurafsky, D. Distant supervision for relation extraction without labeled data. In Proceedings of the 47th Annual Meeting of the Association for Computational Linguistics and the 4th International Joint Conference on Natural Language Processing of the AFNLP, Singapore, 2–7 August 2009; pp. 1003–1011.
58. Zheng, S.C.; Hao, Y.X.; Lu, D.Y.; Bao, H.Y.; Xu, J.M.; Hao, H.W.; Xu, B. Joint entity and relation extraction based on a hybrid neural network. *Neurocomputing* **2017**, *257*, 59–66. [CrossRef]
59. Li, P.F.; Mao, K.Z. Knowledge-oriented convolutional neural network for causal relation extraction from natural language texts. *Expert Syst. Appl.* **2019**, *115*, 512–523. [CrossRef]
60. Ru, C.S.; Tang, J.T.; Li, S.S.; Xie, S.X.; Wang, T. Using semantic similarity to reduce wrong labels in distant supervision for relation extraction. *Inform. Process. Manag.* **2018**, *54*, 593–608. [CrossRef]
61. Zeng, D.J.; Liu, K.; Chen, Y.; Zhao, J. Distant supervision for relation extraction via piecewise convolutional neural networks. In Proceedings of the 2015 Conference on Empirical Methods in Natural Language Processing, Lisbon, Portugal, 17–21 September 2015; pp. 1753–1762.
62. Ren, X.; Wu, Z.Q.; He, W.Q.; Qu, M.; Voss, C.R.; Ji, H.; Abdelzaher, T.F.; Han, J.W. CoType: Joint Extraction of Typed Entities and Relations with Knowledge Bases. In Proceedings of the 26th International Conference on World Wide Web (Www'17), Perth, Australia, 3–7 April 2017; pp. 1015–1024.
63. Huang, Y.Y.; Wang, W.Y. Deep residual learning for weakly-supervised relation extraction. In Proceedings of the 2017 Conference on Empirical Methods in Natural Language Processing, Copenhagen, Denmark, 27 July 2017.
64. Hermann, K.M.; Kocisky, T.; Grefenstette, E.; Espeholt, L.; Kay, W.; Suleyman, M.; Blunsom, P. Teaching Machines to Read and Comprehend. *Adv. Neural Inf. Process. Syst.* **2015**, *28*, 1693–1701.
65. Chorowski, J.; Bahdanau, D.; Serdyuk, D.; Cho, K.; Bengio, Y. Attention-Based Models for Speech Recognition. *Adv. Neural Inf. Process. Syst.* **2015**, *28*, 577–585.
66. Bahdanau, D.; Cho, K.; Bengio, Y. Neural machine translation by jointly learning to align and translate. In Proceedings of the 54th Annual Meeting of the Association for Computational Linguistics, Berlin, Germany, 7–12 August 2016.
67. Zhou, P.; Shi, W.; Tian, J.; Qi, Z.Y.; Xu, B. Attention-Based Bidirectional Long Short-Term Memory Networks for Relation Classification. In Proceedings of the 54th Annual Meeting of the Association for Computational Linguistics, Berlin, Germany, 7–12 August 2016; pp. 207–212.
68. Zgank, A.; Kacic, Z. Predicting the Acoustic Confusability between Words for a Speech Recognition System using Levenshtein Distance. *Elektron Elektrotech.* **2012**, *18*, 81–84. [CrossRef]
69. Yan, Z.Q.; Wu, Q.; Ren, M.; Liu, J.Q.; Liu, S.W.; Qiu, S. Locally private Jaccard similarity estimation. *Concurr. Comput. Pract. Exp.* **2019**, *31*, e4889. [CrossRef]
70. Ye, J. Improved cosine similarity measures of simplified neutrosophic sets for medical diagnoses. *Artif. Intell. Med.* **2015**, *63*, 171–179. [CrossRef]
71. Hong, T.P.; Lin, C.W.; Yang, K.T.; Wang, S.L. Using *TF-IDF* to hide sensitive itemsets. *Appl. Intell.* **2013**, *38*, 502–510. [CrossRef]
72. Alammary, A.S. Arabic Questions Classification Using Modified *TF-IDF*. *IEEE Access* **2021**, *9*, 95109–95122. [CrossRef]
73. Larabi-Marie-Sainte, S.; Alnamlah, B.S.; Alkassim, N.F.; Alshathry, S.Y. A new framework for Arabic recitation using speech recognition and the Jaro Winkler algorithm. *Kuwait J. Sci.* **2022**, *49*. [CrossRef]
74. Volz, J.; Bizer, C.; Gaedke, M.; Kobilarov, G. Discovering and Maintaining Links on the Web of Data. *Lect. Notes Comput. Sci.* **2009**, *5823*, 650–665.
75. Vrandecic, D.; Krotzsch, M. Wikidata: A Free Collaborative Knowledgebase. *Commun. ACM* **2014**, *57*, 78–85. [CrossRef]
76. Niu, X.; Rong, S.; Wang, H.F.; Yu, Y. An effective rule miner for instance matching in a web of data. In Proceedings of the 21st ACM International Conference on Information and Knowledge Management, Maui, HI, USA, 29 October 2012; pp. 1085–1094.
77. Zhang, D.K.; Yin, J.; Zhu, X.Q.; Zhang, C.Q. Network Representation Learning: A Survey. *IEEE Trans. Big Data* **2020**, *6*, 3–28. [CrossRef]
78. Grover, A.; Leskovec, J. node2vec: Scalable Feature Learning for Networks. In Proceedings of the 22nd Acm Sigkdd International Conference on Knowledge Discovery and Data Mining, Kdd'16, San Francisco, CA, USA, 13 August 2016; pp. 855–864.
79. Li, G.H.; Luo, J.W.; Wang, D.C.; Liang, C.; Xiao, Q.; Ding, P.J.; Chen, H.L. Potential circRNA-disease association prediction using DeepWalk and network consistency projection. *J. Biomed. Inform.* **2020**, *112*, 103624. [CrossRef] [PubMed]
80. Bordes, A.; Usunier, N.; Garcia Duran, A.; Weston, J.; Yakhnenko, O. Translating embeddings for modeling multi-relational data. In Proceedings of the 26th International Conference on Neural Information Processing Systems, Red Hook, NY, USA, 5–10 December 2013; pp. 2787–2795.

81. Ji, G.L.; Liu, K.; He, S.Z.; Zhao, J. Knowledge graph completion with adaptive sparse transfer matrix. In Proceedings of the 30th Association for the Advancement of Artificial Intelligence Conference on Artificial Intelligence, Phoenix, AZ, USA, 12–17 February 2016; pp. 985–991.
82. Lin, Y.K.; Liu, Z.Y.; Sun, M.S.; Liu, Y.; Zhu, X. Learning entity and relation embeddings for knowledge graph completion. In Proceedings of the Twenty-Ninth AAAI Conference on Artificial Intelligence, Austin, TX, USA, 25–30 January 2015; pp. 2181–2187.
83. Zhu, J.Z.; Jia, Y.T.; Xu, J.; Qiao, J.Z.; Cheng, X.Q. Modeling the Correlations of Relations for Knowledge Graph Embedding. *J. Comput. Sci. Technol.* **2018**, *33*, 323–334. [CrossRef]
84. Li, S.Y.; Pan, R.; Luo, H.Y.; Liu, X.; Zhao, G.S. Adaptive cross-contextual word embedding for word polysemy with unsupervised topic modeling. *Knowl.-Based. Syst.* **2021**, *218*, 106827. [CrossRef]
85. Wang, B.L.; Sun, Y.Y.; Chu, Y.H.; Yang, Z.H.; Lin, H.F. Global-locality preserving projection for word embedding. *Int. J. Mach. Learn. Cybern.* **2022**, *13*, 2943–2956. [CrossRef]
86. Pan, F.Y.; Li, S.K.; Ao, X.; He, Q. Relation Reconstructive Binarization of word embeddings. *Front. Comput. Sci.* **2022**, *16*, 162307. [CrossRef]
87. Wang, X.Z.; Zhang, H.; Liu, Y. Sentence Vector Model Based on Implicit Word Vector Expression. *IEEE Access* **2018**, *6*, 17455–17463. [CrossRef]
88. Basirat, A.; Nivre, J. Real-valued syntactic word vectors. *J. Exp. Theor. Artif. Intell.* **2020**, *32*, 557–579. [CrossRef]
89. Rumelhart, D.E.; Hinton, G.E.; Williams, R.J. Learning internal representations by error propagation. *Read. Cogn. Sci.* **1988**, *323*, 318–362.
90. Santos, R.; Murrieta-Flores, P.; Calado, P.; Martins, B. Toponym matching through deep neural networks. *Int. J. Geogr. Inf. Sci.* **2018**, *32*, 324–348. [CrossRef]
91. Levenshtein, V.I. Binary codes capable of correcting deletions, insertions, and reversals. *Sov. Phys. Dokl.* **1966**, *10*, 707–710.
92. Mikolov, T.; Sutskever, I.; Chen, K.; Corrado, G.; Dean, J. Distributed representations of words and phrases and their compositionality. In Proceedings of the 26th International Conference on Neural Information Processing Systems, Lake Tahoe, NV, USA, 5–10 December 2013; pp. 3111–3119.
93. Zhou, C.H.; Wang, H.; Wang, C.S.; Hou, Z.Q.; Zheng, Z.M.; Shen, S.Z.; Cheng, Q.M.; Feng, Z.Q.; Wang, X.B.; Lv, H.R.; et al. Prospects for the research on geoscience knowledge graph in the Big Data Era. *Sci. China Earth Sci.* **2021**, *64*, 1105–1114. [CrossRef]
94. Pillai, S.G.; Soon, L.K.; Haw, S.C. Comparing DBpedia, Wikidata, and YAGO for Web Information Retrieval. *Lect. Note Netw. Syst.* **2019**, *67*, 525–535.
95. Farber, M.; Bartscherer, F.; Menne, C.; Rettinger, A. Linked Data Quality of DBpedia, Freebase, OpenCyc, Wikidata, and YAGO. *Semant. Web* **2018**, *9*, 77–129. [CrossRef]
96. Das, A.; Mitra, A.; Bhagat, S.N.; Paul, S. Issues and Concepts of Graph Database and a Comparative Analysis on list of Graph Database tools. In Proceedings of the 2020 International Conference on Computer Communication and Informatics (ICCCI), Coimbatore, India, 22–24 January 2020; pp. 353–358.
97. Vigo, M.; Matentzoglu, N.; Jay, C.; Stevens, R. Comparing ontology authoring workflows with Protege: In the laboratory, in the tutorial and in the 'wild'. *J. Web Semant.* **2019**, *57*, 100473. [CrossRef]
98. Chen, C.M.; Hu, Z.G.; Liu, S.B.; Tseng, H. Emerging trends in regenerative medicine: A scientometric analysis in CiteSpace. *Expert Opin. Biol. Ther.* **2012**, *12*, 593–608. [CrossRef]
99. Gong, F.; Wang, M.; Wang, H.F.; Wang, S.; Liu, M.Y. SMR: Medical Knowledge Graph Embedding for Safe Medicine Recommendation. *Big Data Res.* **2021**, *23*, 100174. [CrossRef]
100. Dong, X.; Gabrilovich, E.; Heitz, G.; Horn, W.; Lao, N.; Murphy, K.; Strohmann, T.; Sun, S.; Zhang, W. Knowledge vault: A web-scale approach to probabilistic knowledge fusion. In Proceedings of the 20th ACM SIGKDD International Conference on Knowledge Discovery and Data Mining, New York, NY, USA, 24–27 August 2014; pp. 601–610.
101. Hao, Y.C.; Zhang, Y.Z.; Liu, K.; He, S.Z.; Liu, Z.Y.; Wu, H.; Zhao, J. An End-to-End Model for Question Answering over Knowledge Base with Cross-Attention Combining Global Knowledge. In Proceedings of the 55th Annual Meeting of the Association for Computational Linguistics (Acl 2017), Vancouver, BC, Canada, 30 July–4 August 2017; pp. 221–231.
102. Gajendran, S.; Manjula, D.; Sugumaran, V. Character level and word level embedding with bidirectional LSTM—Dynamic recurrent neural network for biomedical named entity recognition from literature. *J. Biomed. Inform.* **2020**, *112*, 103609. [CrossRef]
103. Nguyen, T.H.; Grishman, R. Relation Extraction: Perspective from Convolutional Neural Networks. In Proceedings of the 1st Workshop on Vector Space Modeling for Natural Language Processing, Denver, CO, USA, 31 May 2015; pp. 39–48.
104. Zhang, D.X.; Wang, D. Relation Classification via Recurrent Neural Network. *arXiv* **2015**, arXiv:1508.01006v1.
105. Wang, Z.H.; Wang, D.; Li, Q. Keyword Extraction from Scientific Research Projects Based on SRP-*TF-IDF*. *Chin. J. Electron.* **2021**, *30*, 652–657.
106. Sun, G.; Lv, H.Z.; Wang, D.Y.; Fan, X.P.; Zuo, Y.; Xiao, Y.F.; Liu, X.; Xiang, W.Q.; Guo, Z.Y. Visualization Analysis for Business Performance of Chinese Listed Companies Based on Gephi. *Comput. Mater. Contin.* **2020**, *63*, 959–977.
107. Wutiepu, W.; Yang, Y.C.; Han, S.J.; Guo, Y.F.; Nie, S.J.; Liu, C.; Fan, W.L. Zircon U-Pb age, Hf isotope, and geochemistry of Late Permian to Triassic igneous rocks from the Jiapigou gold ore belt, NE China: Petrogenesis and tectonic implications. *Geol. J.* **2020**, *55*, 501–516. [CrossRef]

108. Hovy, E.; Lin, C. Automated text summarization and the SUMMARIST system. In Proceedings of the TIPSTER '98, Baltimore, MD, USA, 13–15 October 1998; pp. 197–214.
109. Piantadosi, S.T. Zipf's word frequency law in natural language: A critical review and future directions. *Psychon. B Rev.* **2014**, *21*, 1112–1130. [CrossRef]
110. Ye, T.Z.; Lv, Z.C.; Pang, Z.S.; Zhang, D.H.; Liu, S.Y.; Wang, Q.M.; Liu, J.J.; Cheng, Z.Z.; Li, C.L.; Xiao, K.Y.; et al. *Metallogenic Prognosis Theories and Methods in Exploration Areas*, 1st ed.; Geological Publishing House: Beijing, China, 2014; pp. 1–703. (In Chinese)

minerals

MDPI

Article

Multitarget Intelligent Recognition of Petrographic Thin Section Images Based on Faster RCNN

Hanyu Wang [1,2,3], Wei Cao [1,2,3,*], Yongzhang Zhou [1,2,3,*], Pengpeng Yu [1,2,3] and Wei Yang [3,4]

1 School of Earth Science and Engineering, Sun Yat-sen University, Zhuhai 519000, China; wanghy327@mail2.sysu.edu.cn (H.W.); yupp3@mail.sysu.edu.cn (P.Y.)
2 Centre for Earth Environment and Resources, Sun Yat-sen University, Zhuhai 519000, China
3 Guangdong Provincial Key Lab of Geological Process and Mineral Resources, Zhuhai 519000, China; yangw58@mail.sysu.edu.cn
4 Guangdong Avi Technology Research Institute, Guangzhou 510000, China
* Correspondence: caow25@mail.sysu.edu.cn (W.C.); zhouyz@mail.sysu.edu.cn (Y.Z.)

Abstract: The optical features of mineral composition and texture in petrographic thin sections are an important basis for rock identification and rock evolution analysis. However, the efficiency and accuracy of human visual interpretation of petrographic thin section images have depended on the experience of experts for a long time. The application of image-based computer vision and deep-learning algorithms to the intelligent analysis of the optical properties of mineral composition and texture in petrographic thin section images (in plane polarizing light) has the potential to significantly improve the efficiency and accuracy of rock identification and classification. This study completed the transition from simple petrographic thin image classification to multitarget detection, to address more complex research tasks and more refined research scales that contain more abundant information, such as spatial, quantitative and category target information. Oolitic texture is an important paleoenvironmental indicator that widely exists in sedimentary records and is related to shallow water hydraulic conditions. We used transfer learning and image data augmentation in this paper to identify the oolitic texture of petrographic thin section images based on the faster region-based convolutional neural network (Faster RCNN) method. In this study, we evaluated the performance of Faster RCNN, a two-stage object detection algorithm, using VGG16 and ResNet50 as backbones for image feature extraction. Our findings indicate that ResNet50 outperformed VGG16 in this regard. Specifically, the Faster RCNN model with ResNet50 as the backbone achieved an average precision (AP) of 92.25% for the ooids test set, demonstrating the accuracy and reliability of this approach for detecting ooids. The experimental results also showed that the uneven distribution of training sample images and the complexity of images both significantly affect detection performance; however, the uneven distribution of training sample images has a greater impact. Our work is preliminary for intelligent recognition of multiple mineral texture targets in petrographic thin section images. We hope that it will inspire further research in this field.

Keywords: Faster RCNN; object detection; intelligent recognition; petrographic thin section; VGG; ResNet; big data mining; computer vision

1. Introduction

In recent years, the introduction of big data and artificial intelligence algorithms has been a major development in geological research fields. Machine learning and deep learning algorithms have been used to uncover hidden relationships in massive structured and unstructured geoscience data [1–6], identify geochemical anomalies [7–12], and predict potential mineral resources [13–17]. Furthermore, with the development of deep learning, image-based computer vision technology has made it possible to intelligently identify and classify rocks by their optical properties [18–22].

An important rock identification method is observing and studying the optical properties of mineral composition and texture in petrographic thin sections under a polarizing microscope [23]. This process requires the professional ability and expert experience of observers; thus, some scholars have used machine learning and deep learning algorithms to classify rocks with promising results [20,24,25]. For instance, Młynarczuk et al. [24] used the nearest neighbor method, K-nearest neighbor algorithm, closest approach mode method, and optimal spherical domain method to classify nine kinds of rock flakes and achieved satisfactory accuracy. Shu et al. [25] compared manually selected features and automatically extracted features based on the k-means method to classify rock images. They found that k-means has greater robustness and flexibility in the classification effect of rock images, albeit requiring a large number of training sets as support.

With advances in deep learning methods in computer vision, rock image classification research methods have shifted toward deep learning. Deep learning models generally adopt mature convolutional neural network models, such as the Visual Geometry Group Network (VGG) [26], UNet [27], Residual Network (ResNet) [28], Dense Convolutional Network (DenseNet) [29], Inception-v3 [30] and Faster Region-based Convolutional Neural Network (Faster RCNN) [31]. These models have demonstrated excellent performance in computer vision and have also achieved good classification results in rock image classification. Classification studies of rock images can be divided into two categories according to the image scale: macroscopic hand specimen rock and microscopic petrographic thin section images. While some scholars have conducted research on rock image classification from a macroscopic perspective [25,32–35], others have studied petrographic thin section image classification from a microscopic perspective [24,36–39]. For instance, Zhang et al. [33] classified granite, phyllite and breccia images based on the Inception-V3 deep learning model and transfer learning method. They used a total of 571 rock images and achieved a satisfactory overall classification effect despite the small size of the training set. Xu and Zhou [39] used the UNet model to classify and recognize pyrite, chalcopyrite, galena and sphalerite in petrographic thin section images. After data enhancement, they obtained a recognition success rate of more than 90% for each mineral. Bai et al. [40] used the VGG model to classify petrographic thin section images of dolomite, andesite, oolitic limestone, granite, lithic sandstone and quartz sandstone, with 1000 samples in each category. They achieved an accuracy rate of 82% on the test set. Polat et al. [37] used the transfer learning method to classify 1200 rock thin sections of 6 kinds of volcanic rocks based on ResNet50 and DenseNet121 models, with an average classification accuracy of more than 98%. These results demonstrate the potential of transfer learning and data enhancement in achieving good results even with small data sets. In these classification studies, deep learning models have been used to classify rocks by learning the features in rock images or petrographic thin section images, which usually contain only one rock type and one target [33,36–38]. To classify multitarget images, the tedious process of standardizing the images into single-target images of the same size and resolution is usually needed.

The above studies of rock and petrographic thin section images have largely focused on mineral composition identification, with few investigations into mineral textural and structural characteristics. In fact, it is not only the mineral composition, but also the mineral textural and structural characteristics that are crucial to determining the rock type and its genesis. Oolitic texture is a kind of texture widely existing in sedimentary rocks. It is composed of small ooids (spherical or ellipsoidal particles composed of a core and a shell surrounding the core) with high porosity and permeability. Therefore, oolitic shoal reservoirs are important targets for oil and gas exploration [41,42]. However, previous studies on intelligent recognition of oolitic texture only focused on whether a petrographic thin section image has oolitic texture or not [40], but did not pay attention to information such as the size, shape, aspect ratio, sorting index, etc., that is bearing in ooids, and is important for reconstructing the paleogeographic environment. Intelligent identification of these minimal units is more challenging than mineral composition identification, as it requires multitarget localization and classification, and thus meets the requirements of multicomponent intelligent recognition in petrographic thin section images.

In this study, the Faster RCNN model was applied to multitarget identification of oolitic texture on petrographic thin section images. The training and test sets were composed of sedimentary rock petrographic thin section images, which contain multiple ooids. This study has two main characteristics: (1) it transitions the research based on petrographic thin section images from simple classification to multitarget detection, making it more consistent with real situations and the research task more complex; (2) the research scale is more refined and contains more abundant information, such as the spatial, quantitative and category target information, which is of reference significance for multitarget mineral intelligent recognition on petrographic thin section images. The present study offers potential benefits to users of the model who can utilize the automatically detected ooids as a reference for their own research on oolitic texture, thereby reducing the time and effort required for manual identification and measurement. Furthermore, this study lays the groundwork for future research on oolitic texture and related fields, including mineralogy, geology and paleontology, which may lead to novel discoveries and insights.

2. Materials and Methods

The microscopic image data of oolitic texture are from the "Micro image data set of some rock-forming minerals, typical metamorphic minerals and oolitic thin sections" of the public database ScienceDB [43]. Each microscopic image contains multiple ooids.

The object detection model can accomplish both localization and classification tasks and, thus, multitarget recognition. The object detection model generates a large number of anchor boxes on the original image by direct or indirect methods and calculates the intersection over union (IOU) between these boxes and the ground truth box, as shown in Figure 1. Positive and negative samples are determined according to the threshold of the IOU ratio, and the positive and negative samples are classified and trained by regression. Therefore, the object detection task is decomposed into classification and regression tasks [44–46].

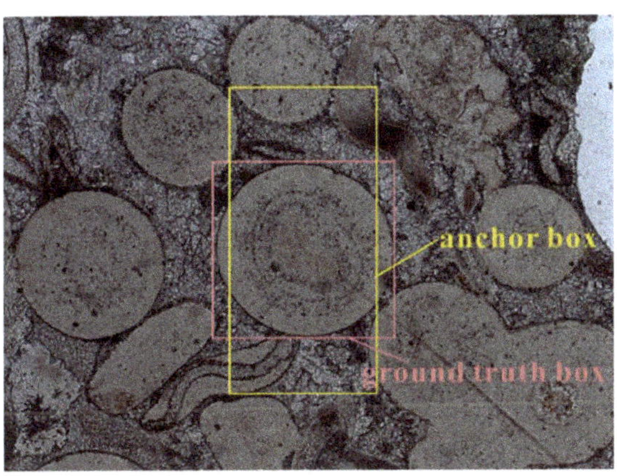

Figure 1. Schematic of the anchor box and ground truth box (The underlying image was obtained from the public database ScienceDB [43]).

Object detection models can be divided into two categories: one-stage and two-stage models. One-stage models are generally faster, but two-stage object detection models have shown higher accuracy and better small target detection capabilities than one-stage models [44]. This study applies the classic Faster RCNN model in the two-stage approach to multitarget identification of oolitic texture in petrographic thin section images due to the varying mineral composition and texture scales.

2.1. Faster RCNN

The Faster RCNN model, as depicted in Figure 2, consists of a feature extraction backbone network, region proposal network (RPN), region of interest (ROI) pooling and classification and regression. The calculation process can be divided into the following four steps [31]:

1. Backbone network: The backbone network comprises a series of convolution, batch normalization, activation function and pooling operations. It is used to extract image features and generate feature maps.
2. Region proposal networks (RPN): A large number of anchor boxes are indirectly generated on an image, and the IOU ratio between each anchor box and the ground truth box is calculated. The anchor boxes are labeled as positive samples and negative samples according to the IOU threshold, and the positive and negative samples are classified and trained by regression. The final output is 300 relatively accurate region proposals (ROI).
3. ROI pooling: The region proposal output in the previous step is projected onto the feature map. The feature maps of these region proposals are different sizes, so it is necessary to standardize them and output feature maps with the same size of 7 × 7, which is convenient to connect the subsequent fully connected layer.
4. Classification and regression: More accurate classification and regression are performed on the feature map output in the previous step.

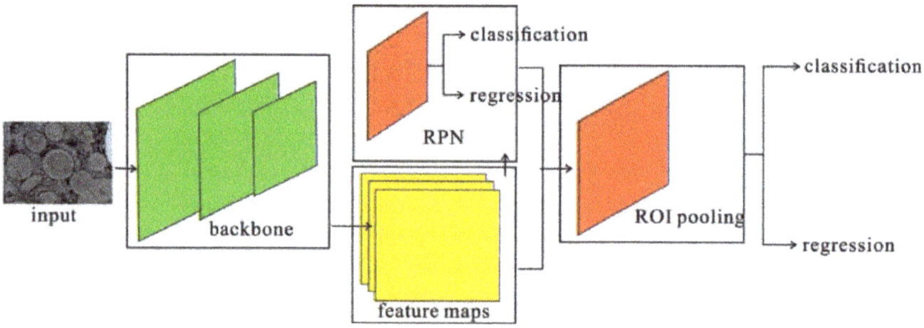

Figure 2. Architecture of the Faster RCNN framework.

2.1.1. VGG16

During the 2014 ImageNet Large Scale Visual Recognition Challenge (ILSVRC) [47], the VGG Lab at the University of Oxford introduced the VGG series network model architecture [26]. The primary innovation of the VGG series is the utilization of small convolution kernel continuous convolution, as opposed to large convolution kernel convolution operation. This approach extends the network depth while ensuring the same receptive field.

The VGG models are characterized by their simple and elegant structure, as well as their impressive performance. The VGG model comprises five VGG-blocks (VGG-block1, VGG-block2, VGG-block3, VGG-block4, VGG-block5), five max pooling layers and three fully connected layers (FC). The varying number of convolutional layers in each VGG-block constitutes VGG-block1, VGG-block2, VGG-block3, VGG-block4 and VGG-block5. VGG11, VGG13, VGG16 and VGG19 are composed of different combinations of VGG-blocks. Among the VGG series, VGG16 is the most classic. In this study, we selected VGG16, which utilizes a 3 × 3 convolution kernel size as the backbone for feature extraction.

Figure 3 illustrates the architecture of VGG16, which comprises five VGG-blocks. VGG-block1 and VGG-block2 each consist of two convolution layers that employ a 3 × 3 convolution kernel size. In contrast, VGG-block3, VGG-block4 and VGG-block5 each contain three convolution layers that also utilize a 3 × 3 convolution kernel size. Subsequently, three fully connected layers are connected to the output.

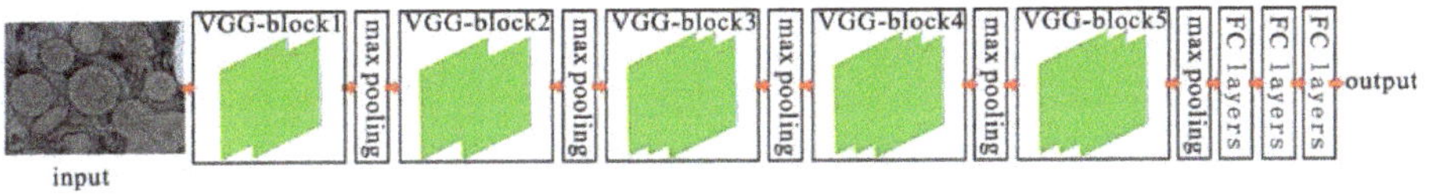

Figure 3. VGG16 Architecture.

2.1.2. ResNet50

The residual network model (ResNet) is a classical convolutional neural network model proposed by He et al. [28]. It emerged in the ILSVRC and won the classification, localization and detection category. The residual block in the network structure is the primary feature that distinguishes ResNet from the previous convolutional neural networks. The introduction of the residual block expands the depth of the convolutional neural network and develops a network structure with dozens or hundreds of layers [28].

In the standard convolutional network, the input is x, and after the convolutional layer, the input is transformed into y. The mapping that this standard convolutional network needs to learn is F, where y = F(x). In the ResNet convolution network, the input is x. After going through the convolution layer, the input x on the shortcut connection (which may need to be transformed) is added, and the output after feature transformation is y. The mapping that this convolution network needs to learn is F, where F(x) = y-x, as the learning mapping is the difference between output and input. Therefore, the network structure corresponding to F(x) is referred to as the residual building block. Compared with the standard convolutional network, the residual network has added shortcut connections, as shown in Figure 4.

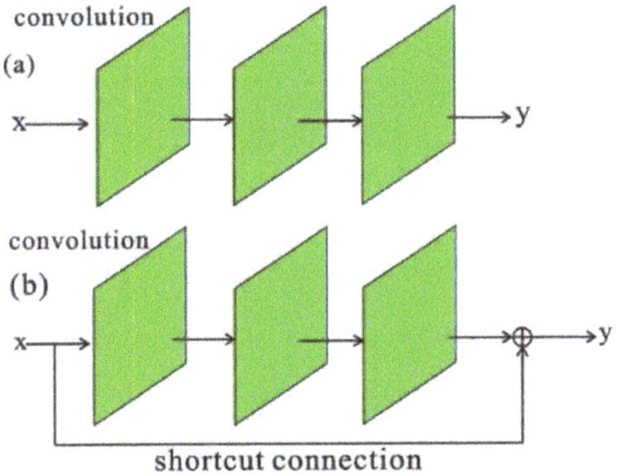

Figure 4. (**a**) Convolution module and (**b**) residual block.

ResNet is stacked with one convolutional layer, four residual modules and one fully connected layer. Common ResNets are ResNet18, ResNet34, ResNet50, ResNet101 and ResNet152. ResNet50 is used in this study, as shown in Figure 5. ResNet50 contains five blocks. Conv1 has only one convolution layer, and the convolution kernel is 7 × 7. Conv2 has 3 residual modules; Conv3 contains 4 residual modules; Conv4 contains 6 residual modules; and Conv5 contains 3 residual modules. Each residual module has a shortcut connection and contains 3 convolution layers; the convolution kernel sizes are 1 × 1, 3 × 3 and 1 × 1. The backbone network of the Faster RCNN model applied in this study is based on ResNet50.

2.1.3. Transfer Learning

Deep learning models have been shown to be heavily reliant on large data sets [48]. When the size of data sets is limited, deep neural networks with a large number of parameters (e.g., ResNet50) cannot be fully trained, which will lead to overfitting and other problems. To avoid overfitting problems and enhance the generalization ability of the model, transfer learning and data augmentation are usually adopted. Transfer learning generally uses the network weights that have been trained on other large data sets as the initial weights of new tasks. The data enhancement method generates additional image data by translation, flipping, cropping and brightness adjustment of the original image, which is then used to supplement the training data set [48,49].

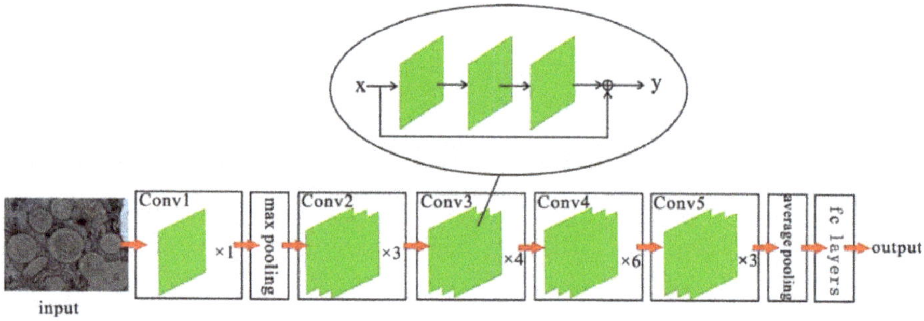

Figure 5. ResNet50 Architecture (the symbol "× number" denotes the number of blocks stacked.).

2.1.4. Average Precision

In the binary decision problem, the classification decision is to classify the samples as positive or negative. Four results are generated by comparing the predicted value with the real value. The decision result of the classifier can be represented by a confusion matrix [50], as shown in Table 1.

Table 1. Confusion matrix table.

	Actual Value: True	Actual Value: False
Predicted values: Positive	TP	FP
Predicted values: Negative	FN	TN

T indicates that the sample is correctly classified, F indicates that the sample is incorrectly classified, P denotes the sample is classified as positive and N denotes the sample is classified as negative.

The meanings of TP, FP, FN and TN are as follows:

TP (true positive): A positive sample is correctly classified as positive.

FP (false positive): A negative sample is incorrectly classified as positive.

FN (false negative): A positive sample is incorrectly classified as negative.

TN (true negative): A negative sample is correctly classified as negative.

Recall: the probability of correct positive samples among all positive samples.

$$\text{Recall} = \frac{TP}{TP + FN} \tag{1}$$

Precision: the probability of positive samples correctly classified among all positive samples returned.

$$\text{Precision} = \frac{TP}{TP + FP} \tag{2}$$

Intersection-Over-Union (IOU): IOU is the ratio of the intersection to the union of the predicted box and the ground truth box. In region proposal networks, samples with an IOU greater than 0.7 are considered positive, while those with an IOU less than 0.3 are considered negative. The IOU threshold is crucial for calculating the average precision

(AP), which is a key metric for evaluating the quality of object detection. In the evaluation of the validation set and test set, an IOU threshold of 0.5 is utilized to calculate the AP.

Average precision: indicates the average accuracy of different recall rates. This index is used to evaluate the performance of ooid detection in this study. The prediction boxes in the test set were sorted according to the confidence from high to low, and the correct rate and recall rate were calculated successively. The Precision-Recall (PR) curve with the recall rate as the abscissa and the correct rate as the ordinate was drawn, and the area enclosed between the rounded PR curve and the coordinate axis was calculated by the integration form as the AP value of this category.

2.2. Experiments

The ooids in the original microscopic images were annotated by LabelImg [51], an open source Python library, and the annotated images were used for training and testing the Faster RCNN. The code has been modified from the source code obtained from GitHub [52]. To improve the computational efficiency and speed up the operation, all the training was completed on a GPU. To accelerate the convergence of the model, the weights that were trained on the PASCAL Visual Object Classes [53] data set were selected as the initial weights. For ablation experiments, VGG16 and ResNet50 were employed as the backbone of the Faster RCNN model. The hardware and software used in the experiment are shown in Table 2 below.

Table 2. Hardware and software configuration.

Hardware/Software	Series/Version
CPU	i7-10700KF@3.8 GHz
GPU	2080ti
DRAM	32 G
SSD	1.5 T
OS	Windows10 Professional
Python	3.7.1
Torch	1.8.1
Torchvision	0.9.1

The performance of a model reaches saturation under ideal conditions as the size of the dataset increases, beyond which it no longer improves. Additionally, the optimal state of the model's performance is achieved with an increase in model complexity, but further increases in complexity lead to a decrease in performance [54]. The size of the dataset and the complexity of the model are closely linked to the model's performance. Complex models require larger datasets to fit the model, and they tend to predict better performance. Warden [55] suggests that for image classification tasks in deep learning, a general rule of thumb is to have 1000 images per classification. However, this requirement can be reduced significantly by using transfer learning and introducing a pre-trained model in a classification task. In the same domain, ten or twenty new categories may suffice. However, if the categories are not in the same field, more training images are necessary.

There are a total of 103 images in the original data set, of which 56 single-polarized microscopic images are selected as experimental data. We chose a random sampling method to reduce human intervention while making it simple and easy to implement. Ninety percent of the images are randomly selected as the training set (81%) and validation set (9%), and the remaining 10% are selected as the test set. The training set images were data enhanced, and the data set was expanded to 315 images by cropping, translation, rotation, flipping, adding random noise and adjusting brightness. Figure 6 shows the comparison between the original image and the image after data enhancement. After the training, the weight with the smallest loss value is selected as the weight of the test set to test and verify its performance.

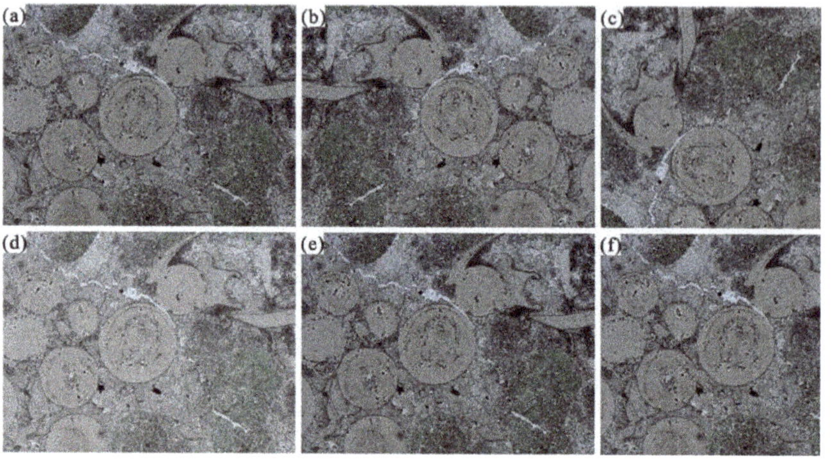

Figure 6. Comparison between the original image and the augmented data. (**a**) is the original image (from the public database ScienceDB [43]); (**b**) is the flipped image; (**c**) is the rotated image; (**d**) is the image after brightness adjustment; (**e**) is the image after adding random noise; (**f**) is the cropped image.

3. Results

The experimental results of this study showed that the multitarget detection of ooids based on Faster RCNN has good performance. In the ablation test of the Faster RCNN model with VGG16 and ResNet50 as backbones, the AP values for VGG16 and ResNet50 in the test set were 87.83% (Figure 7) and 92.25% (Figure 8), respectively. The performance of ResNet50 as the backbone of the Faster RCNN model in the test set was analyzed. Figures 9–14 show the test results of ooid detection in the test set using ResNet50 as the backbone. It was observed that when the features of ooids in the microscopic image were distinct from the background, the detection performance was good, and the ooids were accurately located (Figures 9 and 10). Conversely, when the composition of the image was relatively complex and the difference between the ooids and the background was not obvious, misclassification occurred (Figures 11–13), and the uneven distribution of samples in the training set was found to have a significant impact on the detection performance, with cases of missed detection occurring when the number of similar samples in the training set and the test set was small (Figure 14).

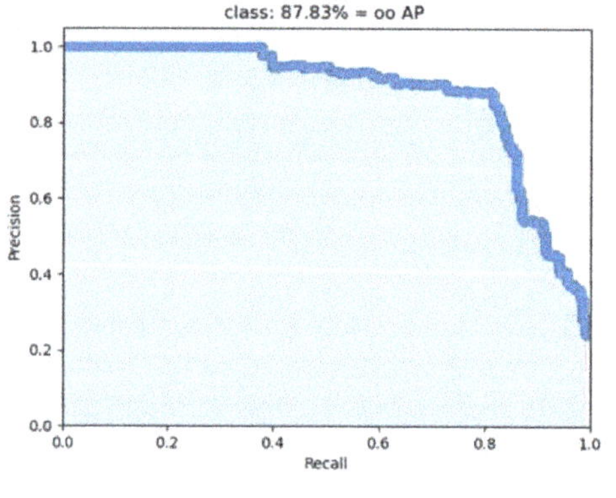

Figure 7. AP of ooids detection using VGG16 as the backbone.

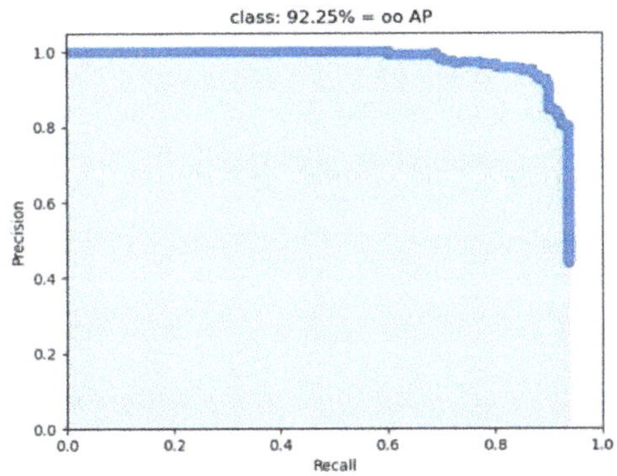

Figure 8. AP of ooids detection using ResNet50 as the backbone.

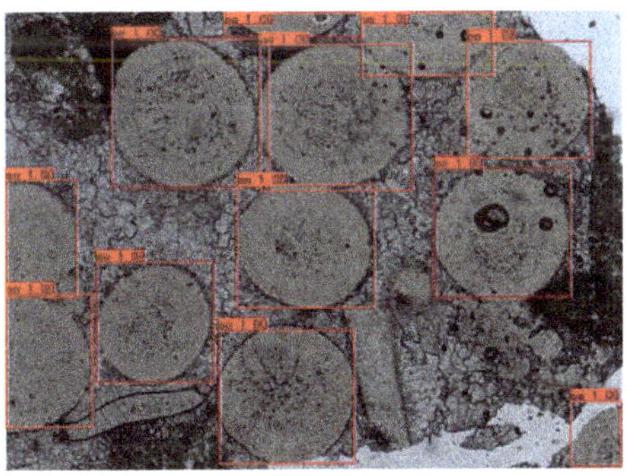

Figure 9. Ooids image with correct detection results.

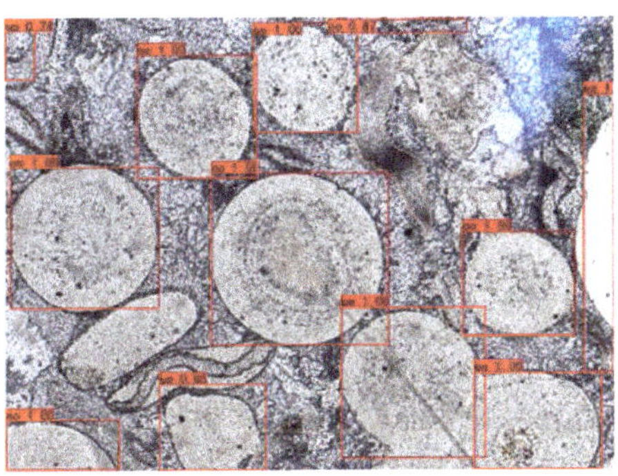

Figure 10. Misclassification of ooids at the image's edges.

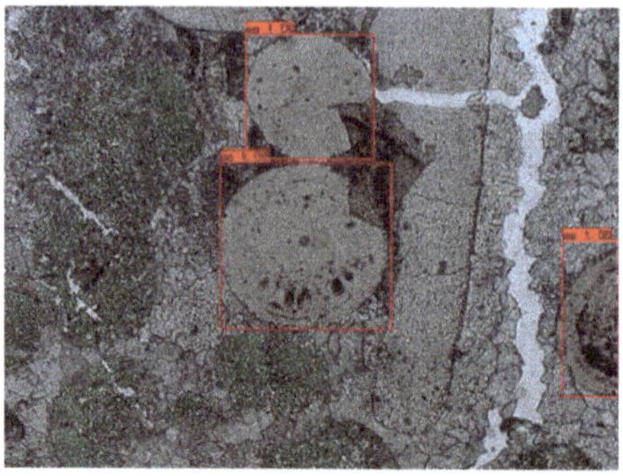

Figure 11. Detection results of ooids with low composition complexity.

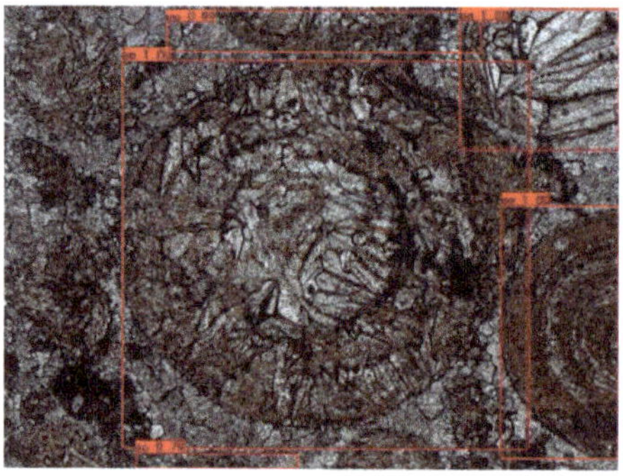

Figure 12. Detection results of ooids with high composition complexity.

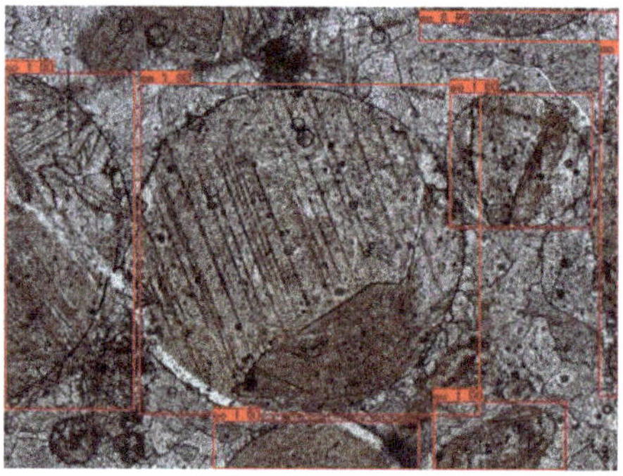

Figure 13. Detection results of ooids with moderate composition complexity.

In Figure 9, the detection model revealed a large number of ooids with moderate size proportions, small differences and uniform distribution, which were significantly distinct from the background. The detection result was highly intuitive, enabling the accurate identification of complete ooids in the image, as well as those partially displayed ooids located at the edge.

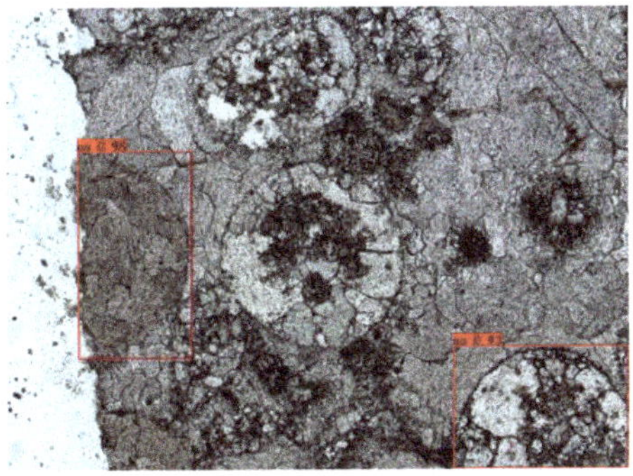

Figure 14. Poor detection results of ooids.

In Figure 10, the model detected a large number of ooids with moderate size proportions, small differences and uniform distribution, which were distinct from the background. The checking model yielded satisfactory results, with complete ooids near the center of the image correctly identified. However, a few misclassifications occurred at the image's edges.

In Figure 11, the composition complexity of components is low, and oolitic features are relatively obvious, while in Figures 12 and 13, the oolitic texture features are more complex, and the size of oolitic particles differs greatly from the background, although not as much as in Figures 9 and 10. The main ooids were correctly detected in all three figures, and large ooids at the edge were localized and detected.

The features of complete ooids in Figure 14 are not as obvious as those in Figures 9–13, and the contrast with the background is not as pronounced, leading to a potential misclassification. Nevertheless, the ooids in Figures 11–13 are also highly similar to the background. Although misclassification may occur in the image edge, the main ooids were still detected and identified without error. Comparing Figures 9–13 revealed that the detectability of ooids is not solely dependent on the distinctiveness of their features and the contrast with the background. A comparison of the original training data set shows that the number of samples with high similarity to the test in Figure 14 in the training set is small, suggesting that the distribution of training samples can have a significant impact on the detection performance.

4. Discussion

The potential causes of misclassification in detection can be attributed to a variety of factors, including the following:

1. The complexity of image components and the lack of distinction between target texture and background features can lead to misclassification in detection. This is exemplified by the petrographic thin section image classification of Bai et al., (2019) [40], where the similarity between oolitic limestone and dolomite thin sections was so high that misclassification occurred. To address this issue, incorporating such samples into the training process may enable the model to learn the differences between them and thus reduce misjudgment.
2. The uneven distribution of training image samples can have a significant impact on detection. This effect was demonstrated by the training set analysis, which revealed that the training set similar to Figures 9–11 accounted for 54%, the training set similar to Figures 12 and 13 accounted for 32%, and the training set similar to Figure 14 accounted for 14%. This skewed distribution of training set samples indicates that the model may not have sufficient learning experience, leading to better performance on simple images than on complex images.
3. The limited size of the training image data sets can lead to insufficient generalization of the model. Data augmentation has been employed to address the issue of small data sets, thus improving the model's generalization capacity [38,39]. After data

augmentation, the overall result was promising, which shows the potential of this method. With the increase in data sets, the generalization ability of the model can be further enhanced, thus improving the detection performance.

When the number of samples is large, deep learning models can be constructed from the very beginning and yield satisfactory classification results. Conversely, when the number of samples is insufficient, transfer learning methods such as appropriate initial weight selection can be employed to expedite the model convergence, and the data set can be augmented and supplemented through data augmentation, ultimately leading to satisfactory classification results. These results suggest that complex deep learning models can be applied to small data sets with the help of appropriate transfer learning and data augmentation.

In this study, a deep learning model was applied to detect multiple targets of ooids in microscopic petrographic thin section images. Object detection was found to be more complex and difficult than classification, with an average precision of 92.25% using ResNet50 as the backbone. Most of the prediction targets had high confidence. However, the uneven distribution of samples in the training set and the component complexity of microscopic image samples both impacted the detection performance, with the former having a greater influence. The results of this experiment suggest that ooid detection methods may provide useful insights for multitarget recognition on microscopic petrological thin section images. Our future work will focus on applying these technologies to intelligently recognize the other mineral textures and structures that reflect the genesis and formation environment of rocks.

5. Conclusions

By analyzing the mineral composition, texture and structural features, petrographic thin section images can be employed to classify rock types, elucidate petrogenesis, invert paleoenvironments and facilitate other relevant studies. In this study, Faster RCNN object detection algorithms were applied to identify multiple objects in petrographic thin section images using oolitic texture images from the public database ScienceDB. Transfer learning was used to accelerate the model convergence, data augmentation was used to enhance the generalization ability of the model, and GPU was utilized to improve the computational efficiency. The following conclusions and understandings are drawn from the experimental research.

1. The AP value of the ooids test set using ResNet50 as the backbone was 92.25%, indicating good overall detection performance. This object detection model was found to be robust and generalizable, as it was able to identify both complete ooids in the middle of the image and partial ooids at the edge.
2. The uneven distribution of samples in the training set and the complex composition of microscopic images affected the detection, with the former having a greater effect. Deep learning was used to learn features from the training set and make predictions on the test set, but the uneven distribution of samples caused the distribution of the learned features to deviate, resulting in missed detection in the prediction process. The complexity of the microscopic image composition, with a small difference between the target and the background, also contributed to misclassification and thus affected the detection performance, although to a lesser extent than the uneven distribution.
3. This study sought to transition from the classification of petrographic thin section images to multitarget detection, incorporating richer content such as spatial, quantitative and categorical target information, as well as more complex tasks. The research scale was further refined, transitioning from rocks to the textures and structures within them, providing a reference for multitarget intelligent recognition on petrographic thin section images.

Author Contributions: Conceptualization, H.W. and Y.Z.; methodology, H.W. and Y.Z.; software, H.W.; validation, H.W. and W.C.; investigation, H.W., W.C. and W.Y.; resources, Y.Z. and W.Y.; data curation, H.W. and W.C.; writing—original draft preparation, H.W. and W.C.; writing—review and editing, Y.Z. and P.Y.; visualization, H.W. and W.C.; supervision, Y.Z.; funding acquisition, Y.Z. All authors have read and agreed to the published version of the manuscript.

Funding: This research was financially supported by National Key Research and Development Program of China (Grant No. 2022YFF0801201), National Natural Science Foundation of China (Grant No. U1911202) and the Key-Area Research and Development Program of Guangdong Province (Grant No. 2020B1111370001).

Data Availability Statement: Data available on request from the authors.

Acknowledgments: We express our utmost gratitude to Fan Xiao and Jin Hong from Sun Yat-sen University. Furthermore, we extend our sincere appreciation to Renguang Zuo, the esteemed guest editor of this special issue, for his exceptional organization and guidance. We would also like to express our appreciation to the two anonymous reviewers who have helped us improve the paper, as well as section managing editor and other editors for their assistance throughout the publication process.

Conflicts of Interest: The authors declare that they have no conflict of interest in relation to this work. We declare that we do not have any commercial or associative interest that represents a conflict of interest in connection with the work submitted.

References

1. Zhou, Y.Z.; Zuo, R.G.; Liu, G.; Yuan, F.; Mao, X.C.; Guo, Y.J.; Xiao, F.; Liao, J.; Liu, Y.P. The great-leap-forward development of mathematical geoscience during 2010–2019: Big data and artificial intelligence algorithm are changing mathematical geoscience. *Bull. Mineral. Petrol. Geochem.* **2021**, *40*, 556–573. [CrossRef]
2. Zhou, Y.Z.; Chen, S.; Zhang, Q.; Xiao, F.; Wang, S.G.; Liu, Y.P.; Jiao, S.T. Advances and prospects of big data and mathematical geoscience. *Acta Petrol. Sin.* **2018**, *34*, 255–263.
3. Liu, C.; Chen, J.P.; Li, S.; Qin, T. Construction of Conceptual Prospecting Model Based on Geological Big Data: A Case Study in Songtao-Huayuan Area, Hunan Province. *Minerals* **2022**, *12*, 669. [CrossRef]
4. Mao, X.C.; Liu, P.; Deng, H.; Liu, Z.K.; Li, L.J.; Wang, Y.S.; Ai, Q.X.; Liu, J.X. A Novel Approach to Three-Dimensional Inference and Modeling of Magma Conduits with Exploration Data: A Case Study from the Jinchuan Ni–Cu Sulfide Deposit, NW China. *Nat. Resour. Res.* **2023**, *32*, 901–928. [CrossRef]
5. Deng, H.; Zhou, S.F.; He, Y.; Lan, Z.D.; Zou, Y.H.; Mao, X.C. Efficient Calibration of Groundwater Contaminant Transport Models Using Bayesian Optimization. *Toxics* **2023**, *11*, 438. [CrossRef]
6. Zeng, L.; Li, T.B.; Huang, H.T.; Zeng, P.; He, Y.X.; Jing, L.H.; Yang, Y.; Jiao, S.T. Identifying Emeishan basalt by supervised learning with Landsat-5 and ASTER data. *Front. Earth Sci.* **2023**, *10*, 1097778. [CrossRef]
7. Wang, J.; Zhou, Y.Z.; Xiao, F. Identification of multi-element geochemical anomalies using unsupervised machine learning algorithms: A case study from Ag-Pb-Zn deposits in north-western Zhejiang, China. *Appl. Geochem.* **2020**, *120*, 104679. [CrossRef]
8. Zuo, R.G. Machine Learning of Mineralization-Related Geochemical Anomalies: A Review of Potential Methods. *Nat. Resour. Res.* **2017**, *26*, 457–464. [CrossRef]
9. Wu, G.P.; Chen, G.X.; Cheng, Q.M.; Zhang, Z.J.; Yang, J. Unsupervised Machine Learning for Lithological Mapping Using Geochemical Data in Covered Areas of Jining, China. *Nat. Resour. Res.* **2021**, *30*, 1053–1068. [CrossRef]
10. Yu, X.T.; Xiao, F.; Zhou, Y.Z.; Wang, Y.; Wang, K.Q. Application of hierarchical clustering, singularity mapping, and Kohonen neural network to identify Ag-Au-Pb-Zn polymetallic mineralization associated geochemical anomaly in Pangxidong district. *J. Geochem. Explor.* **2019**, *203*, 87–95. [CrossRef]
11. Wu, B.C.; Li, X.H.; Yuan, F.; Li, H.; Zhang, M.M. Transfer learning and siamese neural network based identification of geochemical anomalies for mineral exploration: A case study from the Cu-Au deposit in the NW Junggar area of northern Xinjiang Province, China. *J. Geochem. Explor.* **2022**, *232*, 106904. [CrossRef]
12. Li, H.; Li, X.H.; Yuan, F.; Jowitt, S.M.; Zhang, M.M.; Zhou, J.; Zhou, T.F.; Li, X.L.; Ge, C.; Wu, B.C. Convolutional neural network and transfer learning based mineral prospectivity modeling for geochemical exploration of Au mineralization within the Guandian-Zhangbaling area, Anhui Province, China. *Appl. Geochem.* **2020**, *122*, 104747. [CrossRef]
13. Qin, Y.Z.; Liu, L.M. Quantitative 3D Association of Geological Factors and Geophysical Fields with Mineralization and Its Significance for Ore Prediction: An Example from Anqing Orefield, China. *Minerals* **2018**, *8*, 300. [CrossRef]
14. Zuo, R.G.; Kreuzer, O.P.; Wang, J.; Xiong, Y.H.; Zhang, Z.J.; Wang, Z.Y. Uncertainties in GIS-Based Mineral Prospectivity Mapping: Key Types, Potential Impacts and Possible Solutions. *Nat. Resour. Res.* **2021**, *30*, 3059–3079. [CrossRef]
15. Liu, L.M.; Cao, W.; Liu, H.S.; Ord, A.; Qin, Y.Z.; Zhou, F.H.; Bi, C.X. Applying benefits and avoiding pitfalls of 3D computational modeling-based machine learning prediction for exploration targeting: Lessons from two mines in the Tongling-Anqing district, eastern China. *Ore Geol. Rev.* **2022**, *142*, 104712. [CrossRef]

16. Wang, Z.Y.; Yin, Z.; Caers, J.; Zuo, R.G. A Monte Carlo-based framework for risk-return analysis in mineral prospectivity mapping. *Geosci. Front.* **2020**, *11*, 2297–2308. [CrossRef]
17. Lu, Y.; Liu, L.M.; Xu, G.J. Constraints of deep crustal structures on large deposits in the Cloncurry district, Australia: Evidence from spatial analysis. *Ore Geol. Rev.* **2016**, *79*, 316–331. [CrossRef]
18. Jia, L.Q.; Yang, M.; Meng, F.; He, M.Y.; Liu, H.M. Mineral Photos Recognition Based on Feature Fusion and Online Hard Sample Mining. *Minerals* **2021**, *11*, 1354. [CrossRef]
19. Wu, B.K.; Ji, X.H.; He, M.Y.; Yang, M.; Zhang, Z.C.; Chen, Y.; Wang, Y.Z.; Zheng, X.Q. Mineral Identification Based on Multi-Label Image Classification. *Minerals* **2022**, *12*, 1338. [CrossRef]
20. Su, C.; Xu, S.J.; Zhu, K.Y.; Zhang, X.C. Rock classification in petrographic thin section images based on concatenated convolutional neural networks. *Earth Sci. Inform.* **2020**, *13*, 1477–1484. [CrossRef]
21. Ma, H.; Han, G.Q.; Peng, L.; Zhu, L.Y.; Shu, J. Rock thin sections identification based on improved squeeze-and-Excitation Networks model. *Comput. Geosci.* **2021**, *152*, 104780. [CrossRef]
22. Singh, N.; Singh, T.N.; Tiwary, A.; Sarkar, K.M. Textural identification of basaltic rock mass using image processing and neural network. *Comput. Geosci.* **2010**, *14*, 301–310. [CrossRef]
23. Flügel, E.; Munnecke, A. *Microfacies of Carbonate Rocks: Analysis, Interpretation and Application*; Springer: Berlin, Germany, 2010.
24. Mlynarczuk, M.; Gorszczyk, A.; Slipek, B. The application of pattern recognition in the automatic classification of microscopic rock images. *Comput. Geosci.* **2013**, *60*, 126–133. [CrossRef]
25. Shu, L.; McIsaac, K.; Osinski, G.R.; Francis, R. Unsupervised feature learning for autonomous rock image classification. *Comput. Geosci.* **2017**, *106*, 10–17. [CrossRef]
26. Simonyan, K.; Zisserman, A. Very deep convolutional networks for large-scale image recognition. *arXiv* **2014**, arXiv:1409.1556.
27. Ronneberger, O.; Fischer, P.; Brox, T. U-net: Convolutional networks for biomedical image segmentation. In Proceedings of the Medical Image Computing and Computer-Assisted Intervention—MICCAI 2015: 18th International Conference, Part III 18, Munich, Germany, 5–9 October 2015; pp. 234–241.
28. He, K.; Zhang, X.; Ren, S.; Sun, J. Deep residual learning for image recognition. In Proceedings of the IEEE Conference on Computer Vision and Pattern Recognition, Las Vegas, NV, USA, 26 June–1 July 2016; pp. 770–778. [CrossRef]
29. Huang, G.; Liu, Z.; Van Der Maaten, L.; Weinberger, K.Q. Densely connected convolutional networks. In Proceedings of the Proceedings of the IEEE Conference on Computer Vision and Pattern Recognition, Honolulu, HI, USA, 21–26 July 2017; pp. 4700–4708.
30. Szegedy, C.; Vanhoucke, V.; Ioffe, S.; Shlens, J.; Wojna, Z. Rethinking the inception architecture for computer vision. In Proceedings of the IEEE Conference on Computer Vision and Pattern Recognition, Las Vegas, NV, USA, 26 June–1 July 2016; pp. 2818–2826.
31. Ren, S.Q.; He, K.M.; Girshick, R.; Sun, J. Faster R-CNN: Towards Real-Time Object Detection with Region Proposal Networks. *IEEE Trans. Pattern Anal. Mach. Intell.* **2017**, *39*, 1137–1149. [CrossRef]
32. Liu, X.B.; Wang, H.Y.; Jing, H.D.; Shao, A.L.; Wang, L.C. Research on Intelligent Identification of Rock Types Based on Faster R-CNN Method. *IEEE Access* **2020**, *8*, 21804–21812. [CrossRef]
33. Zhang, Y.; Li, M.C.; Han, S. Automatic identification and classification in lithology based on deep learning in rock images. *Acta Petrol. Sin.* **2018**, *34*, 333–342.
34. Liu, C.Z.; Li, M.C.; Zhang, Y.; Han, S.; Zhu, Y.Q. An Enhanced Rock Mineral Recognition Method Integrating a Deep Learning Model and Clustering Algorithm. *Minerals* **2019**, *9*, 516. [CrossRef]
35. Sun, Y.Q.; Chen, J.P.; Yan, P.B.; Zhong, J.; Sun, Y.X.; Jin, X.Y. Lithology Identification of Uranium-Bearing Sand Bodies Using Logging Data Based on a BP Neural Network. *Minerals* **2022**, *12*, 546. [CrossRef]
36. Cheng, G.J.; Guo, W.H. Rock images classification by using deep convolution neural network. *J. Phys. Conf. Ser.* **2017**, *887*, 012089. [CrossRef]
37. Polat, O.; Polat, A.; Ekici, T. Automatic classification of volcanic rocks from thin section images using transfer learning networks. *Neural Comput. Appl.* **2021**, *33*, 11531–11540. [CrossRef]
38. Ran, X.J.; Xue, L.F.; Zhang, Y.Y.; Liu, Z.Y.; Sang, X.J.; He, J.X. Rock Classification from Field Image Patches Analyzed Using a Deep Convolutional Neural Network. *Mathematics* **2019**, *7*, 755. [CrossRef]
39. Xu, S.T.; Zhou, Y.Z. Artificial intelligence identification of ore minerals under microscope based on deep learning algorithm. *Acta Petrol. Sin.* **2018**, *34*, 3244–3252.
40. Bai, L.; Wei, X.; Liu, Y.; WU, C.; CHEN, L. Rock thin section image recognition and classification based on VGG model. *Geol. Bull. China* **2019**, *38*, 2053–2058. [CrossRef]
41. Tan, X.C.; Zhao, L.Z.; Luo, B.; Jiang, X.F.; Cao, J.; Liu, H.; Li, L.; Wu, X.B.; Nie, Y. Comparison of basic features and origins of oolitic shoal reservoirs between carbonate platform interior and platform margin locations in the Lower Triassic Feixianguan Formation of the Sichuan Basin, southwest China. *Petrol. Sci.* **2012**, *9*, 417–428. [CrossRef]
42. Hollis, C.; Lawrence, D.A.; de Periere, M.D.; Al Darmaki, F. Controls on porosity preservation within a Jurassic oolitic reservoir complex, UAE. *Mar. Petrol. Geol.* **2017**, *88*, 888–906. [CrossRef]
43. Zhu, S.S.; Yang, W.Y.; Lu, B.B.; Huang, G.Y.; Hou, G.S.; Wei, S.P.; Zhang, Y.L. Micro image data set of some rock forming minerals, typical metamorphic minerals and oolitic thin sections. *Sci. Data Bank* **2020**. [CrossRef]
44. Jiao, L.; Zhang, F.; Liu, F.; Yang, S.; Li, L.; Feng, Z.; Qu, R. A survey of deep learning-based object detection. *IEEE Access* **2019**, *7*, 128837–128868. [CrossRef]

45. Oksuz, K.; Cam, B.C.; Kalkan, S.; Akbas, E. Imbalance problems in object detection: A review. *IEEE Trans. Pattern Anal. Mach. Intell.* **2020**, *43*, 3388–3415. [CrossRef]
46. Wu, X.W.; Sahoo, D.; Hoi, S.C.H. Recent advances in deep learning for object detection. *Neurocomputing* **2020**, *396*, 39–64. [CrossRef]
47. Russakovsky, O.; Deng, J.; Su, H.; Krause, J.; Satheesh, S.; Ma, S.; Huang, Z.H.; Karpathy, A.; Khosla, A.; Bernstein, M.; et al. ImageNet Large Scale Visual Recognition Challenge. *Int. J. Comput. Vis.* **2015**, *115*, 211–252. [CrossRef]
48. Shorten, C.; Khoshgoftaar, T.M. A survey on Image Data Augmentation for Deep Learning. *J. Big Data* **2019**, *6*, 60. [CrossRef]
49. Kukačka, J.; Golkov, V.; Cremers, D. Regularization for deep learning: A taxonomy. *arXiv* **2017**, arXiv:1710.10686.
50. Davis, J.; Goadrich, M. The relationship between Precision-Recall and ROC curves. In Proceedings of the 23rd International Conference on Machine Learning, Pittsburgh, PA, USA, 25–29 June 2006; pp. 233–240.
51. Tzutalin. Labellmg. Git Code. 2015. Available online: https://github.com/heartexlabs/labelImg (accessed on 10 March 2022).
52. Bubbliiiing. Faster-RCNN-Pytorch. Git Code. 2022. Available online: https://github.com/bubbliiiing/faster-rcnn-pytorch (accessed on 12 March 2022).
53. Everingham, M.; Van Gool, L.; Williams, C.K.I.; Winn, J.; Zisserman, A. The Pascal Visual Object Classes (VOC) Challenge. *Int. J. Comput. Vis.* **2010**, *88*, 303–338. [CrossRef]
54. Zhu, X.X.; Vondrick, C.; Fowlkes, C.C.; Ramanan, D. Do We Need More Training Data? *Int. J. Comput. Vis.* **2016**, *119*, 76–92. [CrossRef]
55. Warden, P. How Many Images Do You Need to Train a Neural Network? Pete Warden's Blog. 2017. Available online: https://petewarden.com/2017/12/14/how-many-images-do-you-need-to-train-a-neural-network (accessed on 10 June 2023).

minerals MDPI

Article

Fast Initial Model Design for Electrical Resistivity Inversion by Using Broad Learning Framework

Tao Tao [1], Peng Han [1,2,*], Xiao-Hui Yang [3], Qiang Zu [1], Kaiyan Hu [1], Shuangling Mo [1], Shuangshuang Li [1], Qiang Luo [4] and Zhanxiang He [1,2]

1 Department of Earth and Space Sciences, Southern University of Science and Technology, Shenzhen 518055, China; 12031167@mail.sustech.edu.cn (T.T.); zuqiang777@gmail.com (Q.Z.); huky@sustech.edu.cn (K.H.); mosl@mail.sustech.edu.cn (S.M.); 12331093@mail.sustech.edu.cn (S.L.); hezx@sustech.edu.cn (Z.H.)
2 Guangdong Provincial Key Laboratory of Geophysical High-Resolution Imaging Technology, Southern University of Science and Technology, Shenzhen 518055, China
3 Plateau Atmosphere and Environment Key Laboratory of Sichuan Province, School of Atmospheric Sciences, Chengdu University of Information Technology, Chengdu 610225, China; yangxh@cuit.edu.cn
4 Guangdong Provincial Geophysical Prospecting Team, Guangzhou 510080, China; luoqiang595@126.com
* Correspondence: hanp@sustech.edu.cn

Abstract: The electrical resistivity method is widely used in near-surface mineral exploration. At present, the deterministic algorithm is commonly employed in three-dimensional (3-D) electrical resistivity inversion to obtain subsurface electrical structures. However, the accuracy and efficiency of deterministic inversion rely on the initial model. In practice, obtaining an initial model that approximates the true subsurface electrical structures remains challenging. To address this issue, we introduce a broad learning (BL) network to determine the initial model and utilize the limited memory quasi-Newton (L-BFGS) algorithm to conduct the 3-D electrical resistivity inversion task. The powerful mapping capability of the BL network enables one to find the model that elucidates the actual observed data. The single-layer BL network makes it efficient and easy to realize, leading to much faster network training compared to that using the deep learning network. Both the synthetic and field experiments suggest that the BL framework could effectively obtain the initial model based on observed data. Furthermore, in comparison to using a homogeneous medium as the initial model, the L-BFGS inversion with the BL framework-designed initial model improves the inversion accuracy of subsurface electrical structures and expedites the convergence speed of the iteration. This study provides an effective approach for fast initial model design in a data-driven manner when the prior information is unavailable. The proposed method can be useful in high-precision imaging of near-surface mineral electrical structures.

Keywords: broad learning; 3-D electrical resistivity inversion; limited memory quasi-Newton; near-surface mineral exploration

1. Introduction

The electrical resistivity method is a well-known geophysical exploration method, holding significant value for various applications, including engineering surveys [1], environmental monitoring [2,3], mineral exploration [4–7], etc. The complex characteristics of underground structures are difficult to directly interpret using observed data imaging (e.g., apparent resistivity pseudosections). Thus, it is crucial to invert observed data to obtain accurate subsurface electrical structures. So far, various inversion techniques have been applied in electrical resistivity imaging studies, including global optimization, deterministic, and probabilistic methods [8]. Global optimization algorithms, such as simulated annealing [9], genetic algorithms [10,11], and particle swarm optimization [12], play a crucial role in solving optimization problems with multiple local optima by thoroughly exploring the entire search space to identify the global optimum [13]. However, the high computational costs associated with global optimization algorithms limit their practical application in 3-D electrical resistivity inversion problems [11,14,15]. Deterministic methods such as the Gauss–Newton method [16], limited memory quasi-Newton method [17], and conjugate gradient method [18–21] are the popular approaches for the 3-D electrical resistivity inversion due to their promising performance in terms of result accuracy, stability, and convergence speed [8,11,22].

In the practical operation, deterministic inversion faces the challenge of selecting a suitable initial model for the inversion task [8,23]. Using an initial model that better represents the true underground electrical structures reduces the probability that inversion results fall into the local minima [24]. The use of a priori information is an effective approach for addressing the selection of the initial model [25,26], yet the deficiency of prior information is a common situation for underground electrical structure prospecting. As a result, a homogeneous initial resistivity model is principally considered in 3-D electrical resistivity inversion [27–31]. Additionally, the determination of the resistivity value for a homogeneous initial model also relies on the empirical selection [32]. Therefore, the selection of an appropriate initial model becomes a crucial issue in 3-D electrical resistivity deterministic inversion.

Machine learning (ML) applied to geophysical problems is a cutting-edge direction [33]. In recent years, ML has been utilized in many geophysics application areas, including pattern recognition in seismic attributes [34], noise removal [35,36], and inversion tasks [8,37–40]. As a representative of the ML, the deep learning (DL) network comes with strong learning and generalization capabilities, enabling excellent performance in electrical resistivity inversion problems [41–43]. However, the training procedures of DL suffer a time-consuming drawback induced by a mass of hyperparameters [44]. Meanwhile, to further improve the inversion effect, Kong et al. utilize the prediction results of DL as the initial model and reference model for the Gauss–Newton method [45]. But, it still has the problem of the high time cost caused by DL, especially in 3-D structural imaging. To overcome this drawback, Chen and Liu proposed the broad learning (BL) approach equipped with a single-layer network [44]. The training time for the deep learning network on the same training set may be hundreds of times longer than that of the BL network [44]. Recently, The BL network has been successfully applied in geophysical inversions, achieving noteworthy results [46–49]. These applications demonstrate that BL possesses the advantages of excellent efficiency and robust mapping capability.

Therefore, in this study, we introduced the BL network into the 3-D electrical resistivity inversion process. Firstly, the BL network is established based on the samples generated by forward calculation. Secondly, the initial model of 3-D electrical resistivity inversion is derived from observed data by using the constructed BL network. Finally, based on the designed initial resistivity model, the limited memory quasi-Newton (L-BFGS) algorithm is employed to undertake the 3-D electrical resistivity inversion task [17,50]. In the subsequent sections, the inversion results of both synthetic models and field applications are presented to verify the feasibility of the proposed method.

2. Materials and Methods

2.1. BL Framework for Initial Resistivity Model Design

To obtain the suitable initial resistivity model in 3-D electrical resistivity deterministic inversion, the BL network is employed to approximate the mapping relationship between the apparent resistivity data $\boldsymbol{x} \in \mathbb{R}^{N_x}$ (N_x is the number of apparent resistivity data) and the resistivity model $\boldsymbol{y} \in \mathbb{R}^{N_y}$ (N_y is the number of the resistivity model grid) as follows:

$$\boldsymbol{y} = \psi(\boldsymbol{x}), \tag{1}$$

where ψ is the functional mapping between apparent resistivity and the initial resistivity model to be simulated by the BL network.

Figure 1 shows the structure of the BL network for designing initial resistivity models. For the BL network training, the input $X \in \mathbb{R}^{N \times N_x}$ denotes the input apparent resistivity dataset, and the output $Y \in \mathbb{R}^{N \times N_y}$ reflects the output resistivity model dataset, where N represents the number of training samples. The following relationship is constructed to represent the target functional mapping ψ:

$$Y = [Z_1, \cdots, Z_n, H_1, \cdots, H_m]W = [Z, H]W, \tag{2}$$

where $Z = [Z_1, \cdots, Z_n]$ is all the mapping features of the input apparent resistivity dataset X, and n is the number of groups of mapping features. Each group of mapping features with p neurons can be obtained by sparse autoencoder. $H = [H_1, \cdots, H_m]$ is the number of the enhancement nodes derived from Z, and m is the number of groups of enhancement nodes. Each group of enhancement nodes is calculated by the nonlinear activation function as follows:

$$H_j = \varphi\left(ZW_j^H + \beta_j^H\right), \tag{3}$$

where W_j^H represents the jth random weight matrix; β_j^H represents the jth random bias vector, and H_j is the jth group of enhancement nodes. In this study, the $tanh(\cdot)$ function is chosen as the nonlinear activation function. W is the only unknown matrix in the BL network, and it can be determined by the pseudoinverse as follows:

$$W = [Z, H]^+ Y = A^+ Y, \tag{4}$$

where $A = [Z, H]$. Particularly, the ridge regression algorithm is applied to solve the pseudoinverse A^+ [44].

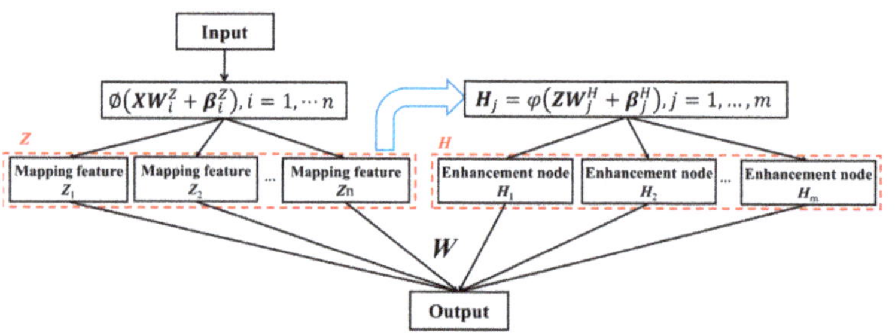

Figure 1. The BL network structure.

The BL network training is completed with the weighting matrix W determined for a given optimal network complexity (namely, a fixed combination of the parameters p, n, and m) [46]. Input the observed apparent resistivity data x into the trained BL network, and the initial resistivity model $\hat{y} \in \mathbb{R}^{N_y}$ can be designed:

$$\hat{y} = [Z', H']W, \tag{5}$$

where Z' is the mapping features of the apparent resistivity data $x \in \mathbb{R}^{N_x}$, and H' is the number of the corresponding enhancement nodes. For more detailed information about the BL network, please refer to [17].

2.2. Generation of Training Dataset

The BL is a supervised learning network that depends on a training dataset to learn the mapping relationship ψ. In this study, the training dataset is created through forward calculation and consists of the resistivity model (network output) and its corresponding apparent resistivity (network input). Specifically, the resistivity model is constructed with a single cuboid anomalous body 'A' embedded in a homogeneous background 'B', as depicted in Figure 2a. The size and spatial position of anomalous body 'A' in each resistivity model are randomly generated. The resistivity values of 'A' and 'B' are randomly selected within a range from 1 to 500 Ω·m. Figure 2b illustrates the spatial distribution of observed apparent resistivity data for the resistivity model, calculated using the finite difference method.

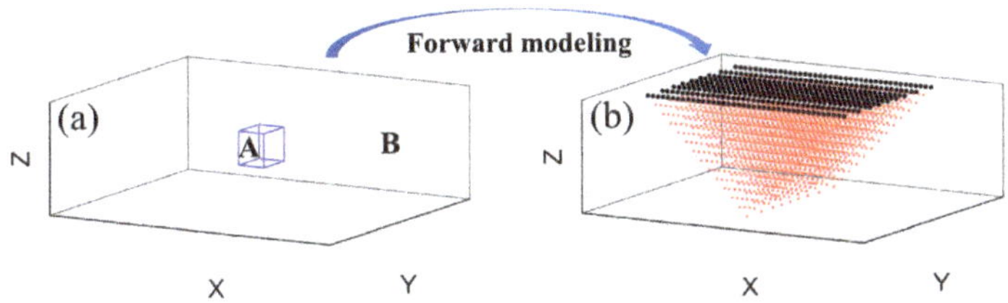

Figure 2. Schematic of a pair training sample. (**a**) Output resistivity model sample ('A' represents the regular anomalous body, and 'B' represents the background model). (**b**) Input apparent resistivity data sample (The black dots represent the locations of surface electrodes, while the red dots represent the spatial positions of apparent resistivity data points. Wenner arrays are adopted for visualization).

In the synthetic experiments, nine profiles are employed for data collection, each featuring 40 electrodes spaced 1 m apart. The data for each profile are acquired using the Wenner array, resulting in the collection of 247 data points. The model mesh is divided into 92 (X direction) × 84 (Y direction) × 29 (Z direction). In the survey line area, the minimum grid spacing of the resistivity model in the lateral direction is set to half of the electrode spacing. Beyond this region, the lateral grid spacing gradually increases with a factor of 1.3. For the first three layers in the vertical direction, the grid spacing is 0.1 m, and then it increases with a factor of 1.2.

2.3. Select of Training Dataset Size

The choice of training dataset size in the training phase depends on various factors, such as the complexity of the resistivity model to be predicted and the desired degree of precision [8]. Generally, a larger training dataset can help improve the performance and generalization ability of the network [8]. However, collecting an excessive number of training samples significantly increases the computation time consumption while providing minimal improvements in results [37]. Therefore, it is meaningful to choose the appropriate number of training samples.

To determine the appropriate training dataset size, we trained the BL using different training dataset sizes and evaluated the results using a validation dataset consisting of 500 samples that were not involved in the training process. Table 1 provides mean absolute percentage error (*MAPE*) at different training dataset sizes. As the size of the training dataset increases, the *MAPE* values of the training dataset and validation dataset gradually decrease. When the training dataset size is 12,000, both the training dataset and validation dataset have the minimum *MAPE* values, and the *MAPE* of the validation dataset does not decrease further compared to a training dataset size of 11,000. Therefore, to reduce the time spent on collecting training samples, we utilized the training dataset consisting of 12,000 samples to train the BL network.

Table 1. Training of BL network with different sizes of training datasets.

Number of Training Dataset	6000	8000	10,000	11,000	12,000
MAPE of the training dataset (%)	1.26	1.16	1.07	1.02	0.99
MAPE of the validation dataset (%)	1.06	1.04	1.03	1.02	1.02

2.4. Choose of BL Network Complexity

The complexity of the BL network is determined by the following three key factors: the number of feature nodes (p); the number of mapping groups (n); and the corresponding enhancement nodes (m). When these parameters are set to small values, the network structure becomes simple, which may not be able to efficiently extract features and result in poor mapping capability. On the other hand, if the network structure is too complex, it can lead to overfitting the training dataset [48]. Therefore, selecting an appropriate network complexity is crucial.

In this study, we select the appropriate network complexity by conducting an exhaustive search on these three parameters (p, n, m) within the range [50:50:200], [40:20:100], and [100:100:500], using 12,000 training samples and 500 validation samples. All samples are generated in the manner described in Section 2.2, and the validation dataset is not included in the training process. To assess the predictive accuracy of different network complexities, the deviation between the predicted resistivity model and the true one is calculated by the mean absolute percentage error (*MAPE*). When the p, n, and m are set to 200, 80, and 100, respectively, both the training and validation datasets exhibit the lowest *MAPE* values. Therefore, this study considers these three values as the optimal network complexity for the BL.

2.5. L-BFGS Inversion with Designed Initial Model

To circumvent the challenge of selecting the initial resistivity model in 3-D electrical resistivity deterministic inversion, we adopt the output result $\hat{y}$ obtained from the BL framework as the initial resistivity model for the inversion task by using the L-BFGS algorithm, and the corresponding objective function is as follows:

$$\Gamma(y) = (x - f(y))^{\mathrm{T}} C_x^{-1} (x - f(y)) + \mu \left(y - y_{ref}\right)^{\mathrm{T}} C_y^{-1} \left(y - y_{ref}\right), \quad (6)$$

where x denotes the apparent resistivity data used for the inversion; y is the inverted resistivity model at each iteration, with the initial resistivity model being the output result $\hat{y}$ of the BL framework; $f(\cdot)$ is the forward calculation operator, and C_x is the data covariance matrix, which takes into account the relative importance of the observations [51]. To avoid the inversion process from focusing on fitting large values in the observed data, the observed data are employed to construct the data covariance matrix. The superscript T represents the transpose operation; y_{ref} is the homogeneous reference model whose resistivity value is given by the mode of the designed initial model vector $\hat{y}$; C_y is the model covariance matrix; and μ is the regularization parameter. For the details of the L-BFGS algorithm, readers can refer to [17,52,53]. The inversion process using the designed initial model is concisely outlined as follows:

1. Set fitting tolerance error ξ and a maximum number of iterations *Num*. Input apparent resistivity data x and the designed initial resistivity model $\hat{y}$;
2. Calculate the apparent resistivity data x^{cal} of the current kth iteration resistivity model y_k by forward modeling: $x^{cal} = f(y_k)$;
3. Calculate the partial derivative $\nabla\Gamma(y_k, x)$ of Equation (6), and calculate the search direction p_k. Obtain the appropriate step α_k by line search and update resistivity model $y_{k+1} = y_k + \alpha_k p_k$;
4. Compute the misfit between observed data x and calculated data x^{cal}. If misfit $< \xi$ or $k > Num$ inversion stop; otherwise, set $k = k + 1$ and go to step 2.

3. Results

In this section, four synthetic examples and a field experiment are used to assess the performance of 3-D electrical resistivity inversion based on the initial model designed by a BL framework. In addition to qualitatively evaluating the inversion results through visual judgment, we perform the quantitative evaluation on the final inversion results using the $MAPE = \frac{1}{N_y}\sum_{i=1}^{N_y}\left|\frac{y_i - \hat{y}_i}{y_i}\right|$ (where $\hat{y}$ represents the final inversion result) and assess the data misfit using the normalized root-mean-error $NRMSE = \sqrt{\frac{1}{N_x}\sum_{i=1}^{N_x}\left(\frac{x_i - \hat{x}_i}{x_i}\right)^2}$ (where $\hat{x}$ represents the apparent resistivity data obtained through the forward calculation of the inversion model). In this study, the initial resistivity model design is conducted on a workstation equipped with a 3.80 GHz Intel Xeon W-2235 CPU.

3.1. Synthetic Experiment 1

Synthetic experiment 1 tests the performance of the 3-D electrical resistivity inversion with the designed initial model using a single regular anomalous body model (Figure 3a). It consists of a 10 Ω·m rectangular prism embedded in a 100 Ω·m homogeneous medium. The dimensions of the rectangular prism are 4 m (length) × 8 m (width) × 2 m (height), with its top located at a depth of 3 m. The spatial heterogeneity of the synthetic resistivity model is determined by multiple profiles aligned in parallel along the y-direction (Figure 3a, black dots). Figure 3b shows the initial resistivity model designed by the BL framework, with training and prediction times of 207 s and 0.5 s, respectively. The designed initial model approximately captures the structure of the anomalous body.

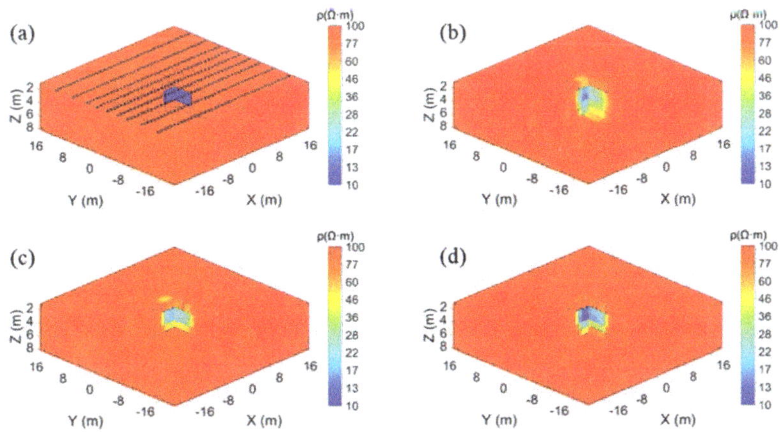

Figure 3. Synthetic model and inversion results. (**a**) The synthetic model. The black dots represent the positions of surface electrodes; nine profiles are deployed. (**b**) The initial resistivity model is designed by the BL framework. (**c**,**d**) The L-BFGS inversion results using the homogeneous initial model and the designed initial model, respectively.

Figure 3c illustrates the inversion result of the L-BFGS algorithm using the homogeneous initial model (inversion time approximately 115 min). It delineates the shape of the anomalous body, but there are redundant structures at the top interface of the anomalous body, which can be misinterpreted as the anomalous extending to the surface. Figure 3d shows the L-BFGS inversion result using the initial resistivity model designed by the BL framework (inversion time approximately 110 min). Comparing Figure 3c with Figure 3d, it is evident that Figure 3d effectively suppresses redundant structures on the upper boundary of the anomalous body, leading to a more accurate delineation of the anomalous body boundary. Furthermore, the *MAPE* values of the L-BFGS inversion results for those two initial models are 2.62% (homogeneous initial model) and 2.35% (designed initial model), respectively. It indicates that the inversion result using the proposed method is more accurate compared to the traditional method.

The *NRMSE* value (0.31%) of the first iteration of the proposed method corresponds to the *NRMSE* value of the designed initial model. However, after 100 iterations, the data misfit (*NRMSE* = 0.11%) using the traditional method is smaller than directly using the output of the BL network (the designed initial model). It indicates that the designed initial model fits the data less effectively than the inversion results of the homogeneous initial model. We, thereby, used the outputting model of the trained BL network as the initial model for further L-BFGS inversion.

Figure 4 displays the *NRMSE* curves of the L-BFGS method utilizing two different initial models. The *NRMSE* value achieved by the proposed method is consistently lower than that obtained by the traditional method at the same iteration number. For example, after 35 iterations of the L-BFGS method utilizing the designed initial model, the *NRMSE* value is smaller than the *NRMSE* value obtained by the traditional method after 100 iterations. This test demonstrates the capacity of the initial model crafted by a BL network to enhance the convergence speed of deterministic inversion.

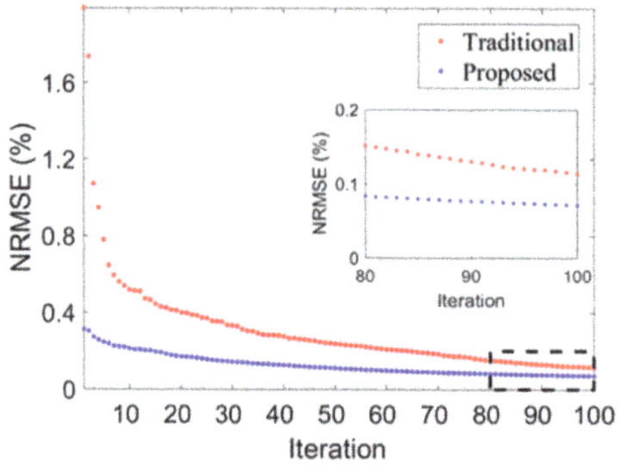

Figure 4. *NRMSE* curves of L-BFGS inversion shown in Figure 3 using the homogeneous initial model and the designed initial model, respectively. The red dots indicate the homogeneous medium as the initial resistivity model (Traditional). The blue dots indicate the initial resistivity model designed by the BL framework (Proposed).

3.2. Synthetic Experiment 2

In many cases, the distribution of the electrical property of the underlying geology exhibits an irregular structure. Hence, this section utilizes an irregular anomaly body model to assess the proposed method. The background and anomaly resistivity values are set at 50 Ω·m and 100 Ω·m, respectively. The irregular anomaly body is positioned approximately as a parallelogram on the horizontal plane and is situated at a depth ranging from 4 m to 7 m underground. For a clearer visual representation of the irregular anomaly body's shape, Figure 5a depicts the true model slices at various depths ([Z = 4 m, 5 m, 6 m, 7 m]). Figure 5b displays the initial model slices generated by the BL network. The total training and prediction time of the BL network is within 4 min. While the resistivity values of the anomaly body may not precisely match the true value of 100 Ω·m, the shape of the anomaly body closely resembles that of the true model. Figure 5c illustrates slices of the L-BFGS inversion result using the homogeneous initial model (inversion time of approximately 129 min). It inadequately delineates both the value and shape of the anomaly body and the accuracy of the inversion result degrades with an increase in the depth of the anomaly body. Furthermore, the structural details of the anomaly body are not clearly discernible at depths between 6 m and 7 m. Figure 5d exhibits slices of the L-BFGS inversion result utilizing the initial model designed by the BL network (inversion time of approximately 102 min). An evident high-resistivity anomalous body is discerned at depths ranging from 3 m to 7 m, corresponding to the actual anomaly body area. The *MAPE* values of the L-BFGS inversion results for the two initial models are 1.91% (homogeneous initial model) and 1.28% (designed initial model), respectively. Consequently, the subsurface resistivity structure presented in Figure 5d is deemed more realistic in comparison to the L-BFGS inversion result using the homogeneous initial model (Figure 5c). The results suggest that utilizing the designed initial model can improve the accuracy of electrical resistivity inversion.

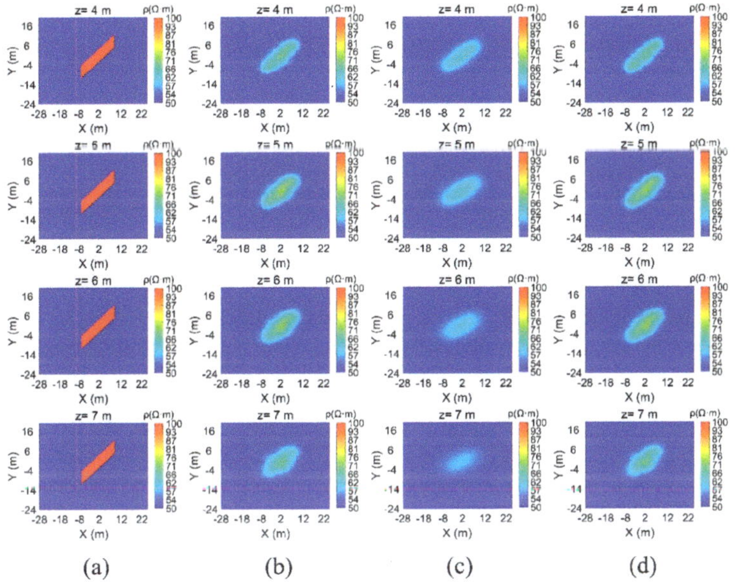

Figure 5. Synthetic model slices, designed initial model slices, and inversion results slices. By row, from top to bottom: the slices at a depth of 4 m, 5 m, 6 m, and 7 m, respectively. By the column, from left to right: (**a**) The synthetic model slices; (**b**) The initial model slices are designed by the BL framework; (**c**) The inversion result slices of L-BFGS using the homogeneous initial model; (**d**) The inversion result slices of L-BFGS inversion using the designed initial model. Note: The resistivity model in the training dataset contains only one regular anomaly body, and the resistivity values of both the anomaly and the model background are random numbers between 1 and 500 Ω·m.

The *NRMSE* curves of the L-BFGS inversion for two different initial models are shown in Figure 6, and the *NRMSE* value for inversion using the designed initial model decreases more rapidly than that of the homogeneous initial model inversion. Specifically, at 13 inversion iterations, the *NRMSE* value using the designed initial model is smaller than the *NRMSE* value at 100 inversion iterations using the homogeneous initial model.

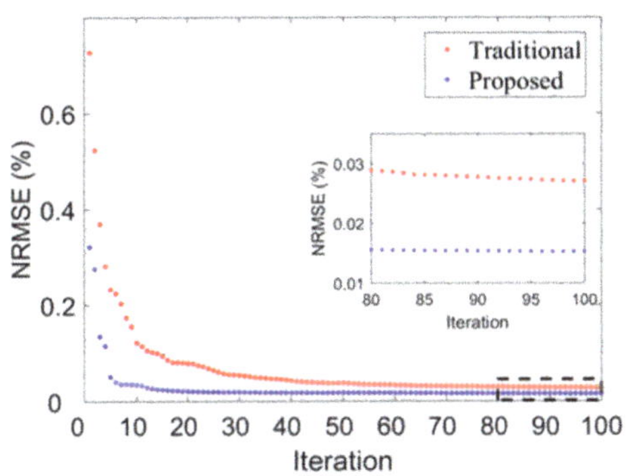

Figure 6. *NRMSE* curves of L-BFGS inversion shown in Figure 5 using the homogeneous initial model and the designed initial model, respectively. The red dots indicate the homogeneous medium as the initial resistivity model (Traditional). The blue dots indicate the initial resistivity model designed by the BL framework (Proposed).

This synthetic example illustrates that the BL network, trained on single regular model samples, can effectively design the initial model for irregular underground electrical structures. Moreover, employing the designed initial model in electrical resistivity inversion can enhance the convergence speed of the inversion process.

3.3. Synthetic Experiment 3

In practice, subsurface electrical structures may contain multiple anomalous bodies. Therefore, we consider a complex synthetic model to test the generalization of the proposed method. The synthetic model consists of three 10 Ω·m conductive rectangular prisms and three 400 Ω·m resistive rectangular prisms embedded in a 100 Ω·m homogeneous medium. All six prisms are located 3 m to 5 m below the surface, as shown in Figure 7a. It is important to note that the resistivity model in the training dataset is a single regular anomaly body, and the model with multiple anomalies is not included. The total training and prediction time of the BL network is within 4 min (Training (218 s), Prediction (0.6 s)). Figure 7b shows the initial resistivity model designed by the BL framework. It roughly delineates the electrical structures of the true model, showing that the training dataset only contains a single anomalous body and can also perform initial model design for electrical structures with multiple anomalous bodies.

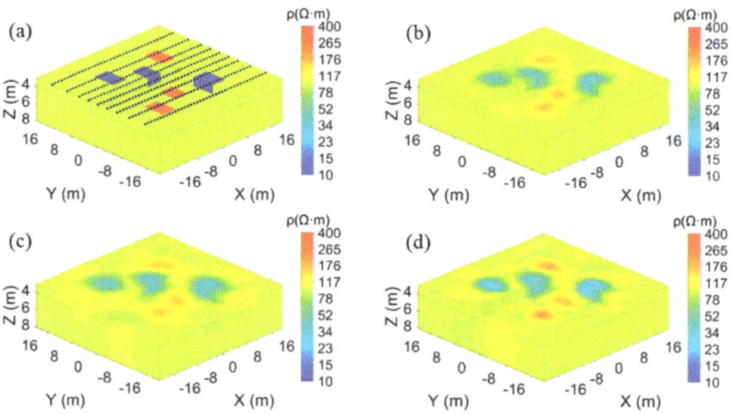

Figure 7. Synthetic model and inversion results. (**a**) The synthetic model. (**b**) The initial model. (**c**,**d**) The L-BFGS inversion results using the homogeneous initial model and the designed initial model, respectively. Note: The resistivity model in the training dataset contains only one regular anomaly body, and the resistivity values of both the anomaly and the model background are random numbers between 1 and 500 Ω·m.

Figure 7c,d displays the L-BFGS inversion results using the homogeneous initial model (inversion time of approximately 116 min) and the designed initial model (inversion time of approximately 93 min), respectively. Compared to Figure 7c, Figure 7d shows better recovery of the values and shape of the anomalous bodies. Furthermore, the *MAPE* values for the L-BFGS inversion results using the homogeneous initial model (Traditional) and the designed initial model (Proposed) are 4.97% and 4.30%, respectively. This indicates that the inversion result obtained using the proposed method is closer to the true model compared to the traditional method.

We also show the *NRMSE* curves of the L-BFGS inversion for two different initial resistivity models in Figure 8. The results indicate that the *NRMSE* values using the proposed method are much smaller than those of the traditional method under the same iteration number.

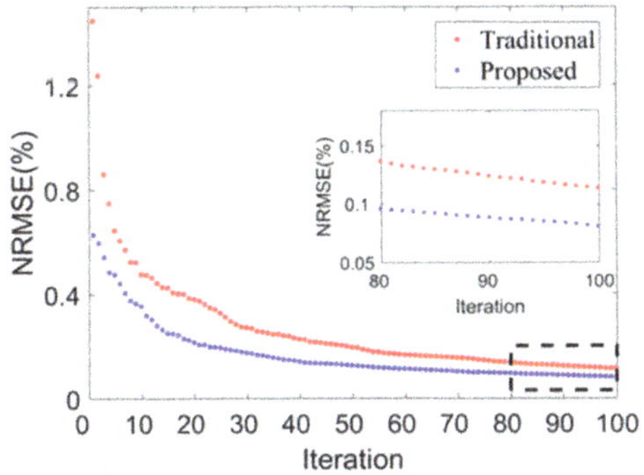

Figure 8. *NRMSE* curves of the L-BFGS inversion shown in Figure 7 using the homogeneous initial model (Traditional) and the designed initial model (Proposed), respectively. The red dots indicate the homogeneous medium as the initial resistivity model (Traditional). The blue dots indicate the initial resistivity model designed by the BL framework (Proposed).

3.4. Synthetic Experiment 4

In this section, a complex synthetic model, as illustrated in Figure 9a, is employed to evaluate the proposed method. The synthetic model comprises a 100 Ω·m rectangular prism and a 300 Ω·m irregular resistive anomalous body embedded in a 200 Ω·m homogeneous half-space. The depth of the anomaly bodies ranges from 3 m to 5 m, with the high-conductive body surrounded by the high-resistive body on three faces. The initial model designed through the Broad Learning (BL) framework is illustrated in Figure 9b, and the corresponding training and prediction times are 200 s and 0.5 s, respectively. It approximately depicts structures of high and low resistivity anomalies that resemble the characteristics of the actual model. The L-BFGS inversion results using the homogeneous initial model and the designed initial model are presented in Figure 9c (inversion time of approximately 122 min) and Figure 9d (inversion time of approximately 106 min), respectively. In comparison with the inversion result in Figure 9c (*MAPE* 1.11%), the result obtained with the designed initial model (Figure 9d) (*MAPE* 0.81%) more accurately characterizes the value and shape of anomaly bodies.

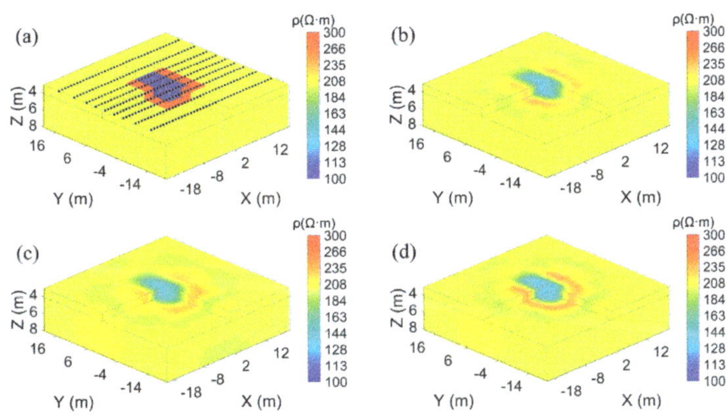

Figure 9. Synthetic model, designed initial model, and inversion results. (**a**) Synthetic model. (**b**) The initial model is designed by the BL framework. (**c**) The L-BFGS inversion result using the homogeneous initial model. (**d**) The L-BFGS inversion result using the designed initial model. Note: The resistivity model in the training dataset contains only one regular anomaly body, and the resistivity values of both the anomaly and the model background are random numbers between 1 and 500 Ω·m.

Figure 10 illustrates the *NRMSE* curves of the L-BFGS method with the two different initial models. The *NRMSE* value from the inversion using the designed initial model is smaller than the *NRMSE* value of the homogeneous initial model inversion at the same number of iterations. Specifically, at 54 iterations, the *NRMSE* value using the designed initial model is smaller than the *NRMSE* value at 100 iterations of the homogeneous initial model inversion.

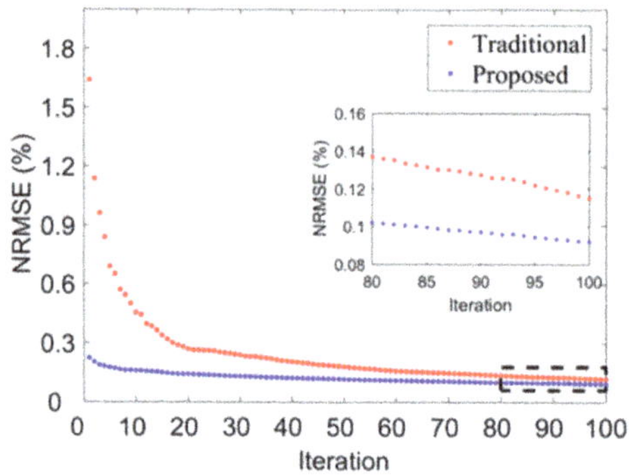

Figure 10. *NRMSE* curves of the L-BFGS inversion shown in Figure 9 using the homogeneous initial model (Traditional) and the designed initial model (Proposed), respectively. The red dots indicate the homogeneous medium as the initial resistivity model (Traditional). The blue dots indicate the initial resistivity model designed by the BL framework (Proposed).

3.5. Field Experiment

In addition, we conduct a field experiment in a farmland area south of Shenzhen, China. In the experimental area, we carried out three profiles with 1.5 m profile spacing, as shown in Figure 11a. For each profile, 20 electrodes are laid out with an electrode interval of 0.4 m. We bury an empty foam box (high resistivity) (Figure 11c) of 0.3 m (length) $\times$ 0.27 m (width) $\times$ 0.2 m (height) under the No. 5 electrode of profile P1. Its top interface is 0.1 m below the ground surface. In addition, we bury an iron box (high conductivity) (Figure 11b) with a size of 0.42 m (length) $\times$ 0.3 m (width) $\times$ 0.28 m (height) under No. 10 and No. 11 electrodes of profile P2; the top interface of the iron box is approximately 0.05 m from the surface, and the long side is parallel to the survey profile. The standard Wenner configuration is used for the survey. The model mesh is divided into 67 $\times$ 51 $\times$ 23. Setting the electrode spacing to less than half the size of the targets could enhance the spatial resolution of electrical resistivity exploration.

A high-resistivity anomalous body in the No. 5 electrode region of profile P1 and a low-resistivity anomalous body in the middle of profile P2 can be observed in the initial resistivity model designed by the BL framework (Figure 12a). These two resistivity anomalous bodies correspond to the foam box and the iron box, respectively. However, there are redundant structures at the bottom of the anomalous bodies.

The L-BFGS inversion results after 10 iterations, employing the homogeneous initial model (inversion time approximately 24 min) and the designed initial model (inversion time approximately 22 min), are presented in Figures 12b and 12c, respectively. Compared with the inversion result in Figure 12b, the result in Figure 12c shows a clearer characterization of the foam box and iron box. The *NRMSE* curves in Figure 12d also indicate that the L-BFGS inversion using the designed initial resistivity model fits the observed apparent resistivity data better than the homogeneous initial model. This field experiment demonstrates that higher inversion accuracy can be obtained based on the designed initial resistivity model using the BL framework rather than directly using L-BFGS inversion.

Figure 11. Schematic of the field area. (**a**) A photo shows the three profiles of the field site. P1, P2, and P3 represent profile 1, profile 2, and profile 3, respectively. (**b**) The photo shows the iron box under profile P2. (**c**) The photo shows the foam box under profile P1.

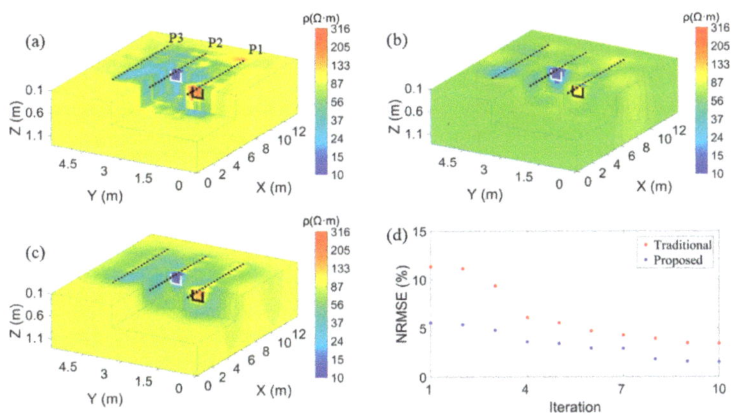

Figure 12. Inversion results and iteration *NRMSE* curves. The white rectangle indicates the location of the iron box, and the black rectangle indicates the location of the foam box. (**a**) The initial model. (**b**) The L-BFGS inversion result using the homogeneous initial model. (**c**) The L-BFGS inversion result using the designed initial model. (**d**) The *NRMSE* curves of the L-BFGS inversion using the homogeneous initial model (Traditional) and the designed initial model (Proposed), respectively. Note: The resistivity model in the training dataset contains only one regular anomaly body, and the resistivity values of both the anomaly and the model background are random numbers between 1 and 500 Ω·m.

4. Discussion

4.1. Generalization Ability of BL Network

Synthetic experiments showcase the generalization ability of the BL network in characterizing both multiple anomalous electrical bodies and irregular anomalous bodies. In these experiments, all earth model samples used for training the BL network are of resistivity values within the range of (1, 500) Ω·m. However, in practical operation, the resistivity values of earth models can surpass this range. To assess the generalization capability of the

BL network, we test it with five earth models having resistivity values exceeding 500 Ω·m. The five synthetic models are depicted in the left column of Figure 13, and the specific parameter values of the models are detailed in Table 2. The right column of Figure 13 displays the predicted results using the BL network trained before (the samples within the range of (1, 500) Ω·m). The BL network can still effectively delineate the structural characteristics of the true subsurface model. Notably, the BL network can design an accurate initial model, even when the resistivity values of the models exceed the training dataset.

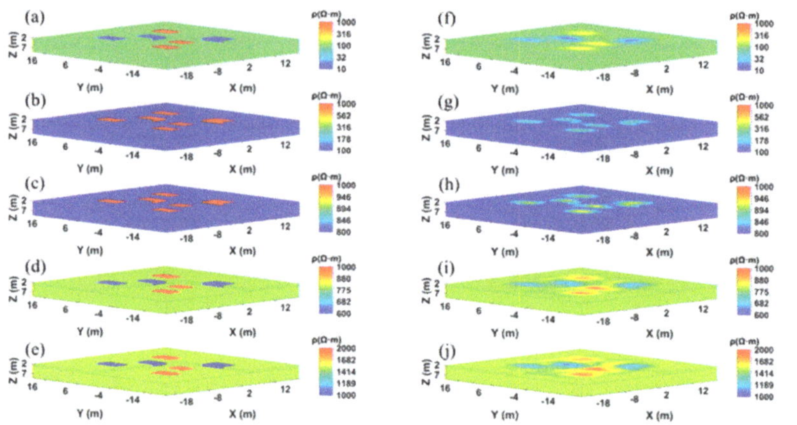

Figure 13. Synthetic models and initial models. (**a–e**) The synthetic models. The resistivity values corresponding to the true models are presented in Table 2. (**f–j**) The initial model designed by the BL network. Note: The resistivity model in the training dataset contains only one regular anomaly body, and the resistivity values of both the anomaly and the model background are random numbers between 1 and 500 Ω·m.

Table 2. The resistivity values of synthetic models.

Synthetic Model	Low Resistivity	High Resistivity	Background Resistivity	*MAPE* (%)
(a)	10 Ω·m	1000 Ω·m	100 Ω·m	4.34
(b)		1000 Ω·m	100 Ω·m	3.12
(c)		1000 Ω·m	800 Ω·m	0.43
(d)	600 Ω·m	1000 Ω·m	800 Ω·m	0.51
(e)	1000 Ω·m	2000 Ω·m	1500 Ω·m	0.98

4.2. Noise Resistance Test

In practice, observation data are often contaminated by noise. To assess the impact of noise on the proposed method, we add 20% Gaussian random noise ($noise_i = x_i \times 20\% \times r_i$; x_i represents ith observed apparent resistivity data, and r_i represents the Gaussian random number) to the apparent resistivity data of Experiment 1. Figure 14a shows the initial resistivity model designed by the BL framework, which contains a noticeable false anomaly in the background. Figure 14b shows the L-BFGS inversion result using the initial resistivity model designed by the BL framework. Compared with the result in Figure 14a, the redundant structures in Figure 14b are significantly reduced, and the anomalous body is closer to the true model. As shown in Figure 14a, although the input data are contaminated and changed by the noise, the BL network can still provide useful information about the anomalous body. Compared with the result shown in Figure 3b, the mapping accuracy decreased when relatively strong noise (20%) existed. Therefore, we use the output result of the BL framework as the initial resistivity model for deterministic inversion to further improve the accuracy of the inversion result.

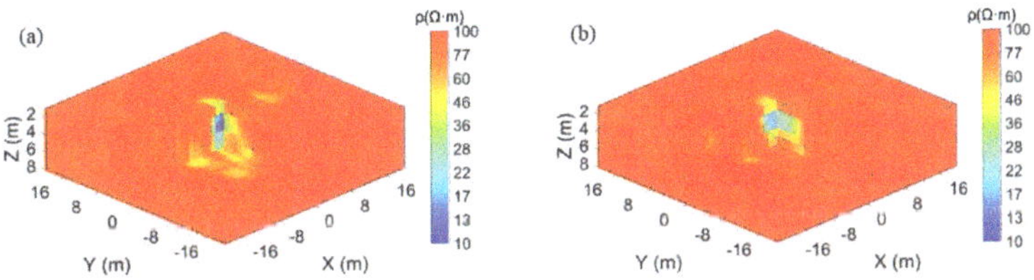

Figure 14. Initial resistivity model and inversion result. (**a**) The initial resistivity model is designed by the BL framework. (**b**) The L-BFGS inversion result uses the initial resistivity model designed by the BL framework.

The impact of noise is an important issue for field data. Noise within the field data diminishes the precision of inversion results, consequently impeding the accurate identification of authentic anomalies. To enhance the accuracy of imaging results, we plan to integrate denoising methods with the proposed inversion technique in future work.

4.3. Checkerboard Test

To further verify the reliability of the field experiment results, we employ the chessboard test to analyze the resolution of the inversion results. The checkerboard model is depicted in Figure 15a. The background resistivity value is 50 Ω·m, with the low-resistance anomaly at 10 Ω·m and the high-resistivity anomaly at 100 Ω·m. The sizes of the low-resistance anomaly and the high-resistance anomaly are both 0.4 m (length) × 0.4 m (width) × 0.2 m (height). The intervals between the anomaly bodies in the X direction and Y direction are 1 m and 0.8 m, respectively. Figure 15b presents the initial model designed through the BL network. In the whole region, the initial model designed by the BL network could approximately represent the pattern of the anomaly, although in regions lacking measuring points some redundant structures existed. Figure 15c depicts the inversion result obtained through the L-BFGS algorithm using a homogeneous initial model. Below the survey line, the inversion result reproduces the structure of the anomaly bodies. However, beyond the coverage area of the survey line, the inversion result fails to depict the anomaly bodies. Figure 15d shows the inversion result obtained through the L-BFGS algorithm using the designed initial model. Compared Figure 15c with Figure 15d, it is evident that Figure 15d gives a more accurate delineation of the anomaly bodies. The *NRMSE* curves of the L-BFGS inversion for two different initial models are depicted in Figure 16. The *NRMSE* value for inversion using the designed initial model decreases more rapidly than that of the homogeneous initial model inversion. Specifically, at 27 inversion iterations, the *NRMSE* value using the designed initial model is smaller than that at 100 inversion iterations using the homogeneous initial model.

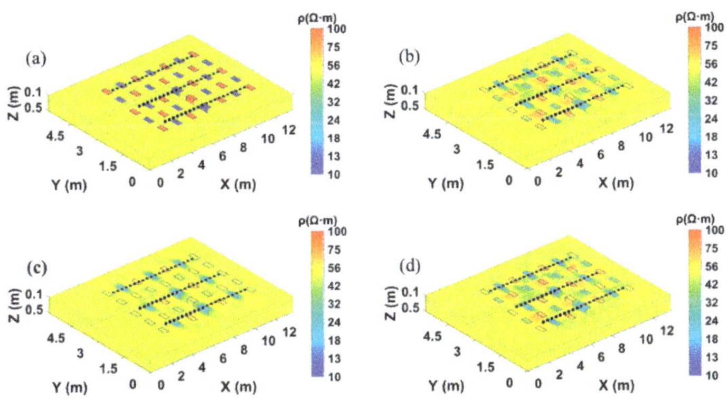

Figure 15. Synthetic model and inversion results. The black dots represent the locations of the electrodes on the surface. (**a**) The synthetic model. The black dots represent the positions of surface electrodes; nine

profiles are deployed. (**b**) The initial resistivity model is designed by the BL framework. (**c**,**d**) The L-BFGS inversion results using the homogeneous initial model and the designed initial model, respectively. Note: The resistivity model in the training dataset contains only one regular anomaly body, and the resistivity values of both the anomaly and the model background are random numbers between 1 and 500 Ω·m.

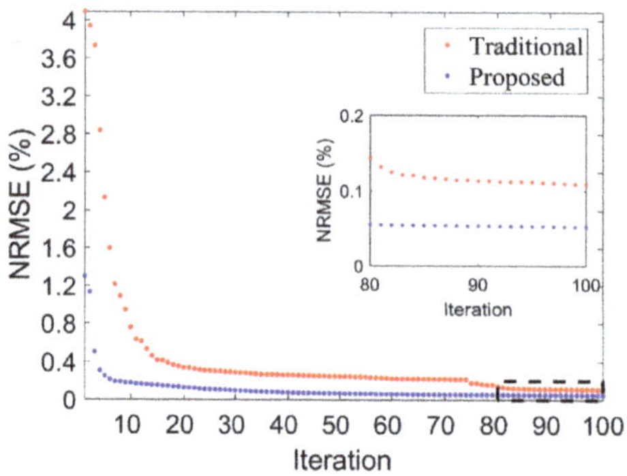

Figure 16. *NRMSE* curves of L-BFGS inversion shown in Figure 15 using the homogeneous initial model and the designed initial model, respectively. The red dots indicate the homogeneous medium as the initial resistivity model (Traditional). The blue dots indicate the initial resistivity model designed by the BL framework (Proposed).

Different arrays demonstrate varying sensitivities to underground structures, and the choice of observation arrays should align with specific problem requirements, such as signal-to-noise ratio and horizontal or vertical resolutions. In this study, the Wenner array, selected for its high signal-to-noise ratio, is employed to test the proposed method. The BL network trained in this article is only suitable for the given observation strategy. In the practical operation, the parameters, including observation arrays, the number of observation data, survey line spacing, and electrode spacing to generate the training dataset, should be set the same as field observation. For measurement data with different observation arrays or containing multiple observation arrays, a new training set needs to be constructed for network training.

5. Conclusions

The initial model selection is crucial and challenging in 3-D electrical resistivity deterministic inversion. In this study, we introduce the Broad Learning (BL) framework for fast initial model design and employ the L-BFGS algorithm for the 3-D electrical resistivity inversion task. The BL framework is constructed based on the training dataset consisting of only one regular anomaly body, yet it could provide a reliable and effective initial resistivity model, even though the underground electrical structure may contain multiple anomalous bodies. Synthetic and field experiments verify the effectiveness of the BL framework in designing the initial resistivity model. Furthermore, in comparison to the homogeneous initial resistivity model, the L-BFGS inversion using the initial resistivity model designed by the BL framework can increase the convergence speed and improve the accuracy of the inversion result. This study provides a useful tool for the initial resistivity model design in deterministic inversion and can be helpful in high-precision imaging of electrical structures.

Metallic minerals typically exhibit lower resistivities compared to non-metallic minerals or surrounding rocks. This contrast in resistivity values enables the identification of mineralized zones through electrical resistivity exploration. The proposed inversion method in this study facilitates the high-precision imaging of subsurface electrical structures, enabling geologists and geophysicists to better delineate and understand the distribution of mineralized zones.

Author Contributions: Conceptualization, T.T., X.-H.Y. and P.H.; methodology, T.T., X.-H.Y. and P.H.; investigation, T.T., X.-H.Y. and P.H.; data curation, Q.Z., S.M. and S.L.; writing—original draft preparation, T.T.; writing—review and editing, T.T., X.-H.Y., K.H. and P.H.; visualization, T.T., X.-H.Y., K.H. and P.H.; supervision, Q.L., Z.H. and P.H.; project administration, P.H.; funding acquisition, P.H. All authors have read and agreed to the published version of the manuscript.

Funding: This research is partly supported by the National Key Research and Development Program of China (Grant No. 2022YFF0800601), Science and Technology Program of Shenzhen (Grant No. JCYJ20210324104602006), Guangdong Provincial Key Laboratory of Geophysical High-resolution Imaging Technology (Grant No. 2022B1212010002).

Data Availability Statement: The associated data and codes used by this study are available at the link https://github.com/TaoTao-geophysics/BL-network-design-initial-model accessible from 8 January 2024.

Acknowledgments: We extend our sincere gratitude to C. L. Philip Chen and Zhulin Liu for generously providing the broad learning code. We also wish to express our appreciation to Rui Wang, Hongyan Chen, Zhiyi Zeng, and Peijie Wang for their valuable discussions throughout this study.

Conflicts of Interest: The authors declare no conflicts of interest.

References

1. Cardarelli, E.; Cercato, M.; Di Filippo, G. Geophysical investigation for the rehabilitation of a flood control embankment. *Near Surf. Geophys.* **2010**, *8*, 287–296. [CrossRef]
2. Ibraheem, I.M.; Tezkan, B.; Bergers, R. Integrated interpretation of magnetic and ERT data to characterize a landfill in the north-west of Cologne, Germany. *Pure Appl. Geophys.* **2021**, *178*, 2127–2148. [CrossRef]
3. Yu, N.; Liu, H.; Feng, X.; Li, T.; Du, B.; Wang, C.; Wang, W.; Kong, W. Advancing CO_2 Storage Monitoring via Cross-Borehole Apparent Resistivity Imaging Simulation. *IEEE Trans. Geosci. Remote Sens.* **2023**, *61*, 1–12. [CrossRef]
4. Ali, M.A.H.; Mewafy, F.M.; Qian, W.; Alshehri, F.; Ahmed, M.S.; Saleem, H.A. Integration of Electrical Resistivity Tomography and Induced Polarization for Characterization and Mapping of (Pb-Zn-Ag) Sulfide Deposits. *Minerals* **2023**, *13*, 986. [CrossRef]
5. Uhlemann, S.; Chambers, J.; Falck, W.E.; Tirado Alonso, A.; Fernández González, J.L.; de Gea, A.E. Applying electrical resistivity tomography in ornamental stone mining: Challenges and solutions. *Minerals* **2018**, *8*, 491. [CrossRef]
6. Shin, Y.; Shin, S.; Cho, S.-J.; Son, J.-S. Application of 3D Electrical Resistivity Tomography in the Yeoncheon Titanomagnetite Deposit, South Korea. *Minerals* **2021**, *11*, 563. [CrossRef]
7. Su, Z.; Revil, A.; Ghorbani, A.; Zhang, X.; Zhao, X.; Richard, J. Combining Electrical Resistivity, Induced Polarization, and Self-Potential for a Better Detection of Ore Bodies. *Minerals* **2023**, *14*, 12. [CrossRef]
8. Vu, M.; Jardani, A. Convolutional neural networks with SegNet architecture applied to three-dimensional tomography of subsurface electrical resistivity: CNN-3D-ERT. *Geophys. J. Int.* **2021**, *225*, 1319–1331. [CrossRef]
9. Sharma, S.P. VFSARES—A very fast simulated annealing FORTRAN program for interpretation of 1-D DC resistivity sounding data from various electrode arrays. *Comput. Geosci.* **2012**, *42*, 177–188. [CrossRef]
10. Başokur, A.T.; Akca, I. Object-based model verification by a genetic algorithm approach: Application in archeological targets. *J. Appl. Geophys.* **2011**, *74*, 167–174. [CrossRef]
11. Liu, B.; Li, S.; Nie, L.; Wang, J.; Zhang, Q. 3D resistivity inversion using an improved Genetic Algorithm based on control method of mutation direction. *J. Appl. Geophys.* **2012**, *87*, 1–8. [CrossRef]
12. Barboza, F.M.; Medeiros, W.E.; Santana, J.M. A user-driven feedback approach for 2D direct current resistivity inversion based on particle swarm optimization Feedback inversion using PSO. *Geophysics* **2019**, *84*, E105–E124. [CrossRef]
13. Sosa, A.; Velasco, A.A.; Velazquez, L.; Argaez, M.; Romero, R. Constrained optimization framework for joint inversion of geophysical data sets. *Geophys. J. Int.* **2013**, *195*, 1745–1762. [CrossRef]
14. Aleardi, M.; Vinciguerra, A.; Hojat, A. A convolutional neural network approach to electrical resistivity tomography. *J. Appl. Geophys.* **2021**, *193*, 104434. [CrossRef]
15. Alyousuf, T.; Li, Y. Inversion using adaptive physics-based neural network: Application to magnetotelluric inversion. *Geophys. Prospect.* **2022**, *70*, 1252–1272. [CrossRef]
16. Pidlisecky, A.; Haber, E.; Knight, R. RESINVM3D: A 3D resistivity inversion package. *Geophysics* **2007**, *72*, H1–H10. [CrossRef]
17. Wu, P.; Tan, H.; Tao, T.; Ma, H.; Ding, Z.; Xu, L. Three-dimensional joint inversion of the resistivity method and ambient noise method with cross-gradient constraints. *Chin. J. Geophys.* **2020**, *63*, 3912–3930.
18. Wu, X.; Xu, G. Study on 3-D resistivity inversion using conjugate gradient method. *Chin. J. Geophys.* **2000**, *43*, 450–458.
19. Peng, M.; Tan, H.; Moorkamp, M. Structure-coupled 3-D imaging of magnetotelluric and wide-angle seismic reflection/refraction data with interfaces. *J. Geophys. Res. Solid Earth* **2019**, *124*, 10309–10330. [CrossRef]

20. Kong, W.; Tan, H.; Lin, C.; Unsworth, M.; Lee, B.; Peng, M.; Wang, M.; Tong, T. Three-Dimensional Inversion of Magnetotelluric Data for a Resistivity Model with Arbitrary Anisotropy. *J. Geophys. Res. Solid Earth* **2021**, *126*, e2020JB020562. [CrossRef]
21. Xu, L.; Tan, H.; Wu, P.; Peng, M.; Wang, S.; Jiang, M. Electrical characteristics of the crust in the south part of Longmenshan fault zone: Evidence from magnetotelluric inversion with velocity structure constraints. *Chin. J. Geophys.* **2022**, *65*, 3434–3450.
22. Ma, H.; Guo, Y.; Wu, P.; Tan, H. 3-D joint inversion of multi-array data set in the resistivity method based on MPI parallel algorithm. *Chin. J. Geophys.* **2018**, *61*, 5052–5065.
23. Wilson, B.; Singh, A.; Sethi, A. Appraisal of Resistivity Inversion Models with Convolutional Variational Encoder–Decoder Network. *IEEE Trans. Geosci. Remote Sens.* **2022**, *60*, 1–10. [CrossRef]
24. Yuval, D.; Oldenburg, W. DC resistivity and IP methods in acid mine drainage problems: Results from the Copper Cliff mine tailings impoundments. *J. Appl. Geophys.* **1996**, *34*, 187–198. [CrossRef]
25. Wunderlich, T.; Fischer, P.; Wilken, D.; Hadler, H.; Erkul, E.; Mecking, R.; Günther, T.; Heinzelmann, M.; Vött, A.; Rabbel, W. Constraining electric resistivity tomography by direct push electric conductivity logs and vibracores: An exemplary study of the Fiume Morto silted riverbed (Ostia Antica, western Italy). *Geophysics* **2018**, *83*, B87–B103. [CrossRef]
26. Pidlisecky, A.; Knight, R.; Haber, E. Cone-based electrical resistivity tomography. *Geophysics* **2006**, *71*, G157–G167. [CrossRef]
27. Loke, M.H.; Barker, R.D. Rapid least-squares inversion of apparent resistivity pseudosections by a quasi-Newton method1. *Geophys. Prospect.* **1996**, *44*, 131–152. [CrossRef]
28. Günther, T.; Rücker, C.; Spitzer, K. Three-dimensional modelling and inversion of DC resistivity data incorporating topography—II. Inversion. *Geophys. J. Int.* **2006**, *166*, 506–517. [CrossRef]
29. Wagner, F.; Mollaret, C.; Günther, T.; Kemna, A.; Hauck, C. Quantitative imaging of water, ice and air in permafrost systems through petrophysical joint inversion of seismic refraction and electrical resistivity data. *Geophys. J. Int.* **2019**, *219*, 1866–1875. [CrossRef]
30. Palacios, A.; Ledo, J.J.; Linde, N.; Luquot, L.; Bellmunt, F.; Folch, A.; Marcuello, A.; Queralt, P.; Pezard, P.A.; Martínez, L. Time-lapse cross-hole electrical resistivity tomography (CHERT) for monitoring seawater intrusion dynamics in a Mediterranean aquifer. *Hydrol. Earth Syst. Sci.* **2020**, *24*, 2121–2139. [CrossRef]
31. Goebel, M.; Knight, R.; Kang, S. Enhancing the resolving ability of electrical resistivity tomography for imaging saltwater intrusion through improvements in inversion methods: A laboratory and numerical study. *Geophysics* **2021**, *86*, WB101–WB115. [CrossRef]
32. Slezak, K.; Jozwiak, W.; Nowozynski, K.; Orynski, S.; Brasse, H. 3-D studies of MT data in the Central Polish Basin: Influence of inversion parameters, model space and transfer function selection. *J. Appl. Geophys.* **2019**, *161*, 26–36. [CrossRef]
33. Mousavi, S.M.; Beroza, G.C. Deep-learning seismology. *Science* **2022**, *377*, eabm4470. [CrossRef] [PubMed]
34. Qi, J.; Zhang, J.; Lyu, B.; Marfurt, K.J. Seismic Geometric Nonparallelism Attributes. *IEEE Geosci. Remote Sens. Lett.* **2020**, *19*, 1–5. [CrossRef]
35. Wu, S.; Huang, Q.; Zhao, L. De-noising of transient electromagnetic data based on the long short-term memory-autoencoder. *Geophys. J. Int.* **2021**, *224*, 669–681. [CrossRef]
36. Xue, J.; Huang, Q.; Wu, S.; Nagao, T. LSTM-Autoencoder Network for the Detection of Seismic Electric Signals. *IEEE Trans. Geosci. Remote Sens.* **2022**, *60*, 1–12. [CrossRef]
37. Wu, S.; Huang, Q.; Zhao, L. Convolutional neural network inversion of airborne transient electromagnetic data. *Geophys. Prospect.* **2021**, *69*, 1761–1772. [CrossRef]
38. Wu, S.; Huang, Q.; Zhao, L. Instantaneous inversion of airborne electromagnetic data based on deep learning. *Geophys. Res. Lett.* **2022**, *49*, e2021GL097165. [CrossRef]
39. Wu, S.; Huang, Q.; Zhao, L. Fast Bayesian Inversion of Airborne Electromagnetic Data Based on the Invertible Neural Network. *IEEE Trans. Geosci. Remote Sens.* **2023**, *61*, 1–11. [CrossRef]
40. Zhang, C.; Li, J.; Yu, H.; Liu, B. Autoencoded Elastic Wave-Equation Traveltime Inversion: Toward Reliable Near-Surface Tomogram. *IEEE Trans. Geosci. Remote Sens.* **2023**, *61*, 1–13. [CrossRef]
41. Liu, B.; Guo, Q.; Li, S.; Liu, B.; Ren, Y.; Pang, Y.; Guo, X.; Liu, L.; Jiang, P. Deep learning inversion of electrical resistivity data. *IEEE Trans. Geosci. Remote Sens.* **2020**, *58*, 5715–5728. [CrossRef]
42. Liu, B.; Pang, Y.; Jiang, P.; Liu, Z.; Liu, B.; Zhang, Y.; Cai, Y.; Liu, J. Physics-Driven Deep Learning Inversion for Direct Current Resistivity Survey Data. *IEEE Trans. Geosci. Remote Sens.* **2023**, *61*, 1–11. [CrossRef]
43. Li, Y.; Yang, D. Electrical imaging of hydraulic fracturing fluid using steel-cased wells and a deep-learning method Electrical hydraulic fracturing imaging. *Geophysics* **2021**, *86*, E315–E332. [CrossRef]
44. Chen, C.P.; Liu, Z. Broad learning system: An effective and efficient incremental learning system without the need for deep architecture. *IEEE Trans. Neural Netw. Learn. Syst.* **2017**, *29*, 10–24. [CrossRef]
45. Kong, S.; Oh, J.; Yoon, D.; Ryu, D.-W.; Kwon, H.-S. Integrating Deep Learning and Deterministic Inversion for Enhancing Fault Detection in Electrical Resistivity Surveys. *Appl. Sci.* **2023**, *13*, 6250. [CrossRef]
46. Yang, X.-H.; Han, P.; Yang, Z.; Miao, M.; Sun, Y.-C.; Chen, X. Broad Learning Framework for Search Space Design in Rayleigh Wave Inversion. *IEEE Trans. Geosci. Remote Sens.* **2022**, *60*, 1–17. [CrossRef]
47. Yang, X.-H.; Han, P.; Yang, Z.; Chen, X. Two-stage broad learning inversion framework for shear-wave velocity estimation. *Geophysics* **2023**, *88*, WA219–WA237. [CrossRef]

48. Hu, K.; Ren, H.; Huang, Q.; Zeng, L.; Butler, K.E.; Jougnot, D.; Linde, N.; Holliger, K. Water Table and Permeability Estimation from Multi-Channel Seismoelectric Spectral Ratios. *J. Geophys. Res. Solid Earth* **2023**, *128*, e2022JB025505. [CrossRef]
49. Zhou, Y.; Li, X.; Tang, Q.; Kuok, S.C.; Fei, K.; Gao, L. An Assimilating Model Using Broad Learning System for Incorporating Multi-Source Precipitation Data with Environmental Factors Over Southeast China. *Earth Space Sci.* **2022**, *9*, e2021EA002043. [CrossRef]
50. Byrd, R.H.; Lu, P.; Nocedal, J.; Zhu, C. A limited memory algorithm for bound constrained optimization. *SIAM J. Sci. Comput.* **1995**, *16*, 1190–1208. [CrossRef]
51. Colombo, D.; Turkoglu, E.; Li, W.; Sandoval-Curiel, E.; Rovetta, D. Physics-driven deep-learning inversion with application to transient electromagnetics. *Geophysics* **2021**, *86*, E209–E224. [CrossRef]
52. Nocedal, J. Updating quasi-Newton matrices with limited storage. *Math. Comput.* **1980**, *35*, 773–782. [CrossRef]
53. Wang, T.; Wang, K.-P.; Tan, H.-D. Forward modeling and inversion of tensor CSAMT in 3D anisotropic media. *Appl. Geophys.* **2017**, *14*, 590–605. [CrossRef]

minerals

MDPI

Article

Applying Data-Driven-Based Logistic Function and Prediction-Area Plot to Map Mineral Prospectivity in the Qinling Orogenic Belt, Central China

Hongyang Bai [1,2], Yuan Cao [3], Heng Zhang [3], Wenfeng Wang [1,2,4,*], Chaojun Jiang [3] and Yongguo Yang [1,2]

1 School of Resources and Geosciences, China University of Mining and Technology, Xuzhou 221116, China
2 Key Laboratory of Coalbed Methane Resources and Reservoir Formation Process of the Ministry of Education, China University of Mining and Technology, Xuzhou 221000, China
3 The Second Geology Prospecting Institute of Henan Bureau of Geology and Mineral Exploration and Development, Zhengzhou 451464, China
4 School of Geology and Mining Engineering, Xinjiang University, Urumqi 830047, China
* Correspondence: wangwenfeng@cumt.edu.cn

Abstract: This study combines data-driven-based logistic functions with prediction–area (P–A) plot for delineating target areas of orogenic Au deposits in the eastern margin of the Qinling metallogenic belt, central China. First, appropriate geological and geochemical factors were identified, optimized, and transformed into a series of fuzzy numbers with a range of 0–1 through a data-driven-based logistic function in order to determine the evidence layer for prospecting orogenic Au. In addition, the P–A plot was derived on the above evidence layers and their corresponding fuzzy overlay layers to pick out a proper prediction scheme, in the process of which acidic magmatic activity proved to be the most important factor of ore-controlling. Moreover, to further prove the advantages of this method, a traditional linear knowledge-driven approach was carried out for comparative purposes. Finally, based on concentration–area (C–A) fractal theory, the fractal thresholds were determined and a mineral prospecting map was generated. The obtained prediction map consisted of high, medium, low, and weak metallogenic potential areas, accounting for 2.5%, 16.1%, 38.4%, and 43% of the study area, containing 2, 3, 1, and 0 of the 6 known mine occurrences contained, respectively. The P–A plot indicated that the result predicted 83% of Au deposits with 17% of the area, confirming the joint application of the data-driven-based logistic function and P–A plot to be a simple, effective, and low-cost method for mineral prospectivity mapping, that can be a guidance for further work in the study area.

Keywords: mineral prospectivity mapping; logistic function; prediction-area; concentration-area; orogenic Au

1. Introduction

Mineral prospectivity mapping (MPM) is a comprehensive area of study, during which geological engineers identify metallogenic target areas with known information and data in the study area to guide further exploration. It is essentially a classification technique [1], by which the study area can be divided into areas with high, moderate, and low favorability of mineralization [2,3]. The process of MPM generally includes the understanding of the metallogenic model, the identification of ore-controlling factors, data treatment or transformation, and finally the integration of these results to describe the ore-forming target area [4]. Simply, its objective is to portray the smallest area usually containing the most mineral deposits. In the above process, however, two key difficulties remain. One is to convert evidence layers with different orders of magnitude values into the same space and integrate them [5]; the other is to determine a group of reasonable thresholds to demarcate the study area [2].

Recent decades have witnessed researches on the aforementioned issues by many scholars. For example, Behera et al. [6] applied C–A fractal and fuzzy analytic hierarchy processes to predict geochemical anomalies and mineral prospects in Sonakhan Greenstone Belt in central India. Yousefi et al. [1,7] compared the effectiveness of Boolean logic, a geometric mean model, and expected value model in metallogenic prediction. Carranza [8] employed data-driven evidence confidence functions to predict the mineralization of gold deposits in Baguio District, Philippines. Knox-Robinson [2] used vector fuzzy logic to integrate spatial data and study the metallogenic prospect of gold deposits in Kalgoorlie Terrane orogenic belt, Western Australia. Carranza [9] proposed a Wildcat model based on logic function.

All the methods described above generally fall into three categories [5,10,11]. The most common one is the knowledge-driven method that mainly assesses mineralization evidence based on the knowledge of geological experts and assigns different weights to each mineralization factor. It is generally suitable for study areas with low exploration levels [9], in which considerable mineralization data are absent. Among them, fuzzy sets and fuzzy logistics, proposed by Zadeh [12], have been widely applied and proven to be of great value in MPM by many geologists [2,11,13–20]. However, this method has the weakness of suffering from experts' bias, thus resulting in different experts often having different opinions [1]. Another approach is the traditional data-driven method that can better establish the relationship between known mines and various evidence layers for metallogenic prediction [21–26]. Nevertheless, its deficiency is that it requires a training batch of known mines to establish this connection, rendering it infeasible in areas with low exploration levels. As a result, it would be a deviation from the fact that unknown deposits have a lesser chance of being involved in such training. In addition, there are approaches (or systems) of integration [27] or hybridization of the two aforementioned methods [28–33]. However, most of these still cannot overcome the above dilemma or remain in development.

However, there are very few ways to overcome the bias of metallogenic prediction due to personal knowledge limitations or preferences. In order to solve the above defects, many geologists began to try empirical function for MPM [8,9,16,28,34]. In these methods, a data-driven approach based on logistic functions proposed by Yousefi et al. [7,35,36] and Yousefi and Carranza [1], in which weights can be assigned to the evidence layers without experts' bias and the known mines, is able to alleviate the aforementioned drawbacks. At the same time, with the development of computers, big data and artificial intelligence technology have been applied to MPM more and more frequently due to their advancement and economy. For example, Liu et al. [37] applied convolutional neural networks in prospecting predictions of the Zhaojikou Pb–Zn ore deposit in Anhui Province, China; Zuo and Xu [38] proved the advantages of graphical deep learning models compared with convolutional neural networks in metallogenic prediction; Zhou et al. [39] summarized the research progress of big data technology in Earth science and considered it to be an important technology for future breakthroughs in this field.

In this study, the MPM of orogenic Au was implemented by combining logistic function with P–A plot based on the analysis of regional geological background and ore-controlling factors, especially a considerable amount of stream sediment geochemical data. The prospectivity map shows the delineated middle–high favorable areas, accounting for 17% of the study area and encompassing five out of the six known mines, would be an ideal prediction output

2. Geological Setting

The study area lies in the southwest of Henan province, central China, covering an area of 1001 km^2. It is located tectonically at the junction between the southern margin of the North China Plate and the northern margin of the Yangtze Plate (the central and eastern part of the Qinling structural belt), with the Jingziguan–Shigang complex syncline generating the Xixia fault depression basin [40,41]. Crossing four metallogenic zones (north

Qinling orogenic belt, south Qinling fold belt, Xixia–Tongbai polymetallic metallogenic belt, and the Danjiangkou gold–silver–vanadium–antimony metallogenic zone), the study area is characterized by complex geological conditions, frequent magmatic activities, and significant structural deformation (regional) metamorphic effects. Consequently, the metallogenic geological conditions are superior, with vanadium and gold as the main metals and marbles and limestones as the main non-metallic minerals, resulting in an area with rich reserves and prospecting potential [41].

The Au deposits in the region are mainly the orogenic belt type, whose metallogenic geological conditions mainly include ore-bearing rocks, ore-controlling structures, and magma activities [41–45]. However, because the ore-bearing rocks could not be more complex and diverse, and a variety of rocks can generate ore-bearing rocks when other mineralization conditions are in place, only ore-controlling structures and magmatic activities are considered in this study, as per common engineering practice. Magma, as the main heat source, plays an important role in the formation of gold deposits. It can not only provide a corresponding motive force for element migration and aggregation, but also supply the matter source [46] (Craw et al., 2006). As Mao et al. [47] and Chen et al. [48] mentioned, most gold deposits are related to Mesozoic granites in the Xiaoqinling–Xiong'ershan region. At the same time, the tectonic belt plays a crucial role in controlling mineralization, and the known gold spots are mainly distributed along the fault zone from the northwest to southeast. The location and simple geological maps of the study area are shown in Figure 1.

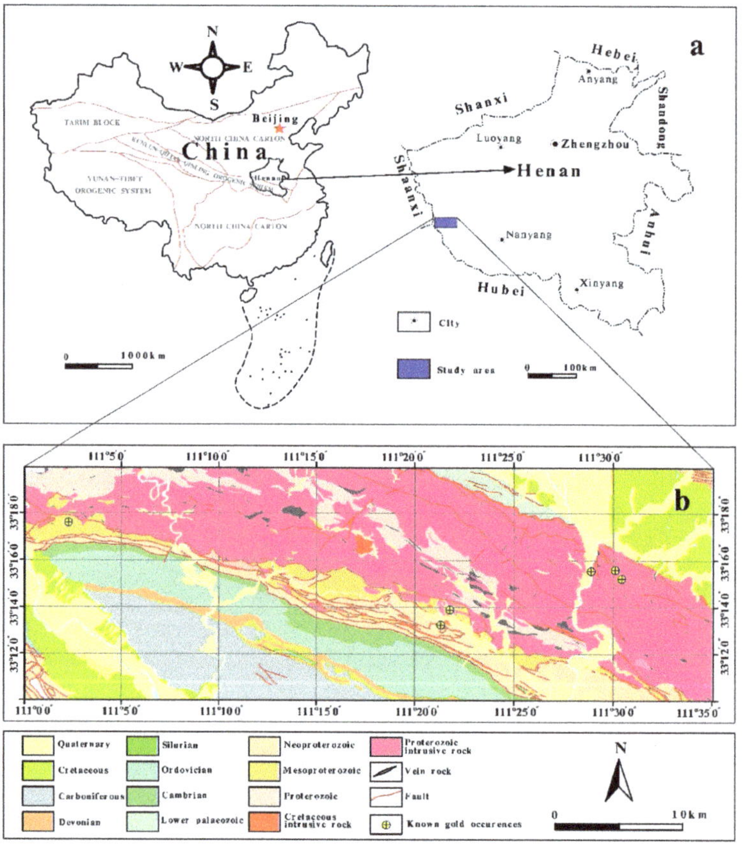

Figure 1. (**a**) Location of the study area, and (**b**) simple geological map of the study area (After [49]).

3. Data Set

3.1. Data Sources

In this study, we used the 1:50,000 regional geological map and 1:50,000 stream sediment survey data prepared by the Second Geology Prospecting Institute of Henan Bureau of Geology and Mineral Exploration and Development. In total, 4036 samples (Figure 2) were collected, processed and analyzed, with a sampling density of 4–6 points per square kilometer in about 1000 square kilometers in the study area. Each of the original samples weighed more than 150 g with a particle size of less than 60 mesh (<216 μm).

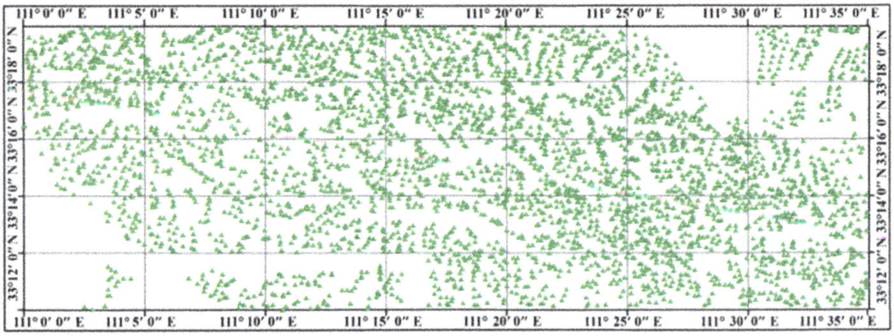

Figure 2. Location map of stream sediment samples (after [49]).

The concentrations of Au, Ag, As, Sb, Cu, Pb, Zn, Mo, W, and Cd were analyzed by using graphite furnace atomic absorption spectrometry (Au), atom emission spectrometry (Ag), atomic fluorescence spectrometry (As and Sb), and inductively coupled plasma optical emission spectrometry (Cu, Pb, Zn, Mo, W, and Cd). The detection limits were 0.3 ppb for Au, 0.02 ppm for Ag, 0.2 ppm for As, 0.04 ppm for Sb,1 ppm for Cu, 1 ppm for Pb, 5 ppm for Zn, 0.3 ppm for Mo, 0.3 ppm for W, and 0.04 ppm for Cd.

Tight quality control was maintained at every stage of the process, in accordance with [50,51].

3.2. Data Preprocessing

On the basis of [52], we determined a proper cell with 200 m × 200 m for all of the evidence maps. Subsequently, the geological data and geochemical data were processed to obtain evidence values in each cell. In this way, the study area contained a total of 25,024 cells.

On the basis of the 1:50,000 geology survey, the main geological information related to the mineralization, such as magmata and faults, was extracted as the evidence layers. Meanwhile, based on the measurement of the 1:50,000 stream sediment survey, stage factor analysis (SFA), as proposed by Yousefi et al. [7], was performed to assess the multiple elements closely related to gold mineralization in the study area. Subsequently, the corresponding multi-element factors were grid-processed to determine their value in each cell.

3.2.1. Geological Evidence, Main Heat Sources, and Faults

In this study, the granite vein and the Yanshanian granite porphyry were extracted as the evidence layers and the distance from the intrusion contact was used as the indicator criterion. However, the further from the intrusion, the less the possibility of mineralization; thus, the inverse square of the distance from intrusive was taken as the evidence value in each cell (Figure 3a). It is generally accepted that faults are important channels for the movement of geological fluids [53]. Without faults, there could be little migration and enrichment of elements; as a result, it would be impossible to generate gold deposits as well. Consequently, we took the proximity to fault as the evidence value, whose method of acquisition was similar to that of the heat source (Figure 3b).

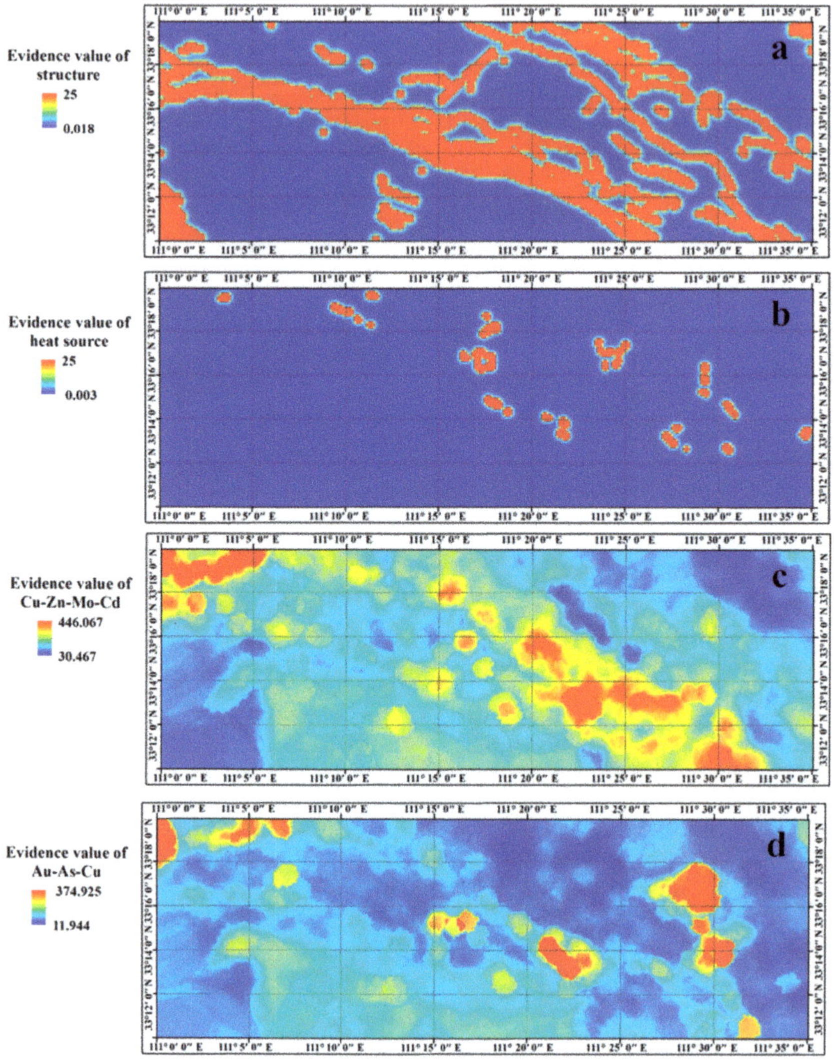

Figure 3. Evidence values of: (**a**) structure; (**b**) heat source; (**c**) Cu–Zn–Mo–Cr factor; and (**d**) Au–As–Pb factor.

3.2.2. Geochemical Evidence

Stage factor analysis, based on the principal component analysis developed by Yousefi et al. [7] as an optimized method, was performed on 4036 samples of 10 elements to obtain multi-element geochemical anomaly factors, using the statistical product and service solutions (SPSS) platform.

From the results of the staged factor analysis (Table 1), two indicator factors (Figure 3c,d) were obtained to reflect the presence of orogenic gold deposits: F1 (Cu–Zn–Mo–Cd) and F2 (Au–As–Pb). Elements with high factor loading values (greater than 0.6 with bold) in the third and fourth stages of the two factors could be used as indicator factors based on geochemical criteria to define exploration targets. In terms of F1 (Cu–Zn–Mo–Cd), these are obviously sulfurophile elements that often form chalcopyrite, sphalerite, molybdenite, and greenockite in nature, respectively. As we all know, S is usually considered to be the mineralizer of Au or an important component of Au-loaded minerals due to the extremely weak activity of Au; sometimes, the source of Au can be transformed into the source of S. In terms of F2 (Au–As–Pb), these are typical low-temperature elements with similar geochemical characteristics; thus, As and Pb are often found in the leading edges of gold belts. Generally, Au, As, and Pb have a strong migration ability in ore-forming fluids, and are most likely to migrate or accumulate along those secondary fracture or ductile shear zones.

Table 1. Rotated factor matrix of staged factor analysis and data of samples from the study area. Loadings in bold represent the selected elements based on a threshold of 0.6 (the absolute threshold value) for each stage.

First Main Phase						Second Main Phase			
First Stage			Second Stage			Third Stage		Fourth Stage	
Element	F1	F2	Element	F1	F2	Element	F1	Element	F2
Au	−0.1	**0.726**	Au	−0.07	**0.719**	Cu	**0.698**	Au	**0.681**
Ag	0.48	0.592	As	0.153	**0.803**	Zn	**0.797**	As	**0.819**
As	0.118	**0.785**	Cu	**0.719**	−0.029	Mo	**0.758**	Pb	**0.783**
Sb	0.184	0.304	Pb	0.291	**0.713**	Cd	**0.838**		
Cu	**0.723**	−0.002	Zn	0.72	0.384				
Pb	0.257	**0.685**	Mo	**0.771**	0.072				
Zn	**0.692**	0.354	Cd	**0.812**	0.16				
Mo	**0.761**	0.1							
W	0.266	0.235							
Cd	**0.803**	0.195							

4. Methodology

4.1. Logistic Function

Different evidence layers have values of diverse orders of magnitude; therefore, they are incomparable and cannot be superimposed to generate a prospecting map. In this context, logistic function was proposed by Bishop (2006) [54], with which discrete data of different orders of magnitude can be converted to continuous values of 0–1, thus avoiding arbitrary division of evidence value by traditional knowledge-based methods. In addition, Yousefi and Carranza [5] proposed an optimized method based on logic function, in which the parameters can be calculated in a data-driven way, so that the converted value can perfectly fall between 0–1.

Therefore, we used an S-shape logistic function to transform the evidence values to obtain fuzzy numbers with a range of 0–1, which could be definite proxies of the evidence values.

$$F_{EV} = \frac{1}{1 + e^{-s(EV-i)}} \tag{1}$$

The S–shaped logistic function, as shown in Equation (1) [5], where EV and F_{EV} are the evidence value and fuzzy score, respectively, while s and I are unknown parameters, has been used by a number of geologists as an efficient tool for MPM, most of whom have obtained anticipative results. However, as Yousefi and Nykänen [5] mentioned, the selection of the s and I parameters in logistic function is often subjective, because different experts often have different preferences. Moreover, some subjectively selected parameters cannot achieve continuous fuzzy numbers. Thus, such experience-based parameters sometimes yield different evidence values with similar or concentrated fuzzy numbers, resulting in the fuzzy score being a poor representative of the source data.

Therefore, Yousefi and Nykänen [5] suggested defining the maximum and minimum values of F_{EV} as close to 1(0.99) and close to 0(0.01), which represent the most and least important evidence value. With known values of EV_{max} and EV_{min} (the maximum and minimum values of EV) as well as F_{Evmax} and F_{Evmin} (the maximum and minimum values of F_{EV}, respectively), Equations (2) and (3) can easily be solved to find s and i. In this way, a data-driven fuzzy score is obtained, which avoids the distortion of conversion caused by experts' subjective bias.

$$\begin{cases} F_{Evmax} = \frac{1}{1+e^{-s(EV_{max}-i)}} \\ F_{Evmin} = \frac{1}{1+e^{-s(EV_{min}-i)}} \end{cases} \quad (2)$$

$$\begin{cases} s = \frac{9.2}{EV_{max}-EV_{min}} \\ i = \frac{EV_{max}+EV_{min}}{2} \end{cases} \quad (3)$$

4.2. Prediction-Area Plot

The P–A plot, a method proposed by Yousefi and Carranza [55], can not only assign weight to each evidence layer, but also evaluate its prediction ability, so as to select a more appropriate evidence layer for MPM. Specifically, it consists of two curves, one corresponding to the left axis, which represents the proportion of the number of mines predicted by the evidence layer to the total, and the other corresponding to the right axis, which represents the proportion of the area account for the total area. There is an intersection point between the two curves, and the higher the intersection point is, the stronger the predictive ability of the evidence layer is. In other words, the higher the intersections of the evidence layer on the P–A plot, the stronger the intrinsic association with the mineral deposit.

5. Results

5.1. Data Transformation

For orogenic gold deposits, faults usually play a role as an activity space or even a storage space for ore-containing hydrotherms [45,56,57]. It is generally believed that the further from the fault, the lower the degree of profitability of the mineralization; thus, the inverse square of the proximity to the fault can be applied as the evidence value. When the evidence value is 0 (i.e., the cell contains the fault itself), its inverse square does not exist; therefore, we manually assign it the maximum value. Based on the GIS platform, the proximity to fault of each cell is computed, the maximum and minimum of whose inverse square are calculated to be 25 and 0.018, respectively. Using Equations (2) and (3), the corresponding s and i values were acquired as 0.3683 and 12.5090. Proximity to the intrusion was used as the data set for the heat source. The further from the intrusive contact, the lower the degree of profitability of mineralization; therefore, we applied the inverse square as the evidence value as well. Due to the nonuniformity of sampling points, the ordinary kriging interpolation [58,59] was conducted on the two multi-element factors (Cu–Zn–Mo–Cd factor and Au–As–Pb factor) to obtain the evidence value of each cell, verifying that they conformed to the normal distribution.

After obtaining the evidence values of the above layers, Equation (1) was utilized to obtain the fuzzy score in each cell (the maximum value was 0.99, and the minimum value was 0.01), which represented the favorable degree of mineralization. The parameters used in the conversion process are shown in Table 2 and the obtained fuzzy score layers are depicted in Figure 4.

Table 2. Parameter values calculated for each evidence layer.

Evidential Layer	s	i
Structure	0.3683	12.5090
Heat source	0.3680	12.5015
Cu-Zn-Mo-Cd-	0.0221	238.2672
Au-As-Pb	0.0253	193.4347

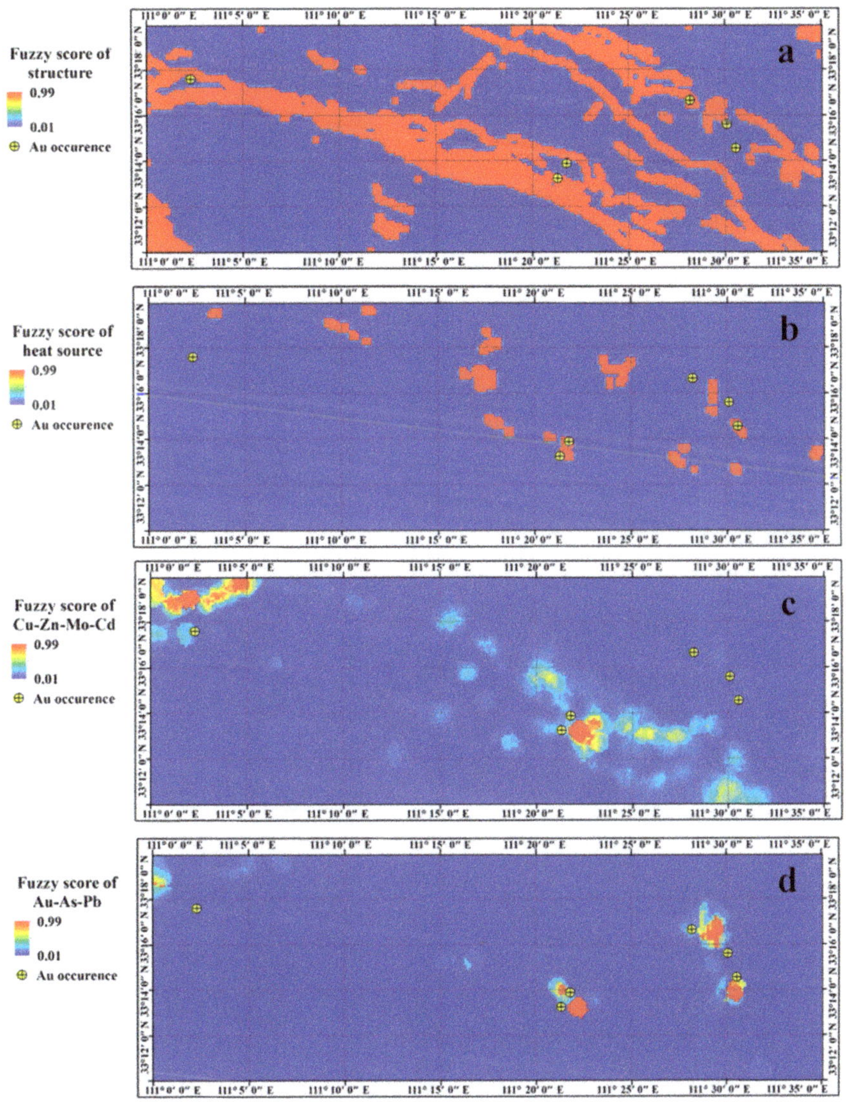

Figure 4. Fuzzy score of: (**a**) structure; (**b**) heat source; (**c**) Cu–Zn–Mo–Cd factor; and (**d**) Au–As–Pb factor.

In order to confirm whether the fuzzy score could be a good representation of the mineralization favorability, the obtained fuzzy score layers (Figure 4a–d) were compared with the evidence value layers (Figure 3a–d) for a comparable purpose. The comparison revealed that both of them were highly similar and the former exhibited a slight convergence compared with the latter, although only in areas where the evidence values had an unobvious anomaly; therefore, it proved better to identify the real anomalies.

This is consistent with the research of Bishop [58] and Yousefi et al. [7] that non-linear transformations (such as the logistic function) yield an optimal decision boundary between different classes of a variable for classification, thus boosting stronger discrimination between anomaly and background values. Therefore, we believe that the data-driven logistic function used in this study is appropriate due to its ability of transforming evidence values of different magnitudes to fuzzy scores with a 0–1 range and retaining the relative importance.

5.2. Evaluation of Fuzzy Evidence Layers

Before conducting the fuzzy overlay, we performed a P–A plot, which could evaluate the mineralization advantage of each layer objectively, to quantify its prediction ability. This method employs the ratio of predicted gold deposits to total deposits, versus the ratio of the accumulated area to total area as indicators based on the known gold deposits [10].

The P–A plot of the evidence layer is shown in Figure 5. The intersection of the two curves represents the prediction ability of the evidence layer. The higher the intersection, the stronger the prediction ability, and the closer it is to mineralization.

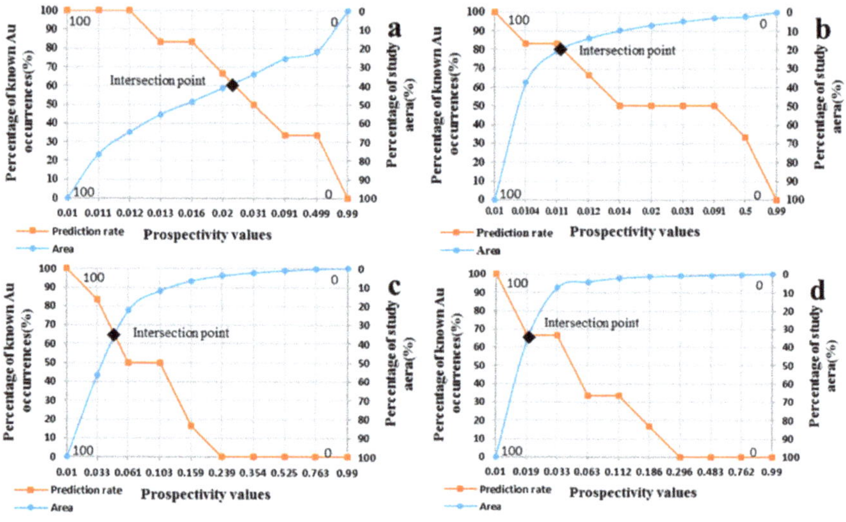

Figure 5. Prediction area (P–A) plot of each layer: (**a**) structure; (**b**) heat source; (**c**) Cu–Zn–Mo–Cr factor; and (**d**) Au–As–Pb factor.

5.3. Integration

As can be seen from Table 3, the prediction capability of each fuzzy evidence layer was quite different. The best one was the heat source, with a value of 81, which significantly exceeded other layers. In order to better determine the relationship between each fuzzy evidence layer and mineralization, we conducted fuzzy overlay with a γ value of 0.95 [60] to integrate these layers.

Table 3. Prediction ability of each evidence layer.

Evidential Layer	% of Known Au Occurrences	% of Study Area
Figure 4a (structure)	61	39
Figure 4b (heat source)	81	19
Figure 4c (Cu–Zn–Mo–Cr)	65	35
Figure 4d (Au–As–Pb)	67	33

The fuzzy evidence layers with prediction ability greater than or equal to 81, 67, 65, and 61 were integrated separately to obtain three overlay maps (Figure 6) that were then estimated by P–A plot, and their evaluation results are shown in Table 4.

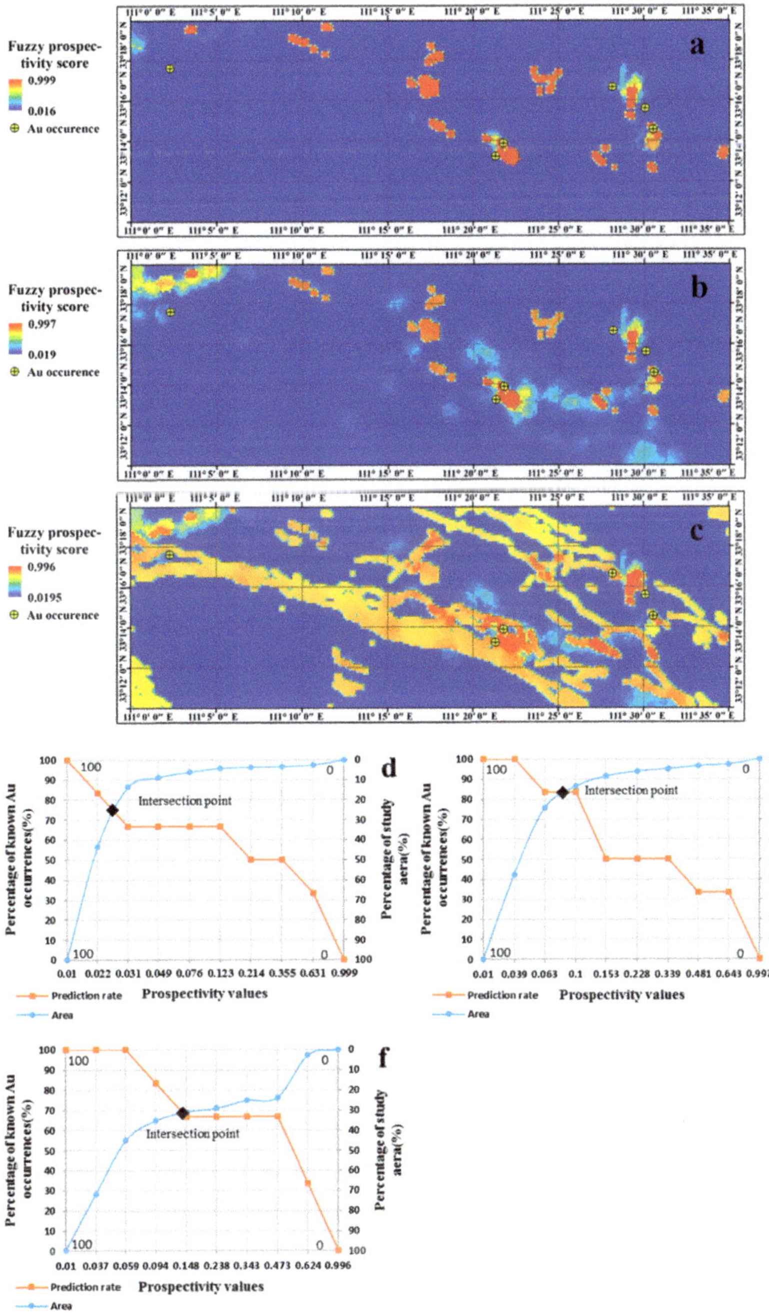

Figure 6. Perspectivity scores: (a) integrated with Figure 4b,d; (b) integrated with Figure 4b–c; and (c) integrated with Figures 4b–d and 3a; (d) P–A plots of Figure 6a; (e) P–A plots of Figure 6b; (f) P–A plots of Figure 6c.

Table 4. Prediction ability of each integrated layer.

Fuzzy Prospectivity Map	% of Known Au Occurrences	% of Study Area
Figure 6a	75	25
Figure 6b	83	17
Figure 6c	68	32

Although there are many similarities between Figure 6a,b, the latter has a slightly stronger predictive value (Figure 6d,e, and Table 4). However, we also noticed that the corresponding forecasting capacity from Figure 6d to 6f possessed a trend of increasing first and then declining. Among them, Figure 6e reached a maximum with a prediction rate of 84%; at this point, the evidence layers for integration were Figure 4b–d, with prediction rates of 81%, 65%, and 67%, respectively. This phenomenon is strange, because several layers with lower prediction ability have a better result when integrated. In spite of this, it is in line with a previous study [55]. What is interesting is that the prediction rate of Figure 6c dropped significantly when faults were added for integration. This may be caused by the multi-phase superposition of tectonic movements in this area, leading to the development of a large number of faults, whereas there is no magmatic hydrothermal activity in some faults.

In summary, it has been proven that the heat sources, Cu–Zn–Mo–Cd factor, and Au–As–Pb factor are the most important aspects in the mineralization process. Consequently, Figure 6b could be used as the ideal perspectivity map in the study area.

5.4. Fuzzy Prospectivity Score Obtained Linearly (A Comparison)

Carranza (2008) [22] suggested that two or more methods should be conducted in MPM to determine a proper metallogenic target area. Thus, in order to further demonstrate the superiority of logistic functions, we employed the knowledge-driven method to divide the evidence layers on the four evidence layers of Figure 3 into 10 classes at specific intervals, and assigned a mineralization favorability weight to each class linearly based on expert judgment for comparison purposes (Figure 7).

Subsequently, fuzzy gamma (γ = 0.95) operation was performed, and the fuzzy prospectivity score and matching P–A plot were obtained (Figure 8a,b). According to Figure 8b, the intersection value was 77, obviously lower than that of Figure 6b (83). The above results indicate that that weighting the evidence layer using the logistic function not only avoids subjective judgment, but also yields a higher prediction rate compared with the traditional discrete linear method. This is consistent with the findings of Yousefi and Carranza [1].

5.5. To Determine the Thresholds

As Mandelbrot [61–63] and Carlson [64] mentioned, in many cases, ore deposits are characterized by aggregation and fractal distribution. Therefore, in order to determine the high, moderate, low and weak areas of the mineralization more accurately, we employed the C–A method to de-fuzzify Figure 6b and obtain a prospectivity map. This method was proposed by Cheng [65], and has been used and approved by a number of geologists [66–72]. It applies the logarithm of the concentration and the logarithm of the corresponding area greater or equal to concentration as the X-axis and Y-axis, respectively, which can reflect the fractal characteristics inherent in the prediction map. According to this method, in the log-log graph, the consistent slope represents a fractal dimension, and the concentration value corresponding to the fractal point could be used as the threshold for differentiating the favorable, unfavorable, and intermediate areas of mineralization [5].

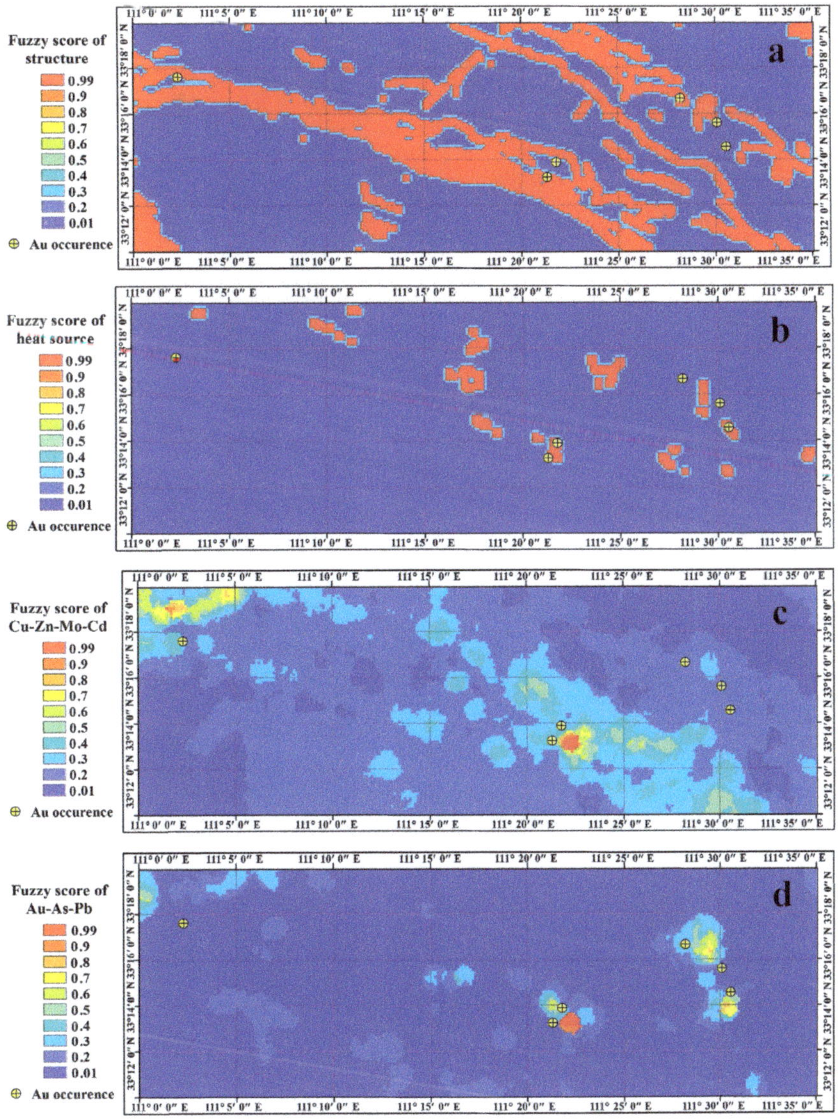

Figure 7. Fuzzy score obtained using the knowledge-driven approach: (**a**) structure; (**b**) heat source; (**c**) Cu–Zn–Mo–Cr factor; and (**d**) Au–As–Pb factor.

In this study, the logarithm of the fuzzy score and the logarithm of the cumulative area were taken as the X-axis and Y-axis, respectively (Figure 9a). Three inflection points were obtained (−1.41, −1.11, and −0.19) and the corresponding fuzzy prediction values (0.039, 0.078, and 0.643) were acquired, which then were used to divide the study area into four parts (Figure 9b). The resulting high-potential area accounting for 2.5% of the study area, containing two known Au occurrences, the moderate-potential area accounting for 16.1% of the study area with three, the low potential area accounting for 38.4% of the study area with one Au occurrence, and the very low-potential area accounting for 43% of the study area with no Au occurrences, would be an ideal metallogenic prediction map.

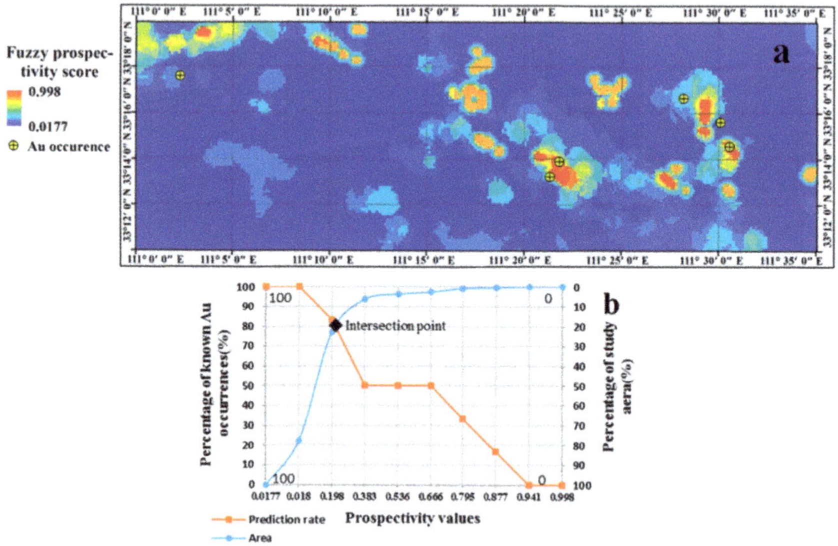

Figure 8. (**a**) Fuzzy prospectivity score integrating by Figure 6–d; (**b**) P–A plot of Figure 8a.

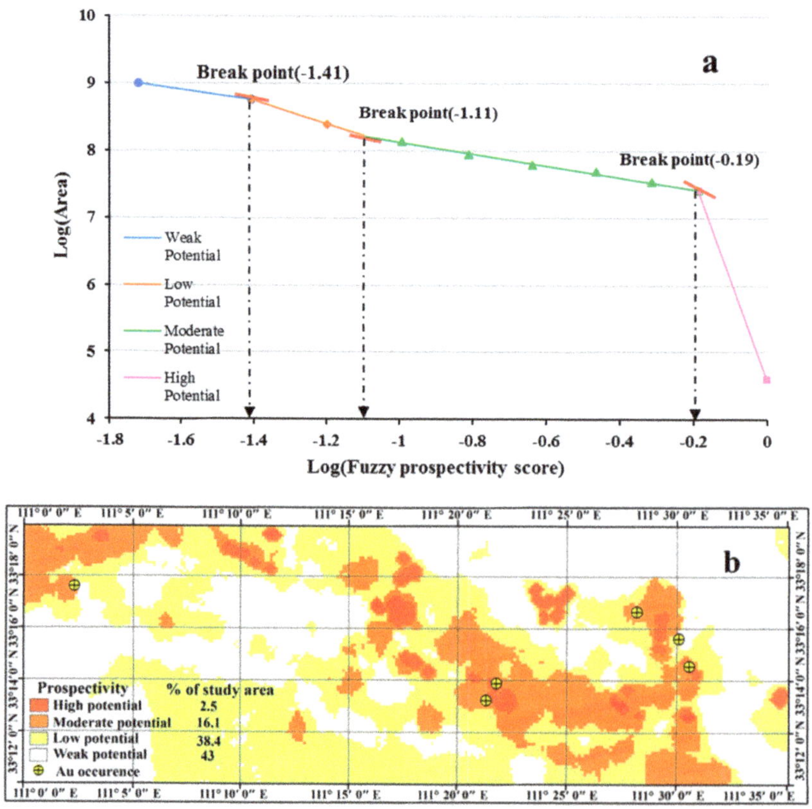

Figure 9. (**a**) Concentration–Area (C–A) model, and (**b**) prospectivity map generated by Figure 6b.

Although there was a known Au occurrence located in the low potential area, we noticed that it was very close to the moderate potential area. This may be attributed to the substitution of point position for area projection. This Au occurrence was, actually, an alteration zone about 0.76 km in length with a trend near east–west, with only its center projected on the horizontal. Therefore, it can be seen from the prospectivity map that the gold occurrence is situated less than two cells away from the favorable metallogenic area (400). In reality, the gold occurrence was partially contained by the moderate-potential areas.

6. Discussion

Previous studies have shown that Mesozoic magmatic activity in the study area may be an important source of heat and provenance for mineralization, despite its limited distribution and small outcrop area. The relatively weak activities are mainly shallow–ultra-shallow acid rocks, which are distributed along faults. Compared with common acid rocks, Au, Ag, As, Sb, Bi, Cu, and Mo are enriched in the rock mass, and their enrichment coefficients are all greater than 3.0, which may be an important source of Au in the deposit. In addition, Au was not only contained in magmatic hydrothermal fluid itself, but also probably extracted from the sedimentary strata of the wall rock, and then transported in the fracture and ductile shear zone.

The main faults in the study area are distributed in the northwest-trending direction, among which the Jianhuaizhai–Huangfengya fault, characterized by rock and ore controlling as well as unconformable forming, is the largest that has developed in the Cambrian, Sinian, and Proterozoic strata. The northeast-trending secondary faults and ductile shear zones are well developed in the study area, intersecting the northwest-trending structural line diagonally, which caused the strata and early faults to be disjointed. On the one hand, these faults can provide channels for hydrothermal migration; on the other hand, the intersection of the northwest and the secondary northeast faults is likely to form important ore-bearing spaces. As a result, the minerals in the area are generally produced in the northwest direction, and the mineral spots are mainly located at the intersection of the main structure and the secondary structure.

Different evidence layers cannot be directly compared and integrated when used for mineral prospectivity mapping because of their different dimensions. Data-driven-based logistic function, which can transform the evidence values of different magnitudes into fuzzy values within the range of 0–1 so as to represent the relative importance of mineralization, could be a suitable means. To obtain s and i by solving equations, an essentially data-driven approach is able to factually reflect the relative importance of evidence values and avoid the trial-and-error process that is usually involved in defining these two parameters [5]. There is an assumption in this process that the largest and smallest evidence values are assigned fuzzy scores close to 1 (0.99) and close to 0 (0.01), respectively. This assumption, consistent with knowledge and practical experience, has been accepted by a large number of geologists and applied in a variety of models.

In this study, four fuzzy layers of geology and geochemistry were evaluated by P–A plot. The results demonstrate that the heat source possessed the highest prediction rate, which is consistent with the strong control of orogenic gold deposits in geological hydrothermals. The prediction rate of faults also reached 61%, in a clear positive correlation with mineralization. At the same time, both multi-element geochemistry evidence layers exhibited prediction rates greater than 65%, which was in line with the SFA analysis. It should be noted that, because there are some secondary hidden faults in the region, they cannot be identified and extracted for MPM, resulting in relatively low prediction ability of faults.

When overlying, a fuzzy gamma operation was used with the recommended value of 0.95 [58]. The results revealed that the heat source, Au–As–Pb factor, and Cu–Zn–Mo–Cd factor had relatively high prediction rates, and were then integrated to derive a perspectivity map with a maximum prediction rate of 83%. This was higher than each of the evidence layers individually, or an integration of all of them. However, the prediction ability demonstrated in Figure 6a to 6c has a tendency of first increasing and then declining. We wonder whether there would be a specific prediction rate that elevates the final prediction ability with the evidence layer integrated over it and vice versa.

7. Conclusions

(1) Based on analysis of regional geological and geochemical characteristics, the epigenetic and ultra-epigenetic acid rock (such as the early Cretaceous granitic porphyry), the northwest-trending faults, and the accompanying secondary faults, as well as pathfinder elements (Au, As, Cu, Zn, Pb, Mo, and Cd) were identified and extracted as the main evidence layers in the search for orogenic gold.

(2) The data-driven-based logistic function demonstrated an excellent ability of converting evidence values of different scales into fuzzy scores with a range of 0–1, and the relative importance of the obtained fuzzy scores was approximately in line with the original evidence values. Meanwhile, in reducing the influence of subjective preferences, the data-driven-based logistic function yielded a better prediction effect than that of traditional knowledge-driven methods.

(3) Based on the analysis and application of geochemical big data, the data-driven logistic function and P–A evaluation were jointly applied to predict mineralization. The results showed that the heat source P–A plot had the highest predictive ability (81%), indicating the strong correlation between mineralization and the intermediate acid intrusive rock (vein), which is in line with the general characteristics of orogenic gold deposits.

(4) The mineralization prediction map generated in the study area, in which 83% of Au occurrences were situated in 17% of the area, confirmed the joint application of data-driven-based logistic function and P–A plot to be a simple, effective, and low-cost method for mineral prospectivity mapping that could provide guidance for further research in the study area.

Author Contributions: Conceptualization, H.B. and W.W.; investigation, Y.C., H.Z. and C.J.; writing—original draft preparation, H.B.; writing—review and editing, W.W; methodology, H.B. and Y.Y.; funding acquisition, W.W. and Y.Y. All authors have read and agreed to the published version of the manuscript.

Funding: This study was supported by the National Key R&D Program of China (Nos. 2021YFC2902003), National Natural Science Foundation of China (Nos. U1903207, 41972176 and 41872248), as well as A Project Funded by the Priority Academic Program Development of Jiangsu Higher Education Institutions (PAPD).

Data Availability Statement: Not applicable.

Acknowledgments: The authors would like to thank the editors and anonymous reviewers for their constructive comments.

Conflicts of Interest: The authors declare that they have no conflict of interest.

References

1. Yousefi, M.; Carranza, E.J.M. Fuzzification of continuous-value spatial evidence for mineral prospectivity mapping. *Comput. Geosci.* **2015**, *74*, 97–109. [CrossRef]
2. Knox-Robinso, C.M. Vectorial fuzzy logic: A novel technique for enhanced mineral prospectivity mapping, with reference to the orogenic gold mineralisation potential of the Kalgoorlie Terrane, Western Australian. *Aust. J. Earth Sci.* **2000**, *47*, 929–941. [CrossRef]
3. Abedi, M.; Norouzi, G.H.; Bahroudi, A. Support vector machine for multi-classification of mineral prospectivity areas. *Comput. Geosci.* **2012**, *46*, 272–283. [CrossRef]
4. Kreuzer, O.P.; Yousefi, M.; Nykänen, V. Introduction to the special issue on spatial modelling and analysis of ore-forming processes in mineral exploration targeting. *Ore Geol. Rev.* **2020**, *119*, 103391. [CrossRef]
5. Yousefi, M.; Nykänen, V. Data-driven logistic-based weighting of geochemical and geological evidence layers in mineral prospectivity mapping. *J. Geochem. Explor.* **2016**, *164*, 94–106. [CrossRef]

6. Behera, S.; Panigrahi, M.K.; Pradhan, A. Identification of geochemical anomaly and gold potential mapping in the Sonakhan Greenstone belt, Central India: An integrated concentration-area fractal and fuzzy AHP approach. *Appl. Geochem.* **2019**, *107*, 45–57. [CrossRef]
7. Yousefi, M.; Kamkar-Rouhani, A.; Carranza, E.J.M. Application of staged factor analysis and logistic function to create a fuzzy stream sediment geochemical evidence layer for mineral prospectivity mapping. *Geochem. Explor. Environ. Anal.* **2014**, *14*, 45–58. [CrossRef]
8. Carranza, E.J.M.; Hale, M. Wildcat mapping of gold potential, Baguio district, Philippines. *Trans. Inst. Min. Metall. Sect. B Appl. Earth Sci.* **2002**, *111*, 100–105. [CrossRef]
9. Carranza, E.J.M. Improved wildcat modeling of mineral prospectivity. *Resour. Geol.* **2010**, *60*, 129–149. [CrossRef]
10. Yousefi, M.; Carranza, E.J.M. Data-driven index overlay and boolean logic mineral prospectivity modeling in greenfields exploration. *Nat. Resour. Res.* **2016**, *25*, 3–18. [CrossRef]
11. Du, X.; Zhou, K.; Cui, Y.; Wang, J.; Zhang, N.; Sun, W. Analytical hierarchy process (AHP) and prediction-area (P-A) plot for mineral prospectivity mapping: A case study from the Dananhu metallogenic belt, Xinjiang, NW China. *Arab. J. Geosci.* **2016**, *9*, 298. [CrossRef]
12. Zadeh, L.A. Fuzzy sets. *IEEE Information and Control.* **1965**, *8*, 338–353. [CrossRef]
13. Moon, W.M.; Ping, A.N. Integration of geological, geophysical, and remote sensing data using fuzzy set theory. *Geoinformatics FCE CTU* **1991**, *2*, 171–176. [CrossRef]
14. Cheng, Q.; Agterberg, F.P. Fuzzy weights of evidence and its application in mineral potential mapping. *Nat. Resour. Res.* **1999**, *8*, 27–35. [CrossRef]
15. Carranza, E.J.M.; Hale, M. Geologically-constrained fuzzy mapping of gold mineralization potential, Baguio district, Philippines. *Nat. Resour. Res.* **2001**, *10*, 125–136. [CrossRef]
16. Luo, X.; Dimitrakopoulos, R. Data-driven fuzzy analysis in quantitative mineral resource assessment. *Comput. Geosci.* **2003**, *29*, 3–13. [CrossRef]
17. Nykänen, V.; Groves, D.I.; Ojala, V.J.; Eilu, P.; Gardoll, S.J. Reconnaissancescale conceptual fuzzy-logic prospectivity modeling for iron oxide copper–gold de-posits in the northern Fennoscandian Shield, Finland. *Aust. J. Earth Sci.* **2008**, *55*, 25–38. [CrossRef]
18. Lusty, P.A.J.; Scheib, C.; Gunn, A.G.; Walker, A.S.D. Reconnaissance-scale prospectivity analysis for gold mineralization in the Southern Uplands-Down-Longford Terrane, NorthernIreland. *Nat. Resour. Res.* **2012**, *21*, 359–382. [CrossRef]
19. Abedi, M.; Norouzi, G.H.; Fathianpour, N. Fuzzy outranking approach: A knowledge-driven method for mineral prospectivity mapping. *Int. J. Appl. Earth Obs. Geoinf.* **2013**, *21*, 556–567. [CrossRef]
20. Ford, A.; Miller, J.M.; Mol, A.G. A comparative analysis of weights of evidence, evidential belief functions, and fuzzy logic for mineral potential mapping using incomplete data at the scale of investigation. *Nat. Resour. Res.* **2016**, *25*, 19–33. [CrossRef]
21. Porwal, A.; Carranza, E.J.M.; Hale, M. Artificial neural networks for mineral potential mapping: A case study from Aravalli Province, Western India. *Nat. Resour. Res.* **2003**, *12*, 155–171. [CrossRef]
22. Carranza, E.J.M. Geochemical anomaly and mineral prospectivity mapping in GIS. In *Handbook of Exploration and Environmental Geochemistry*; Elsevier: Amsterdam, The Netherlands, 2008; Volume 11, p. 351.
23. Carranza, E.J.M.; Hale, M.; Faassen, C. Selection of coherent deposit-type locations and their application in data-driven mineral prospectivity mapping. *Ore Geol. Rev.* **2008**, *33*, 536–558. [CrossRef]
24. Zuo, R. Regional exploration targeting model for Gangdese porphyry copper deposits. *Resour. Geol.* **2011**, *61*, 296–303. [CrossRef]
25. Liu, Y.; Cheng, Q.; Xia, Q.; Wang, X. Mineral potential mapping for tungsten polymetallic deposits in the Nanling metallogenic belt, South China. *J. Earth Sci.* **2014**, *25*, 689–700. [CrossRef]
26. Liu, Y.; Cheng, Q.; Xia, Q.; Wang, X. The use of evidential belief functions for mineral potential mapping in the Nanling belt, South China. *Front. Earth Sci.* **2015**, *9*, 342–354. [CrossRef]
27. Yousefi, M.; Kreuzer, O.P.; Nykänen, V.; Hronsky, J.M.A. Exploration information systems – A proposal for the future use of GIS in mineral exploration targeting. *Ore Geol. Rev.* **2019**, *111*, 103005. [CrossRef]
28. Billa, M.; Cassard, D.; Lips, A.L.W.; Bouchot, V.; Touliere, B.; Stein, G.; Guillou-Frottier, L. Predicting gold-rich epithermal and porphyry systems in the central Andes with a continental-scale metallogenic GIS. *Ore Geol. Rev.* **2004**, *25*, 39–67. [CrossRef]
29. Chung, C.F.; Fabbri, A.G. The representation of geoscience information for data integration. *Nonrenewable Resour.* **1993**, *2*, 122–139. [CrossRef]
30. Porwal, A.; Carranza, E.J.M.; Hale, M. A hybrid neuro-fuzzy model for mineral potential mapping. *Math. Geol.* **2004**, *36*, 803–826. [CrossRef]
31. Porwal, A.; Carranza, E.J.M.; Hale, M. A hybrid fuzzy weights-of-evidence model for mineral potential mapping. *Nat. Resour. Res.* **2006**, *15*, 1–14. [CrossRef]
32. Roy, R.; Cassard, D.; Cobbold, P.R.; Rossello, E.A.; Bailly, M.; Lips, A.L.W. Predictive mapping for copper–gold magmatic-hydrothermal systems in NW Argentina: Use of a regional-scale GIS, application of an expert-guided data-driven approach, and comparison with results from a continental-scale GIS. *Ore Geol. Rev.* **2006**, *29*, 260–286. [CrossRef]
33. Carranza, E.J.M.; Ruitenbeek, F.J.A.V.; Hecker, C.; Meijde, M.V.; Meer, F.D.V. Knowledge-guided data-driven evidential belief modeling of mineral prospectivity in Cabo de Gata, SE Spain. *Int. J. Appl. Earth Obs. Geoinf.* **2008**, *10*, 374–387. [CrossRef]
34. Cassard, D.; Billa, M.; Lambert, A.; Picot, J.C.; Husson, Y.; Lasserre, J.L.; Delor, C. Gold predictivity mapping in French Guiana using an expert-guided data-driven approach based on a regional-scale GIS. *Ore Geol. Rev.* **2008**, *34*, 471–500. [CrossRef]

35. Luo, J. Statistical mineral prediction without defining a training area. *Math. Geol.* **1990**, *22*, 253–260. [CrossRef]
36. Yousefi, M.; Kamkar-Rouhani, A.; Carranza, E.J.M. Geochemical mineralization probability index (GMPI): A new approach to generate enhanced stream sediment geochemical evidential map for increasing probability of success in mineral potential mapping. *J. Geochem. Explor.* **2012**, *115*, 24–35. [CrossRef]
37. Yousefi, M.; Carranza, E.J.M.; Kamkar-Rouhani, A. Weighted drainage catchment basin mapping of stream sediment geochemical anomalies for mineral potential mapping. *J. Geochem. Explor.* **2013**, *128*, 88–96. [CrossRef]
38. Liu, Y.; Zhu, L.; Zhou, Y. Application of Convolutional Neural Network in prospecting prediction of ore deposits: Taking the Zhaojikou Pb-Zn ore deposit in Anhui Province as a case. *Acta Petrol. Sin.* **2018**, *34*, 3217–3224.
39. Zuo, R.; Xu, Y. Graph Deep Learning Model for Mapping Mineral Prospectivity. *Math. Geosci.* **2022**. [CrossRef]
40. Zhou, Y.; Chen, S.; Zhang, Q.; Xiao, F.; Wang, S.; Liu, Y.; Jiao, S. Advances and prospects of big data and mathematical geoscience. *Acta Petrol. Sin.* **2018**, *34*, 255–263.
41. Meng, Q.; Zhang, G. Geologic framework and tectonic evolution of the Qinling orogeny, Central China. *Tectonophysics* **2000**, *323*, 183–196. [CrossRef]
42. Bureau of Geology and Mineral Resources of Henan Province. *Regional Geology Records of Henan Province*; Geological Press: Beijing, China, 1989. (In Chinese with English abstract)
43. Kerrich, R.; Goldfarb, R.; Groves Garwin, S.; Jia, Y. The characteristics, origins, and geodynamic settings of supergiant gold metallogenic provinces. *Sci. China Ser. D Earth Sci.* **2000**, *43*, 1–68. [CrossRef]
44. Kerrich, R.; Wyman, D. Geodynamic setting of mesothermal gold deposits: An association with accretionary tectonic regimes. *Geology* **1990**, *18*, 882–885. [CrossRef]
45. Groves, D.I.; Goldfarb, R.J.; Gebre-Mariam, M.; Hagemann, S.G.; Robert, F. Orogenic gold deposits: A proposed classification in the context of their crustal distribution and relationship to other gold deposit types. *Ore Geol. Revs* **1998**, *13*, 7–27. [CrossRef]
46. Groves, D.I.; Goldfarb, R.J.; Knox-Robinson, C.M.; Ojala, J.; Gardoll, S.; Yun, G.Y.; Holyland, P. Late-kinematic timing of orogenic gold deposits and significance for computer-based exploration techniques with emphasis on the Yilgarn Block, Western Australia. *Ore Geol. Rev.* **2000**, *17*, 1–38. [CrossRef]
47. Craw, D.; Begbie, M.; MacKenzie, D. Structural controls on Tertiary orogenic gold mineralization during initiation of a mountain belt, New Zealand. *Miner. Depos.* **2006**, *41*, 645–659. [CrossRef]
48. Mao, J.; Goldfarb, R.J.; Zhang, Z.; Xu, W.; Qiu, Y.; Deng, J. Gold deposits in the Xiaoqinling–Xiong'ershan region, Qinling Mountains, central China. *Miner. Depos.* **2002**, *37*, 306–325. [CrossRef]
49. Chen, Y.; Pirajno, F.; Qi, J. The Shanggong gold deposit, Eastern Qinling Orogen, China: Isotope geochemistry and implications for ore genesis. *J. Asian Earth Sci.* **2008**, *33*, 252–266. [CrossRef]
50. Bai, H.; Cao, Y.; Zhang, H.; Zhang, C.; Hou, S.; Wang, W. Combining fuzzy analytic hierarchy process with concentration–area fractal for mineral prospectivity mapping: A case study involving Qinling orogenic belt in central China. *Appl. Geochem.* **2021**, *126*, 104894. [CrossRef]
51. *DZ/T 0011-2015*; Specification of Geochemical Reconnaissance Survey(1:50,000). China Geological Survey Nanjing Geological Survey Center: Nanjing, China, 2015. (In Chinese)
52. *DZ01304-2006*; The Specification of Testing Quality Management for Geological Laboratories. Ministry of Land and Resources: Beijing, China, 2006. (In Chinese)
53. Hengl, T. Finding the right pixel size. *Comput. Geosci.* **2006**, *32*, 1283–1298. [CrossRef]
54. Bishop, C.M. *Pattern Recognition and Machine Learning*; Springer Science Business Media, LLC: New York, NY, USA, 2006.
55. Pirajno, F. Intracontinental strike-slip faults, associated magmatism, mineral systems and mantle dynamics: Examples from NW China and Altay-Sayan (Siberia). *J. Geodyn.* **2010**, *50*, 325–346. [CrossRef]
56. Yousefi, M.; Carranza, E.J.M. Prediction–area (P–A) plot and C–A fractal analysis to classify and evaluate evidential maps for mineral prospectivity modeling. *Comput. Geosci.* **2015**, *79*, 69–81. [CrossRef]
57. Herbert, S.; Woldai, T.; Carranza, E.J.M.; Ruitenbeek, F.J.A.V. Predictive mapping of prospectivity for orogenic gold in Uganda. *J. Afr. Earth Sci.* **2014**, *99*, 666–693. [CrossRef]
58. Carranza, E.J.M.; Laborte, A.G. Random forest predictive modeling of mineral prospectivity with small number of prospects and data with missing values in Abra (Philippines). *Comput. Geosci.* **2015**, *74*, 60–70. [CrossRef]
59. Meng, Q.; Liu, Z.; Borders, B.E. Assessment of regression kriging for spatial interpolation – Comparisons of seven GIS interpolation methods. *Cartogr. Geogr. Inf. Sci.* **2013**, *40*, 28–39. [CrossRef]
60. Bishop, T.F.A.; Lark, R.M. The geostatistical analysis of experiments at the landscape-scale. *Geoderma* **2006**, *133*, 87–106. [CrossRef]
61. Bonham-carter, G.F. Geographic Information Systems for Geoscientists: Modelling with GIS. Computer Methods. In *The Geosciences*, 13th ed.; Pergamon Press: New York, NY, USA, 1995.
62. Mandelbrot, B.B. *Statistics of Natural Resources and the Law of Pareto*; I.B.M. Research Note NC-146; Watson Research Center: New York, NY, USA, 1962; p. 131.
63. Mandelbrot, B.B. *The Fractal Geometry of Nature*; W.H. Freeman and Company: New York, NY, USA, 1983; p. 468.
64. Mandelbrot, B.B. Self-affine fractals and fractal dimension. *Phys. Scr.* **1985**, *32*, 257–260. [CrossRef]
65. Carlson, C.A. Spatial distribution of ore deposits. *Geology* **1991**, *19*, 111–114. [CrossRef]
66. Cheng, Q.; Agterberg, F.P.; Ballantyneb, S.B. The separation of geochemical anomalies from background by fractal methods. *J. Geochem. Explor. C* **1994**, *51*, 109–130. [CrossRef]

67. Cheng, Q.; Agterberg, F.P.; Bonham-Carter, G.F. Fractal pattern integration for mineral potential estimation. *Nonrenewable Resour.* **1996**, *5*, 117–130. [CrossRef]
68. Xiang, Z.; Gu, X.; Wang, E.; Wang, X.; Zhang, Y.; Wang, Y. Delineation of deep prospecting targets by combining factor and fractal analysis in the Kekeshala skarn Cu deposit, NW China. *J. Geochem. Explor.* **2019**, *198*, 71–81. [CrossRef]
69. Hodkiewicz, P.F.; Weinberg, R.F.; Gardoll, S.J.; Groves, D.I. Complexity gradients in the Yilgarn Craton: Fundamental controls on crustal-scale fluid flow and formation of world-class orogenic-gold deposits. *Aust. J. Earth Sci.* **2005**, *52*, 831–841. [CrossRef]
70. Ford, A.; Blenkinsop, T.G. Combining fractal analysis of mineral deposit clustering with weights of evidence to evaluate patterns of mineralization: Application to copper deposits of the Mount Isa Inlier, NW Queensland, Australia. *Ore Geol. Rev.* **2008**, *33*, 435–450. [CrossRef]
71. Ghasemzadeh, S.; Maghsoudi, A.; Yousefi, M.; Mihalasky, M.J. Stream sediment geochemical data analysis for district-scale mineral exploration targeting: Measuring the performance of the spatial U-statistic and C-A fractal modeling. *Ore Geol. Rev.* **2019**, *113*, 103–115. [CrossRef]
72. Chen, D.; Wei, J.; Wang, W.; Shi, W.; Li, H.; Zhan, X. Comparison of Methods for Determining the Thresholds of Geochemical Anomalies and the Prospecting Direction-A Case of Gold Deposits in the Gouli Exploration Area, Qinghai Province. *Mineral* **2019**, *9*, 368. [CrossRef]

minerals

MDPI

Article

Recursive Feature Elimination and Neural Networks Applied to the Forecast of Mass and Metallurgical Recoveries in A Brazilian Phosphate Mine

Fernanda Gontijo Fernandes Niquini [1,*], André Miranda Brito Branches [1], João Felipe Coimbra Leite Costa [1], Gabriel de Castro Moreira [1], Claudio Luiz Schneider [2], Florence Cristiane de Araújo [3] and Luciano Nunes Capponi [3]

[1] PPGE3M (Programa de Pós-Graduação em Engenharia de Minas, Metalúrgica e de Materiais), Universidade Federal do Rio Grande do Sul, Av. Bento Gonçalves, 9500-Setor 4-Prédio 74-Sala 211, Campus do Vale, Porto Alegre 91509-900, Brazil

[2] Centro de Tecnologia Mineral (CETEM), 900-Ilha da Cidade Universitária, Av. Pedro Calmon, Rio de Janeiro 21941-596, Brazil

[3] Mosaic Fertilizantes P&k Ltd., Rodovia MGC, Km 196.5, Tapira 38185-000, Brazil

* Correspondence: fernanda.gontijo.fn@gmail.com

Abstract: Geometallurgical models are commonly built by combining explanatory variables to obtain the response that requires prediction. This study presented a phosphate plant with three concentration steps: magnetic separation, desliming and flotation, where the yields and recoveries corresponding to each process unit were predicted. These output variables depended on the ore composition and the collector concentration utilized. This paper proposed a solution based on feature engineering to select the best set of explanatory variables and a subset of them able to keep the model as simple as possible but with enough precision and accuracy. After choosing the input variables, two neural network models were developed to simultaneously forecast the seven geometallurgical variables under study: the first, using the best set of variables; and the second, using the reduced set of inputs. The forecasts obtained in both scenarios were compared, and the results showed that the mean squared error and the root mean squared error increase in all output variables evaluated in the test set was smaller than 2.6% when the reduced set was used. The trade-off between simplicity and the quality of the model needs to be addressed when choosing the final neural network to be used in a 3D-block model.

Keywords: geometallurgy; neural networks; recursive feature elimination; phosphate; feature selection

1. Introduction

Geometallurgical models are essential for the mining industry. These models combine geology, geochemistry, mineralogy, mine planning, rock mechanics, and metallurgy [1] to make predictions about mineral processing performance. This performance can be measured in several ways, depending on the goals to be achieved. Comminution energy requirements; ore resistance to impact breakage ($A \times b$); Ball Mill Work Index [2–6] throughput [3,7,8]; particle size in the product—defined as the size at which 80% of the particles are smaller (commonly known as *P*80) [9]; yields and recoveries [6,10–15]; the contaminant grades in tailings [16,17]; and acid consumption [18,19] are variables frequently modeled in geometallurgical studies. The most common input variables used to build geometallurgical models include the mineralogical features; the geological descriptions (containing lithological, textural, alteration, and weathering information); the chemical composition; and the specific gravity of the ore. All of these generate many variables that need to be analyzed and are related to plant performance.

The data collected from the ore vary in importance in terms of their effects on the metallurgical performance of the process. Proper selection of the features is an important step in geometallurgical modeling. The selection of variables, also called feature selection, obtains the subset of input data that contributes the most to determining the dependent variables (target/output) to be modelled. The advantages of the feature selection procedure are:

- Removing redundant and irrelevant variables;
- Decreasing noise in the forecast, improving the efficiency of the model;
- Reducing the risk of overfitting;
- Reducing the computational cost of processing.

The process starts by deciding which variables should be used as inputs in the geometallurgical model. It is prevalent to start with a correlation matrix analysis aimed at identifying the input variables with the strongest relationships with the outputs to be modelled. If the selected variables present a strong correlation coefficient among themselves, this can lead to a collinearity problem [20]. Furthermore, when the number of variables increases, analyzing the correlation matrix can become a complex problem due to the number of cells to interpret. Other methods that are frequently used to reduce the number of variables are listed in Table 1, along with the core idea of the method, disadvantages, and the appropriate references.

Table 1. Methods used to reduce the number of variables.

Method	Core Idea	Disadvantages	Reference
Super Secondary	Linearly combine several input variables into a single variable	- Possible loss of interpretability of the models that are generated; - Requires previous knowledge to determine which variables are to be considered for amalgamation; - High number of samples needed to avoid the attribution of inflated weights caused by poorly conditioned matrices.	[18]
PCA and MAF	Combination of variables	- Difficulty in interpreting the resulting models that use the factors; - Total variability of data is not considered if not all the components are used.	[21,22]
Agglomerative hierarchical cluster analysis	Group variables instead of samples by similarity	- Demands time for interpreting the results.	[23–26]
Stepwise Methods (forward, backward and stepwise)	Find the subset of variables that minimizes the residual sum of squares	- Do not ensure that collinearity problems will not occur in models and that irrelevant variables will not be part of the selected subset.	[27]

Recursive feature elimination (RFE) [28] is another method used to select a subset of features to reduce dimensionality. The idea is to start with a model containing all of the available variables and iteratively remove one or more features in each algorithm step, until only one variable remains. The improvement in the recursive feature elimination, in relation to the backward elimination, is that the RFE allows the user to choose which technique will be used to elaborate the forecast model. It allows the use of support vector regression, random forest, linear regression, and other techniques which contain a measure of the importance of each variable during the modeling. In contrast, the backward elimination was conceived to be used only with linear regression models. Furthermore, the RFE allows the user to remove more than one variable in each iteration, while the backward elimination was initially developed to remove one variable at a time.

The RFE technique was chosen to be applied in this study for three main reasons: first, the relation between the number of samples and the number of variables was not a problem for the database under study. There were 3056 samples of ten input variables (nine chemical components and the collector concentration used in the flotation test), making techniques such as PCA, MAF, and super secondary not advantageous due to the difficulty in interpreting the resulting variables. The second reason was that the application of clustering to choose variables demands more time for interpretation. This study's core objective was to develop a semi-automatic workflow to be applied and quickly updated whenever new samples were introduced into the database. The third reason is that RFE is an improvement when compared to the stepwise methods, more specifically the backward elimination, allowing the use of other algorithms in its core and expanding the modeling options under evaluation. Another advantage of the RFE algorithm is its use combined with k-fold cross-validation [29], allowing the user to evaluate the variability of the results by varying the samples used.

Thus, the workflow proposed in this paper was to carry out a feature selection study using RFE to reduce model complexity and, after the most important features were selected, to elaborate a neural network model to predict the processing plant response for any ore sample. Some examples of neural networks applied to forecast geometallurgical variables, as proposed in this paper, are listed below:

- Niquini and Costa used neural networks to forecast yield and metallurgical recovery of all plant outputs in a gold [15] and a zinc [14] processing plants. The results showed that neural networks are able to generate accurate and precise predictions while satisfying the mass and metallurgical balance constraints;
- Both and Dimitrakopoulos [30] applied neural networks to forecast throughput and compared the results against the throughput predicted by linear regression. The NN showed better results and the predictions obtained with its use can be integrated into production scheduling;
- Gholami et al. [31], used neural networks to forecast four geometallurgical variables in a copper mine, using as inputs particle size distribution, collector and frother concentrations, solids content, pH and mineralogical variables;
- Jorjani et al. [32] created a neural network model to forecast La, Ce, Y and Nd recoveries from an apatite concentrate. The input variables were leaching time, pulp density, agitation rate and acid concentration;
- Nakhaei et al. [33] developed a neural network model to forecast Cu and Mo grades and recoveries using as inputs Cu and Mo grades measured in the flotation circuit as well as some flotation circuit operational variables;
- Srivastava et al. [34] elaborated a neural network model to predict the power in a SAG mill using, as inputs, the mill power observed in the previous time increments and the feed rate in the evaluated time;
- Panda and Tripathy [35] created a neural network to forecast the grade and metallurgical recovery of chromite using operational variables as inputs.

After developing the neural network model, the process response can be included in a 3D block model used for mine planning. The reduced input scenario was compared to the results obtained when the best scenario, using all variables available, was used to build the neural network forecasting model. By doing this, it was possible to determine the impact of removing features from the neural network inputs.

Process Modelling and Geometalurgical Modelling

A general solution for any flowsheet was established as early as 1979 [36], allowing for accurate solutions for very complex and extensive mineral-processing plant flowsheets. The performance of a mineral-processing plant can be predicted using a plant-wide simulator [37]. However, the simulation prediction accuracy depends on several factors related to the unit operation models and the ore model (also known as a geometallurgical model). These two models cannot be completely separated and, in many cases, the unit operation models contain ore model parameters such as comminution parameters for breakage and selection functions. Nevertheless, it is advantageous to understand the problem as if the unit operation models and the ore models are two separate, but not independent, entities.

The reason for this is that, in the case of mine planning, the plant may, and should, be considered as a fixed entity, a group of processes that will perform only as a function of the ore that is fed to it. System data may change as a function of the ore that is fed to the plant, and this includes solids flowrates, water flowrates, and particle-size distributions. If we look at this problem from this perspective, it is clear that the ore model is the main factor determining plant performance and the accuracy of the predictions made.

At the heart of the mineral-processing plant that was considered in this study, is the flotation circuit. Although flotation is widely used to concentrate large tonnages of a variety of ore minerals, it is a challenging process to model because of the large number of micro-processes involved. King [37] gives an excellent review of the state of the art of flotation modeling.

In King's approach, the flotation process requires four models, for the pulp phase, the bubble phase, the froth phase, and the entrained phase. Each of these models requires several parameters, some of which are difficult to measure in practice. Many of the micro-processes involved in flotation can be bundled in flotation rates. The onset of the development of kinetic models is due to Sutherland [38]. Since then, several authors have contributed to understanding flotation kinetics through developing particle-bubble collision efficiency models [39–41]. Clift, Grace and Weber [42] and Schulze [43] contributed with hydrodynamic considerations of bubble motion. Bubble-rise velocity models were developed by Karamanev and Nikolov [44]. The micro-process of bubble loading and its influence in flotation rates were investigated by Tomlinson and Fleming [45], King, Hatton and Hulbert [46] and Bradshaw and O'Connor [47]. Recovery as a function of particle size was modelled by Trahar and Warren [48].

Another important process in flotation is particle detachment, as described by Drzymala [49]. A simple model for the froth phase was developed by Murphy, Zimmerman and Woodburn [50]. Several other authors have made excellent contributions to understanding and modelling each of the processes involved in flotation. The point here is to show that it is not feasible to include every aspect of flotation simulation into a geometallurgical model. For example, Neethling and Cilliers [51] provided a significantly more accurate and realistic model for the froth phase that includes coalescence. However, the computation is complex, and it is not likely that one simulation could be accomplished in a fraction of a second.

While plant-wide simulation is suited for predicting yields and recoveries in the context of mine planning, it would require a detailed characterization of the ore such as, but not limited to, complete mineralogical assembly and liberation by size information for every drill-core sample that is produced in the mine. At this stage in the development of characterization techniques, this would consume too much time and would also be financially unattractive. Process modelling is best employed in the context of flowsheet optimization and for scaling up unit operations. In these applications, it is feasible to obtain fundamental laboratory-scale parameters for a given ore sample. In the context of geometallurgical modelling for mine planning, the approach that is described in this paper is better suited because the number of samples that need to be characterized and tested is large, the flowsheet does not change, or, more accurately, the geometallurgical model is developed for a given flowsheet and it is valid while the flowsheet is not changed, and there is no need for scale-up parameters as all unit operations have already been scaled at the time the plant was designed.

2. Data Processing

2.1. Geological Context

The geological features of the region were studied extensively [52–55] due to the anomalous occurrences of phosphate, titanium, niobium, and rare earth elements. The anomalies are related to the weathering of Cretaceous alkaline intrusive rocks. The weathering layer often reaches several tens of meters in depth and the unaltered rock is only usually observed in cores from deep diamond drilling.

In the study area, anomalous phosphate concentrations occur due to the presence of apatite. Naturally, the alteration is more intense closer to the surface and the apatite there is almost completely leached. The supergenic concentration of this mineral results from the solubilization and leaching of the more unstable components contained in the original rocks, such as mafic minerals and carbonates, mainly pyroxenites.

During the weathering process, at first, carbonates and mafic minerals are leached, while apatite crystals remain, as well as titanium, the latter in the form of perovskite. With the progress of such transformations, apatite is converted into secondary minerals, from the crandalite group, of reduced solubility and with lesser economic value. Meanwhile, the perovskite transforms into anatase, which gradually becomes residually concentrated. This process has led to the current weathering profile of the study area, where the intermediate-upper layer shows a high content of titanium in the form of anatase, and the intermediate-lower portion presents higher grades of phosphatic ore and lower concentrations of titanium.

The lower portion of the weathering profile comprises semi-altered rocks, which may marginally contain economic concentrations of apatite, but predominately comprise mafic minerals, carbonates and magnetite. Variations due to the heterogeneity of the original magma lead to higher or lower occurrences of minerals such as perovskite, phlogopite, and vermiculite.

The samples taken for this study corresponded to the intermediate-lower portion of the profile, which was classified as ore during the geological logging of the diamond-drilled cores.

2.2. Dependent and Independent Variables

The database under study contained 3056 individual geometallurgical samples, with Al_2O_3, BaO, CaO, Fe_2O_3, MgO, Nb_2O_5, P_2O_5, SiO_2, and TiO_2 grades analyzed from vertical diamond drilling. The feature selection study also considered the collector (fatty acid) concentration used in the flotation batch tests as an input variable. This concentration was determined by the laboratory technician, taking into account some sample aspects such as: apatite crystal cleanliness; P_2O_5 grade after desliming and demagnetization; and ore texture. The milling time adopted in the processing, the depressor (corn starch) concentration, the pH, the flotation time, the strength of the magnetic field and the solid concentration in the flotation pulp were kept under constant values to not affect the results; the influence of these variables on the test responses was nil, and they were not considered as input variables.

Mineralogy and rock texture, although very important aspects of a geometallurgical study, were not directly available for this particular study as specific features, but were certainly reflected in the wide assembly of grade variables considered. Table 2 summarizes the ROM information and the collector concentration, showing that the mean grade of P_2O_5 was 9.74%.

Understanding the relationship between independent and dependent variables requires knowledge of the mineral processing plant's flowsheet (Figure 1). All geometallurgical samples were submitted to batch tests that mimicked the flowsheet: starting with a magnetic separation step, followed by desliming and flotation.

Magnetics, limes, tailings, and apatite concentrate are plant products. Therefore, each batch test generated the relative fraction of the initial mass destined for each product, making it possible to calculate the yield corresponding to each plant output. In this paper the term "yield" was used to express the ratio between weight or throughput of a stream and the throughput of the plant feed [56,57]. The term "recovery" was used to express the fraction of apatite recovered in a stream with respect to the feed stream.

Table 2. ROM descriptive statistics.

Variable	N	Mean	Std. Dev.	Minimum	Q1	Median	Q3	Maximum
Al_2O_3(%)	3056	4.42	2.15	0.06	3.15	4.15	5.25	34.57
BaO (%)	3056	0.34	0.64	0.01	0.15	0.24	0.38	27.21
CaO (%)	3056	13.65	4.39	0.02	10.91	13.46	16.15	42.89
Fe_2O_3(%)	3056	27.33	7.21	7.59	22.68	26.50	31.12	68.64
MgO (%)	3056	4.09	2.50	0.01	2.17	3.86	5.61	17.85
Nb_2O_5(%)	3056	0.14	0.26	0.01	0.06	0.11	0.17	12.00
P_2O_5(%)	3056	9.74	3.08	0.36	7.80	9.46	11.34	31.63
SiO_2(%)	3056	23.02	8.05	1.32	17.84	23.08	27.90	88.40
TiO_2(%)	3056	8.45	4.11	0.13	5.74	7.64	10.41	39.31
Collector dosage (g/t)	3056	627.80	751.30	20.00	200.00	350.00	700.00	5000.00

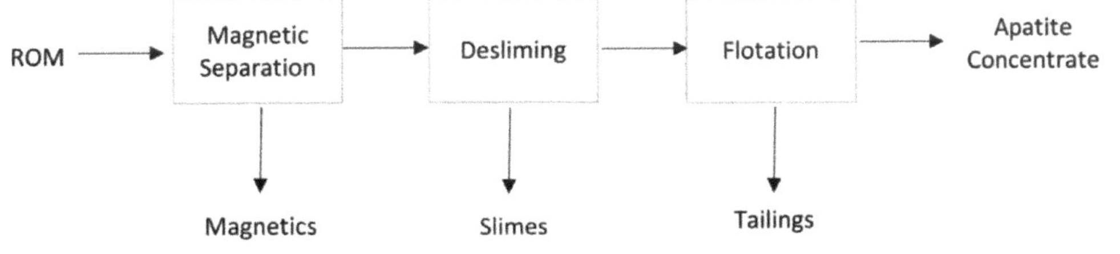

Figure 1. Plant flowsheet.

Let Q_{stream} represent the solids flowrate of a given stream. Let Y_{stream} represent the yield of a given stream, then Equations (1) to (4) follow.

$$Y_{Magnetics} = \frac{Q_{Magnetics}}{Q_{ROM}} \quad (1)$$

$$Y_{Slimes} = \frac{Q_{Slimes}}{Q_{ROM}} \quad (2)$$

$$Y_{Concentrate} = \frac{Q_{Concentrate}}{Q_{ROM}} \quad (3)$$

$$Y_{Tailings} = \frac{Q_{Tailings}}{Q_{ROM}} \quad (4)$$

Thus, four yield values were available (totalling 100%): magnetics, slimes, concentrate, and tailings yield.

Chemical analysis is routinely carried out for the ROM, the flotation feed, and the flotation concentrate streams of the batch tests. The grades in the flotation tailings stream can be calculated from the measured grades in the flotation feed and concentrate streams. The grades in the combined magnetics and slime streams can be calculated using the grades and yields measured in the ROM and in the flotation feed. Let the metallurgical recovery of P_2O_5 in any stream be R_{Stream} and the grade of P_2O_5 in any stream be g_{Stream}. The metallurgical recoveries of P_2O_5 were calculated using the yields and grades as shown in Equations (5) to (7). Considering all output points with chemical information (calculated or analysed), it is possible to determine the P_2O_5 metallurgical recovery with respect to the ROM P_2O_5 content, as presented in Equations (6) to (8). It should be noted that when these three variables are added, the result equals 100%, as should be the case with the yield variables.

$$R_{Slimes+Magnetics} = \left(Y_{Slimes} + Y_{Magnetics}\right) \times \frac{g_{Slimes+Magnetics}}{g_{ROM}} \quad (5)$$

$$R_{Concentrate} = Y_{Concentrate} \times \frac{g_{Concentrate}}{g_{ROM}} \quad (6)$$

$$R_{Tailings} = Y_{Tailings} \times \frac{g_{Tailings}}{g_{ROM}} \quad (7)$$

Therefore, the dependent variables to be modeled are the slimes, magnetics, concentrate, and tailings yields, as well as the metallurgical recoveries in the slime + magnetic, in the flotation concentrate, and the flotation tailings. Table 3 presents the descriptive statistics for all dependent variables under study. On average, 47.81% of all P_2O_5 and 13.74% of all mass fed in the plant are destined for the flotation concentrate.

Table 3. Dependent variables descriptive statistics. Values in %.

Variable (%)	N	Mean	Std. Dev.	Minimum	Q1	Median	Q3	Maximum
Y_{Slimes}	3056	42.85	10.10	4.09	36.07	42.37	49.52	81.62
$Y_{\text{Magnetics}}$	3056	10.79	8.09	0.04	4.74	9.25	14.95	68.35
$Y_{\text{Concentrate}}$	3056	13.74	6.94	0.34	8.47	13.14	18.10	52.55
Y_{Tailings}	3056	32.61	11.86	1.91	23.19	32.17	41.52	68.84
$R_{\text{Slimes+Magnetics}}$	3056	34.81	10.25	1.86	28.02	33.93	41.11	82.51
$R_{\text{Concentrate}}$	3056	47.81	16.28	0.68	36.99	49.98	60.20	92.08
R_{Tailings}	3056	17.38	13.29	0.00	7.20	13.30	24.69	75.10

2.3. Database Partition

Twenty percent of the samples were randomly selected and set apart to be used as a test set to evaluate the model's performance. All models presented were elaborated using the remaining 80% of the samples in the database. The k-fold technique [29] was used to determine the optimum parameters for each proposed model. Then, the models obtained were applied to the test dataset.

3. Materials and Methods

The methodology proposed in this work was based on two main methods: artificial neural network (ANN) and recursive feature elimination, both belonging to the machine learning framework.

As explained before, RFE is a technique used in, so-called, 'Feature Engineering' aimed at identifying the most relevant input variables to predict either one or a set of output variables. The goal is to use the least possible number of variables, thus simplifying the model while maintaining good accuracy in the predictions.

ANNs are computing systems designed to most efficiently map the relationships between input and output variables in complex structures. These two subjects are more extensively described in the following sections, and the overall methodology can be outlined as follows:

(1) Selection of the most relevant input variables using RFE;
(2) The construction of ANN models to predict metallurgical attributes using the input variables selected in the previous step;
(3) Checking the performance of the built model by forecasting the test dataset.

3.1. Recursive Feature Elimination

As already mentioned, one must decide how many variables will be used as inputs in the prediction model. It is common for the geometallurgical dataset to contain data that is either redundant or not relevant to the prediction of the process performance. Recursive feature elimination (RFE) [28] is an algorithm for automatic feature selection that uses a prediction algorithm in its core to rank the inputs by relevance using a predefined metric. The process starts by creating a predictive model with all the information, organizing the information, and measuring model performance. The less critical predictor is removed at each step. The model is rebuilt iteratively until the desired set of inputs is achieved.

Here, the most relevant subset was found using the RFE version with cross-validation (RFECV) [58], partitioning the dataset tenfold, three times (ten-fold validation with three repetitions). Three different machine learning multitarget predictive models were used in its core: linear regression [59], random forest [60], and CatBoost regressor [61], which will be briefly explained below.

Linear regression is one of the most commonly used techniques for building geometallurgical models [15]. Its goal is to find the equation which minimizes the prediction's mean squared error using ordinary least squares. To rank the variables for importance, the recursive feature elimination uses the p-value of each variable, removing the one corresponding to the higher p-value in each algorithm iteration, until only the variable with the smallest p-value remains.

The random forest algorithm establishes cutoffs in the input variables, which help to estimate the output variables under study. The greater the number of times an input variable is selected to best describe the output, the greater its importance in the model's creation. This importance is used inside the RFE to rank the input variables.

The CatBoost algorithm was developed based on the decision trees and gradient boosting theory. The main idea of boosting is to sequentially combine many weak models (small decision trees). Because gradient boosting fits the decision trees sequentially, the fitted trees will learn from former trees' mistakes and reduce the errors. Adding a new function to the existing ones is continued until the selected loss function is no longer minimized [61]. Similar to other decision tree-based algorithms, CatBoost uses feature importance to rank the relevance of all inputs for the model being created.

It should be noted that only techniques which present a measure of variables' importance can be used in RFE. Artificial neural networks, ACE [62], and others which do not provide this kind of information, cannot be used in combination with the recursive feature elimination.

To evaluate the performance of each model built, the following metrics were chosen: mean absolute error (MAE), square root of the mean squared error (RMSE), mean absolute percentage error (MAPE), median absolute percentage error (MdAPE), and the Pearson correlation coefficient. Once this study dealt with multitarget models, the mean value of each variable metric was taken to give a general idea of the model's precision and accuracy. For example, the MAE value was calculated for each of the seven output variables under study. The mean of the seven MAE values was then calculated and this was named the global MAE. The same procedure was used for the other metrics.

The feature selection process was developed in three stages. First, the three predictive algorithm performances were evaluated with RFECV for each subset of relevant inputs in decreasing order until only the most important variable remained, obtaining the global MAE value in each algorithm step. The lowest global MAE value indicated the best predictive algorithm and optimum subset size.

In the second stage, the algorithm selected previously was used inside the RFECV, and the global MAPE for the optimum set was calculated, this being the reference value. All subsets of the variables were analysed (with a lower number of variables than the best case), which provided a global MAPE increase of a maximum of 5%, compared with the reference value. Thus, the smallest subset within the percentage cut-off was chosen as the optimum subset.

In the final third stage, to highlight the importance of the features, an adaptation of the SHAP-values algorithm [63] was used to show the importance of each variable as percentage values.

After the feature selection step, the development of two neural networks started: the first one using the optimum set of variables as inputs (in this case, all available variables) and the second using the variables contained in the optimum subset as inputs. Using this technique to achieve the forecast models was preferred to the CatBoost algorithm, once the neural network results were comparable to the CatBoost results while presenting the advantage of providing the yields and metallurgical balances between the forecasts. Details about the artificial neural network technique are given below.

3.2. Artificial Neural Networks

The artificial neural network technique (ANN) is versatile in geometallurgical model construction. This method can forecast variables with nonlinear relationships with the input parameters, such as the yields and metallurgical recoveries. Additionally, it can simultaneously predict two or more dependent variables using one single network structure. Due to these characteristics and a history of good results [12,14,15], ANN was chosen to predict the geometallurgical responses studied herein. A brief description of the theory related to ANN is given below. For more details about neural networks, their mathematics, and their applications in the mining industry, the reader should refer to Niquini and Costa [14].

Determining which, and how many, variables should be used as inputs is necessary, given that different structures will lead to different results. The independent variables in neural networks are named "input variables" (represented by X), while the dependent variables are named "output variables" (represented by Y). Therefore, the oxide grades in the ROM and the collector dosage inserted in the flotation tests are information that can be used as inputs. The yields and metallurgical recoveries are in the network outputs.

In addition to the input and output layers, there are three other fundamental structures in the ANN: the hidden layer(s), the network weights, and the activation functions (Figure 2).

The user determines the number of hidden nodes and the number of hidden layers. The hidden layer is the intermediary part of the network, where the intermediate calculation to generate the network's final forecast is carried out. Each hidden node is denoted by Z.

The lines that connect the input layer to the hidden layer and the hidden layer to the output layer are named network weights. They determine the nodes with a greater or a lesser influence in each output node forecast: the higher the weight (in absolute terms), the more significant the influence of the related node, and vice-versa. The determination of α and β weight is usually made using the backpropagation algorithm (details in [64]). Equation (8), an adaptation of the notation given by Izenman [64], shows the position of the weights inside the network forecast function.

$$\hat{y}_{i,k} = g_k\left(\sum_{j=0}^{t} \alpha_{kj}\left(f_j\left(\sum_{m=0}^{r} \beta_{jm} x_{i,m}\right)\right)\right) \tag{8}$$

where i refers to the sample number in the database, with $i = 1, \ldots, n$ (where n is the total number of samples); k is the index which represents the output variable under study, with $k = 1, \ldots, s$ (where s is the number of output nodes); j is the index which represents the hidden nodes, with $j = 0, \ldots, t$ (where t is the total number of hidden nodes); and m represents the input nodes, with $m = 0, \ldots, r$ (where r is the total number of input nodes in the network).

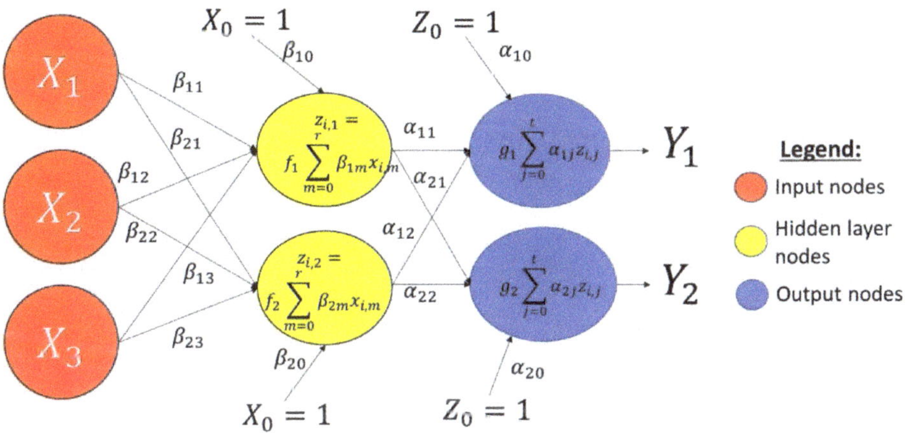

Figure 2. Example of a neural network with three input nodes and two output nodes.

The functions g and f, in Equation (8), and shown in Figure 2, are the activation functions: a set of predetermined functions which are fed with the linear combination of the input nodes (x) and its β weights, in the case of the f function, or fed with the linear combination of the hidden nodes (z) and its α weights, in case of the g function. Of all the activation functions that exist in the literature, the following were used in this study:

Linear function:

$$\varphi(v) = v \tag{9}$$

Sigmoid function:

$$\varphi(v) = \frac{1}{1 + e^{(-v)}} \tag{10}$$

ReLU function:

$$\varphi(v) = max(0, v) \tag{11}$$

Softplus function:

$$\varphi(v) = ln(1 + e^{v}) \tag{12}$$

Tanh function:

$$\varphi(v) = \frac{(e^{v} - e^{-v})}{(e^{v} + e^{-v})} \tag{13}$$

Using Equation (8) and the activation functions mentioned above, it is possible to forecast the geometallurgical variables of interest. The models can predict the processing plant performance in each block of the 3D block model, aiding in routine mine scheduling procedures.

4. Results and Discussion

The first step in elaborating the forecast model for the seven output variables consisting of yields and recoveries, was to determine which input variables would be used in the neural network. Following the methodology proposed, the first stage included finding the best predictive algorithm and subset size. Figure 3 presents the performance of the three algorithms for a different number of features. As can be seen, the CatBoost Regressor algorithm is the best for any quantity of inputs, with an optimal performance being achieved by using 10 features in this scenario. Figure 3 also shows that the global MAE presents a lower rate of decrease from four or more inputs. Hence, it is necessary to analyse whether the size reduction of any chosen subset will significantly affect the model's performance and which variables are selected in each scenario.

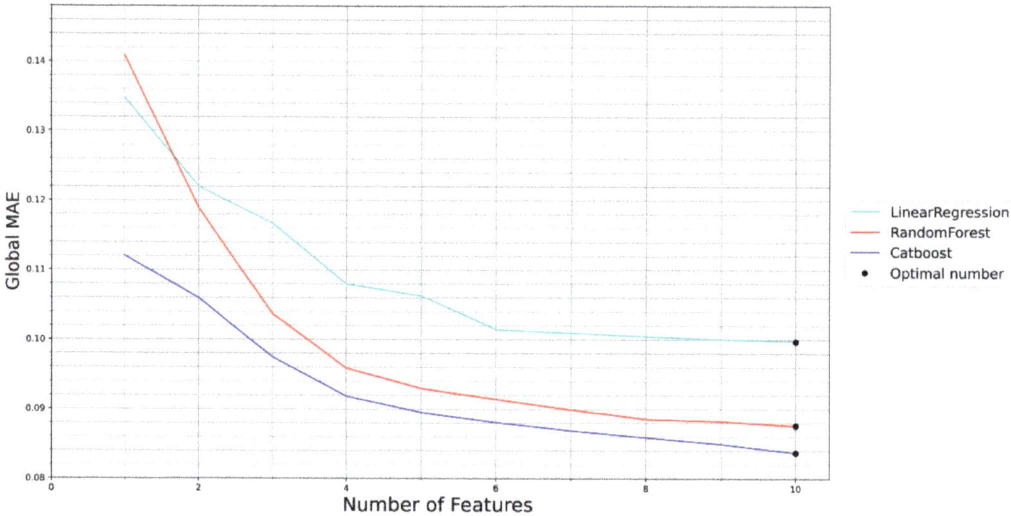

Figure 3. Results from predictive performances for all RFECV core algorithms.

The CatBoost algorithm was applied for the optimum set of variables (10 features) and for the subsets containing four to nine features to reduce model complexity. The results for four, five, and six features are presented due to their importance. A ten-fold cross-validation technique was used, considering different randomly selected seeds. The global MAPE value was calculated for each set size, as presented in Table 4. The increase in the global MAPE value, compared to the best case, is presented in Table 5.

Table 4. Global MAPE values observed for the differing number of input variables.

Metric	Catboost Algortithm with			
	10 Variables	6 Variables	5 Variables	4 Variables
Global MAPE	6.07	6.36	6.50	6.75

By analyzing the results, it is possible to conclude that, by choosing the six most relevant features, the global MAPE increment is lower than 5% when comparing this metric against the global MAPE obtained for the best case. Thus, the scenario using six variables was chosen to continue the study.

Table 5. Increase in global MAPE when subsets of variables were used.

Scenario	Catboost Algortithm with	Increase In Global MAPE (%) Compared Agaist the Best Case
1	6 variables	4.6
2	5 variables	7.1
3	4 variables	11.1

In the next step of the analysis, the CatBoost hyperparameters were calibrated for the two sets of variables: the best one, with all variables available, and the second, with six variables chosen in the reduced scenario. The SHAP-values algorithm was then used to estimate each input variable's importance (at percentage levels) in both scenarios. Figure 4 illustrates the importance of the optimum scenario (10 features), while Figure 5 presents this information for the subset with six variables. The graphics generated by the algorithm can be interpreted as follows: the X axis represents the importance of each feature to the model (the higher the value, the more important the feature), while the Y axis shows the input variables included in the evaluation. The horizontal bar colors are related to the output variables that are to be predicted (the larger is the color bar, more important the feature is to forecast the respective output behavior). More details about the SHAP values technique can be read in [65].

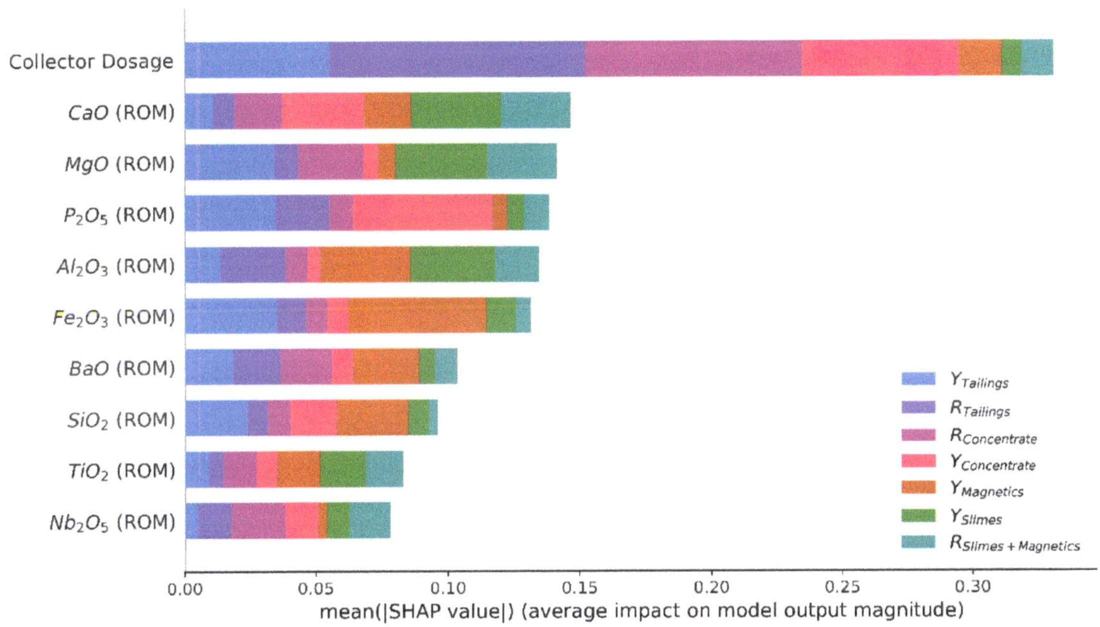

Figure 4. Features importance for the best model (10 inputs).

Amongst the grades, the most important feature to describe the geometallurgical variables was the CaO concentration in the ROM, followed by the MgO concentration in the best scenario and Fe_2O_3 concentration in the sub-optimum scenario. The importance of collector dosage is notorious in both scenarios; being even more important to forecast the flotation concentrate and the tailings recoveries. It is known that the collector dosage cannot influence the desliming and demagnetization processes; thus, the influence of this variable on the yield of slimes, the yield of magnetics and the recovery of slimes + magnetics presented in Figures 4 and 5, are related to a cross-relation with other variables, for example: if the sample has a lot of apatite grains with magnetite, part of the magnetic material will be concentrated in the magnetic separator which will present a high yield, but another part of the apatite grains with magnetite will follow through to flotation. When entering in the flotation step, the laboratory technician will see the apatite grains with magnetite in the sample and will add a higher collector dosage, in order to improve the flotation behavior. Thus, the collector dosage was correlated with the magnetic mass yield not because the first affected the latest, but because the first was calibrated to deal with the characteristic of the sample not concentrated in the magnetic separation step.

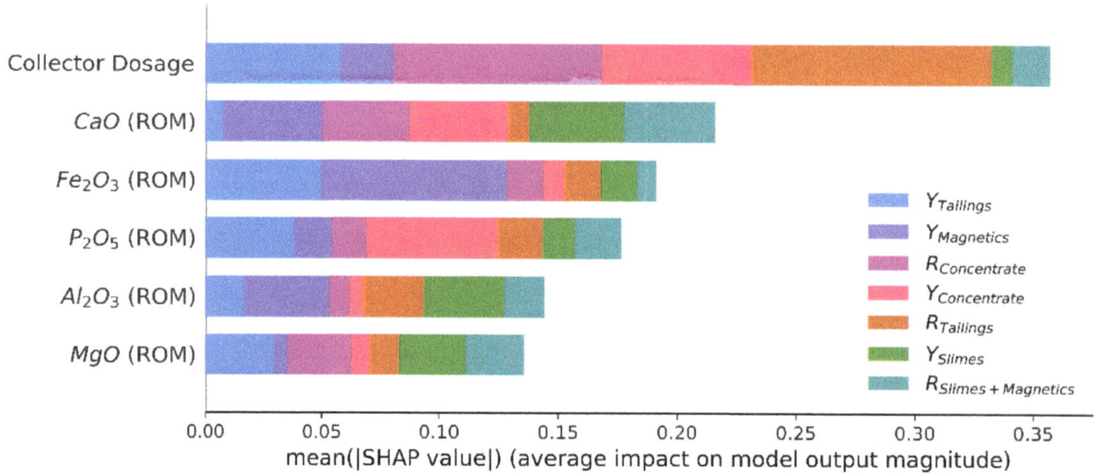

Figure 5. Features importance for the reduced model (six inputs).

BaO, Nb_2O_5, SiO_2 and TiO_2 grades were the variables removed in the reduced scenario. It is known that barite and pyrochlore, the minerals that carry BaO and Nb_2O_5, are present in the deposit and they may affect flotation when their grades are significant; more specifically the barite which competes for the collector of apatite in flotation. Because these minerals are not usually present in high quantities, their influence in the global scenario is reduced, so the grades of BaO and Nb_2O_5 can be removed. The TiO_2 grade was probably selected for removal by the algorithm because it is correlated with the other two variables that were kept in the model (correlation of 0.48 with Fe_2O_3 and 0.46 with MgO). The same occurred with SiO_2, which is correlated with the grades of Fe_2O_3 (0.68), P_2O_5 (0.56) and MgO (0.53).

Next, the neural network's hyperparameters were calibrated, starting with the complete model with the optimum set of variables, presented in Figure 4. Several tests were carried out to identify the best learning rate [66] to adjust this neural network, and the value found was 0.001. One thousand epochs were used. Two hidden layers were used: the first one with eight hidden nodes and a softplus activation function and the second with nine hidden nodes and a linear activation function. The linear activation function was used in the output layer. These sets of hyperparameters were selected because they provided the smallest global MAE for the validation set. The metrics generated by this network are summarized in Table 6. The first value in each cell represents the metric mean of the ten-fold validation, and the second represents the standard error. At the bottom of the table, the test set results obtained when the training and validation sets are combined to create the final model are presented. The MdAPE metric was selected to be analyzed because the outliers (illustrated in Figure 6) inflated the mean absolute percentage error (MAPE), which may lead to an erroneous model interpretation.

Table 6. ANN metrics using 10 input variables.

Variable	Set	Mean ± Std. Error			
		Correlation	MAE	RMSE	MdAPE
$Y_{Concentrate}$	Training	0.854 ± 0.001	2.664 ± 0.013	3.599 ± 0.013	14.938 ± 0.2
Y_{Slimes}		0.608 ± 0.002	6.241 ± 0.021	8.042 ± 0.033	12.297 ± 0.067
$Y_{Magnetics}$		0.855 ± 0.001	3.05 ± 0.013	4.156 ± 0.015	25.445 ± 0.16
$Y_{Tailings}$		0.837 ± 0.001	5.158 ± 0.036	6.534 ± 0.035	13.711 ± 0.077
$R_{Concentrate}$		0.667 ± 0.002	9.296 ± 0.036	12.174 ± 0.044	14.564 ± 0.17
$R_{Slimes+Magnetics}$		0.456 ± 0.002	6.964 ± 0.016	9.149 ± 0.022	16.155 ± 0.048
$R_{Tailings}$		0.688 ± 0.001	6.941 ± 0.031	9.596 ± 0.022	36.768 ± 0.231
$Y_{Concentrate}$	Validation	0.849 ± 0.008	2.708 ± 0.06	3.641 ± 0.082	15.482 ± 0.473
Y_{Slimes}		0.585 ± 0.017	6.35 ± 0.078	8.272 ± 0.134	12.239 ± 0.233
$Y_{Magnetics}$		0.847 ± 0.009	3.115 ± 0.067	4.28 ± 0.135	25.999 ± 0.727
$Y_{Tailings}$		0.83 ± 0.006	5.24 ± 0.077	6.664 ± 0.09	13.909 ± 0.26
$R_{Concentrate}$		0.655 ± 0.014	9.432 ± 0.208	12.291 ± 0.253	14.774 ± 0.416
$R_{Slimes+Magnetics}$		0.435 ± 0.017	7.059 ± 0.136	9.236 ± 0.18	16.348 ± 0.644
$R_{Tailings}$		0.676 ± 0.013	7.065 ± 0.191	9.728 ± 0.191	37.596 ± 1.316
$Y_{Concentrate}$	Test	0.866	2.631	3.601	14.897
Y_{Slimes}		0.583	6.389	8.416	12.449
$Y_{Magnetics}$		0.837	3.276	4.602	27.851
$Y_{Tailings}$		0.833	5.353	6.741	14.263
$R_{Concentrate}$		0.625	9.479	12.843	13.925
$R_{Slimes+Magnetics}$		0.424	7.073	9.276	16.199
$R_{Tailings}$		0.658	7.181	10.420	34.133

The correlation coefficient between the forecasts and the real values for the flotation concentrate yield was 0.866, and the MdAPE found was 14.897 in the test set. For the flotation concentrate P_2O_5 recovery, the correlation coefficient was 0.625, associated with a MdAPE value of 13.925. The slimes yield and the slimes + magnetics P_2O_5 recoveries also showed good MdAPE values (12.449 and 16.199, respectively), although the correlation coefficient between the real values and the forecasts is not very high. The magnetics yields and tailings P_2O_5 recovery were the variables with the highest MdAPE values (27.851 and 34.133, respectively).

After completing the analysis with the optimum scenario, the neural network's calibration was conducted using the six variables presented in Figure 5. Two hidden layers were used: the first with 10 hidden nodes and a sigmoid activation function and the second with 10 hidden nodes and a linear activation function. The linear activation function was used in the output layer. The best learning rate found to adjust this network was 0.002 and one thousand epochs were used. The metrics generated by using this network are summarized in Table 7.

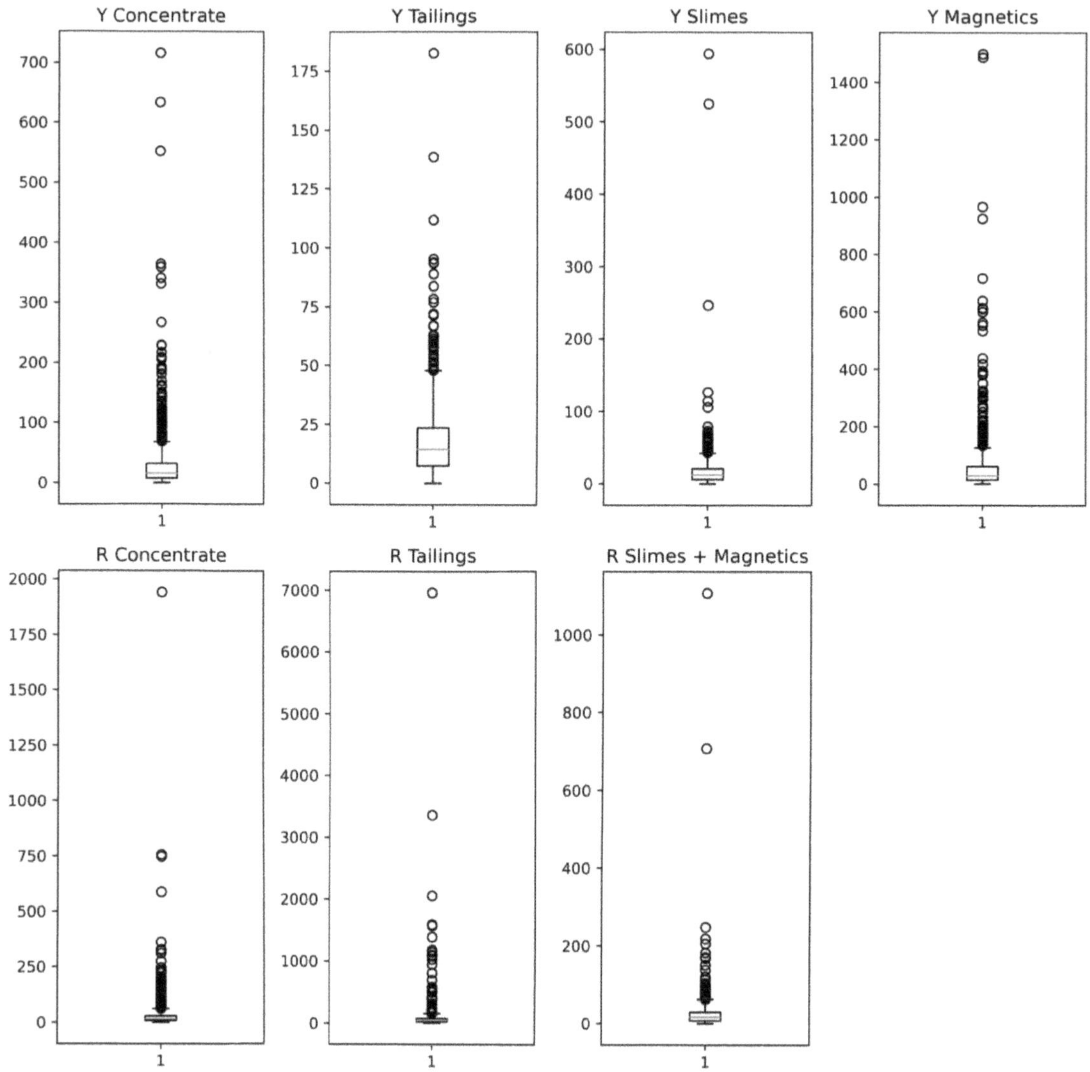

Figure 6. Absolute percentage error (APE) for the test set.

The results presented in Table 7 show that for the reduced set of input variables, the metrics observed were very similar to the ones seen in the optimum scenario, especially for the variables with the most significant economic impact, namely the flotation concentrate yield and P_2O_5 recovery.

Using six and 10 features, a comparison of the two scenarios can be summarized by calculating the relative differences, as shown in Equation (14). Table 8 summarizes the relative differences observed between the Pearson correlation coefficients, MAE, RMSE, and MdAPE mean values.

$$Relative\ Difference = \left(\frac{Metric\ Value_{6features} - Metric\ Value_{10features}}{Metric\ Value_{10features}} \right) * 100\% \quad (14)$$

Table 7. ANN metrics using six input variables.

Variable	Set	Mean ± Std. Error			
		Correlation	MAE	RMSE	MdAPE
$Y_{Concentrate}$	Training	0.855 ± 0.001	2.664 ± 0.025	3.598 ± 0.018	15.122 ± 0.32
Y_{Slimes}		0.599 ± 0.001	6.334 ± 0.022	8.093 ± 0.021	12.523 ± 0.062
$Y_{Magnetics}$		0.846 ± 0.001	3.195 ± 0.019	4.319 ± 0.018	26.523 ± 0.149
$Y_{Tailings}$		0.835 ± 0.001	5.165 ± 0.021	6.528 ± 0.025	13.692 ± 0.072
$R_{Concentrate}$		0.667 ± 0.002	9.32 ± 0.069	12.221 ± 0.06	14.601 ± 0.302
$R_{Slimes+Magnetics}$		0.463 ± 0.002	6.957 ± 0.02	9.121 ± 0.031	16.231 ± 0.082
$R_{Tailings}$		0.692 ± 0.002	6.902 ± 0.055	9.561 ± 0.028	36.608 ± 0.457
$Y_{Concentrate}$	Validation	0.851 ± 0.008	2.706 ± 0.056	3.634 ± 0.075	15.564 ± 0.346
Y_{Slimes}		0.582 ± 0.012	6.412 ± 0.065	8.205 ± 0.092	12.732 ± 0.23
$Y_{Magnetics}$		0.84 ± 0.007	3.225 ± 0.073	4.372 ± 0.098	26.718 ± 0.788
$Y_{Tailings}$		0.829 ± 0.006	5.236 ± 0.071	6.629 ± 0.097	13.855 ± 0.281
$R_{Concentrate}$		0.657 ± 0.012	9.385 ± 0.207	12.32 ± 0.261	14.674 ± 0.404
$R_{Slimes+Magnetics}$		0.447 ± 0.017	7.002 ± 0.135	9.168 ± 0.181	16.463 ± 0.276
$R_{Tailings}$		0.682 ± 0.011	6.985 ± 0.133	9.659 ± 0.174	36.91 ± 0.964
$Y_{Concentrate}$	Test	0.859	2.671	3.650	16.169
Y_{Slimes}		0.586	6.470	8.437	12.656
$Y_{Magnetics}$		0.838	3.312	4.599	28.397
$Y_{Tailings}$		0.827	5.490	6.854	14.806
$R_{Concentrate}$		0.634	9.395	12.637	15.157
$R_{Slimes+Magnetics}$		0.443	7.006	9.191	16.189
$R_{Tailings}$		0.665	7.244	10.233	34.327

Table 8. Relative differences between the sub-optimum scenario and the optimum.

Variable	Set	Relative Difference (%)			
		Correlation	MAE	RMSE	MdAPE
$Y_{Concentrate}$	Training	0.1%	0.0%	0.0%	1.3%
Y_{Slimes}		−1.5%	1.5%	0.6%	1.9%
$Y_{Magnetics}$		−1.1%	4.8%	3.9%	4.2%
$Y_{Tailings}$		−0.2%	0.1%	−0.1%	−0.1%
$R_{Concentrate}$		0.0%	0.3%	0.4%	0.3%
$R_{Slimes+Magnetics}$		1.5%	−0.1%	−0.3%	0.5%
$R_{Tailings}$		0.6%	−0.6%	−0.4%	−0.4%

Table 8. *Cont.*

Variable	Set	Relative Difference (%)			
		Correlation	MAE	RMSE	MdAPE
$Y_{Concentrate}$	Validation	0.2%	−0.1%	−0.2%	0.5%
Y_{Slimes}		−0.5%	1.0%	−0.8%	4.1%
$Y_{Magnetics}$		−0.8%	3.5%	2.1%	2.8%
$Y_{Tailings}$		−0.1%	−0.1%	−0.5%	−0.4%
$R_{Concentrate}$		0.3%	−0.5%	0.2%	−0.7%
$R_{Slimes+Magnetics}$		2.8%	−0.8%	−0.7%	0.7%
$R_{Tailings}$		0.9%	−1.1%	−0.7%	−1.8%
$Y_{Concentrate}$	Test	−0.7%	1.5%	1.4%	8.5%
Y_{Slimes}		0.3%	1.3%	0.3%	1.7%
$Y_{Magnetics}$		0.1%	1.1%	−0.1%	1.9%
$Y_{Tailings}$		−0.7%	2.6%	1.7%	3.8%
$R_{Concentrate}$		1.4%	−0.9%	−1.6%	8.8%
$R_{Slimes+Magnetics}$		4.5%	−1.0%	−0.9%	−0.1%
$R_{Tailings}$		1.1%	0.9%	−1.8%	0.6%

By analysing the relative difference in the test set, it is possible to conclude that when the correlation coefficient decreases, the fall is not marked (smaller than 0.7%), showing that the reduction in the number of input variables did not impact the model's precision. In most cases, an improvement in the correlation coefficient value is noted when the reduced set of variables is used.

Removing four variables from the inputs did not significantly affect the metrics. The relative differences in the MAE and RMSE values showed an increase in errors lower than 2.6% when the reduced set of variables is used instead of the optimum set. The most significant impact observed was in the MdAPE values for the flotation concentrate mass and metallurgical recoveries, showing an increase of 8.5% and 8.8% in the test set, respectively, when the reduced scenario was adopted. The reduced model can be considered as a good option to forecast the metallurgical variables of interest because the MdAPE increase was lower than 4% for all other output variables.

Considering the above results, it can be concluded that it is possible to work with the reduced model without losing much, in terms of model precision and accuracy. There is a gain in simplicity once adding variables in the process can represent a considerable increase in the workload, not only in computer power but also in time spent by the analyst to elaborate, update, and evaluate the models created to estimate the input variables in the block model using kriging or simulation. Analysing through the geostatistical point of view, passing from 10 to six variables to be modelled implies a work-time reduction of 40%, a considerable amount.

5. Conclusions

Reducing the number of variables leads to a lower computational cost. The gain is also felt in the geostatistical workflow, once the number of input variables to be integrated to the resource block model is reduced. However, those benefits are counterbalanced by higher errors in the forecasting of some variables. The trade-off between a small error and a higher labor force is important when applying such models to a complete industrial operation. To help with this decision, it is important to quantify the errors related to the two approaches and compare them, as shown in this paper.

In this case study, the reduction from 10 to six input variables showed that dimension reduction can lead to models capable of returning correlation values close to the ones observed when the complete set of variables is used, with a small loss in the reduced scenario, when it occurs. The increase in MAE and RMSE was smaller than 2.6% for all output variables in the test set when the sub-optimum set of variables was used instead of the set with all 10 features. The four removed variables make sense considering the points discussed in chapter 4: the BaO and Nb_2O_5 are related to the presence of barite and pyrochlore, which only affects the process in high quantities (which is rare); and SiO_2 and TiO_2 grades are correlated with other variables that were maintained in the model, making their presence redundant. Thus, it was decided to select the model elaborated with the reduced number of input variables, due to its quality, and considering the practicality of the whole process, as this is a workflow being suggested for an active mining operation where time response represents a crucial aspect.

We would like to stress that each case study is unique. It is possible to achieve smaller errors in the test set with a reduced set of variables compared with the optimum scenario, especially if small datasets are used, due to overfitting (caused by a high number of variables and a small number of samples). It is also possible to find higher errors that do not justify the dimensionality reduction. Thus, it is important to follow the proposed workflow and analyze the errors in both scenarios to choose the best option for each mine under study.

It is important to mention that the geometallurgical ore model herein described is essentially statistical in nature. Nevertheless, the model is powerful enough to include some of the physical–chemistry aspects of the process. For example, the properly carried out investigation has led to the establishment of a quantitative relationship describing the deleterious effect of the presence of carbonates in the plant's feed, understood as the ratio between CaO and P_2O_5, proving that important geochemical/mineralogical relationships were mathematically predicted, giving an interpretation of the results that makes geological sense.

Another important point that needs to be stressed is related to the estimation of the collector dosage in the block model. In this case study, this explanatory variable was attributed to the blocks by the nearest neighbor estimation method. Thus, mining regions which require a smaller collector dosage can be recognized, and areas which demand a higher collector dosage are also noted.

Final Thoughts about Mineral Processing

Usually, the processing plant's human operator(s) has a well-established objective of a certain number of tonnes of the product (concentrate) with a required grade or quality. Such quality may include different grades (for example, SiO_2 and Fe_2O_3, but most certainly P_2O_5). A conveyor belt equipped with a scale provides the tonnage of the filtered product, so this information is readily available through the control system. At short time intervals, samples are taken and sent for analysis so that, given a short delay, product quality and moisture content are also available to the operator. Given that the plant is operating at a steady pace, the operator may decide to change the processing parameters so that the production objectives are met during his shift. The changes that can be made are relatively small, such as in the reagent dosage, aeration rate, and perhaps also the froth levels, and froth-wash-water flowrate.

The point here is that most of the plant's processing parameters do not change at all, and for the few parameters that do change, so that production requirements can be met, the impacts on the production rate and quality are well-known. Therefore, it is safe to assume that the plant and all its processing parameters can either be changed to meet the required quality or not. Simply put, if the quality can be met, the rock is ore, and if not, the rock is waste. The broad question that is left concerns the metallurgical recovery of P_2O_5 as a function of ore quality. In other words, for the development of the geometallurgical model, the main concern is with the ore model.

Author Contributions: Conceptualization, F.G.F.N. and J.F.C.L.C.; methodology, F.G.F.N. and A.M.B.B.; software, F.G.F.N. and A.M.B.B.; investigation, F.G.F.N.; data curation, F.C.d.A. and L.N.C.; writing—original draft preparation, F.G.F.N. and C.L.S.; writing—review and editing, J.F.C.L.C., C.L.S., G.d.C.M., A.M.B.B., L.N.C. and F.C.d.A.; project administration, J.F.C.L.C., L.N.C. and F.C.d.A.; funding acquisition, J.F.C.L.C. and C.L.S. All authors have read and agreed to the published version of the manuscript.

Funding: This study was funded by Fundação Luiz Engert.

Data Availability Statement: The data is unavailable due to confidentiality.

Acknowledgments: The authors would like to thank the mining company which kindly provided the data used in this study and Carla Grasso for all technical discussions, which made this work possible and the results applicable to the industry.

Conflicts of Interest: The authors declare no conflict of interest.

References

1. Williams, S.R.; Richardson, J.M. *Geometallurgical Mapping: A New Approach that Reduces Technical Risk*; SGS Minerals Services: Truro, UK, 2004; 13p.
2. Lopera, P.A. Geometallurgical Mapping and Mine Modelling—Comminution Studies: La Colosa Case Study, AMIRA P843A. Master's Thesis, University of Tasmania, Hobart, Australia, 2014.
3. Newton, M.J.; Graham, J.M. Spatial modelling and optimisation of geometallurgical indices. In Proceedings of the First AUSIMM International Geometallurgy Conference, Brisbane, QLD, Australia, 5–7 September 2011; pp. 247–261.
4. Keeney, L.; Walters, S.G. A Methodology for geometallurgical mapping and orebody modelling. In Proceedings of the First AUSIMM International Geometallurgy Conference, Brisbane, QLD, Australia, 5–7 September 2011; pp. 217–225.
5. Montoya, P.A.; Keeney, L.; Jahoda, R.; Hunt, J.; Berry, R.; Drews, U.; Chamberlain, V.; Leichliter, S. Techniques applicable to prefeasibility projects—La Colosa case study. In Proceedings of the First AUSIMM International Geometallurgy Conference, Brisbane, QLD, Australia, 5–7 September 2011; pp. 103–111.
6. Sepulveda, E.; Dowd, P.A.; Xu, C.; Addo, E. Multivariate Modelling of Geometallurgical Variables by Projection Pursuit. *Math. Geosci.* **2016**, *49*, 121–143. [CrossRef]
7. Wirfiyata, F.; Mccaffery, K. Applied geo-metallurgical characterisation for life of mine throughput prediction at Batu Hijau. In Proceedings of the International Autogneous Grinding, Semiautogenous Grinding and High Pressure Grinding Roll Technology, Westmount, QC, USA, 25–28 September 2011; pp. 1–22.
8. Johnson, C.L.; Browning, D.A.; Pendock, N.E. Hyperspectral Imaging Applications to Geometallurgy: Utilizing Blast Hole Mineralogy to Predict Au-Cu Recovery and Throughput at the Phoenix Mine, Nevada. *Econ. Geol.* **2019**, *114*, 1481–1494. [CrossRef]
9. Lishchuk, V.; Lund, C.; Ghorbani, Y. Evaluation and comparison of different machine-learning methods to integrate sparse process data into a spatial model in geometallurgy. *Miner. Eng.* **2019**, *134*, 156–165. [CrossRef]
10. Fernandes, F.G.; Cabral, I.E. Regression model utilization to estimate the mass recovery of a phosphate mine in Brazil. *Rem Rev. Esc. Minas* **2016**, *69*, 75–77. [CrossRef]
11. Vieira, M.C.A. Metodologia Para Prever Recuperação de Zinco em Planta de Beneficiamento. Master's Thesis, Programa de Pós-Graduação em Engenharia de Minas, Metalurgia e de Materiais, Universidade Federal do Rio Grande do Sul, Porto Alegre, Brazil, 2016.
12. Niquini, F.G.F.; Costa, J.F.C.L. Comparison between regression models and neural networks Applied to forecast geometallurgical variables. In *Mining Goes Digital, Proceedings of the 39th International Symposium 'Application of Computers and Operations Research in the Mineral Industry' (APCOM 2019), Wroclaw, Poland, 4–6 June 2019*; CRC Press: Boca Raton, FL, USA, 2019; pp. 463–472.
13. Andrade, L.B.; Cabral, I.E.; Costa, J.F.C.L. Modelling and forecasting geometallurgical recovery at a phosphate mine. In *Mining Goes Digital, Proceedings of the 39th International Symposium 'Application of Computers and Operations Research in the Mineral Industry' (APCOM 2019), Wroclaw, Poland, 4–6 June 2019*; CRC Press: Boca Raton, FL, USA, 2019; pp. 551–556.
14. Niquini, F.G.F.; Costa, J.F.C.L. Mass and Metallurgical Balance Forecast for a Zinc Processing Plant Using Artificial Neural Networks. *Nat. Resour. Res.* **2020**, *29*, 3569–3580. [CrossRef]
15. Niquini, F.G.F.; Costa, J.F.C.L. Forecasting mass and metallurgical balance at a gold processing plant using modern multivariate statistics. *REM -Int. Eng. J.* **2020**, *73*, 571–578. [CrossRef]
16. Knight, R.; Olson Hoal, K.; Abraham, A.P.G. Three-dimensional geometallurgical data integration for predicting concentrate quality and tailings composition in a massive sulfide deposit. In Proceedings of the First AUSIMM International Geometallurgy Conference, Brisbane, QLD, Australia, 5–7 September 2011; pp. 227–232.
17. Louwrens, E.; Napier-Munn, T.; Keeney, L. Geometallurgical characterisation of a tailings storage facility—A novel approach. In Proceedings of the Tailings and Mine Waste Management for the 21st Century, Sydney, Australia, 27–28 July 2015; pp. 125–131.
18. Boisvert, J.B.; Rossi, M.E.; Ehrig, K.; Deutsch, C.V. Geometallurgical Modeling at Olympic Dam Mine, South Australia. *Math. Geosci.* **2013**, *45*, 901–925. [CrossRef]

19. Macmillan, E.; Ehrig, K.; Liebezeit, V.P.; Kittler, P.; Lower, C. Use of geometallurgy to predict tailings leach acid consumption at Olympic Dam. In Proceedings of the First AUSIMM International Geometallurgy Conference, Brisbane, QLD, Australia, 5–7 September 2011; pp. 93–102.
20. Hair, J.F.; Black, W.C.; Babin, B.J.; Anderson, R.E.; Tatham, R.L. *Multivariate Data Analysis*, 6th ed.; Pearson University Press: Upper Saddle River, NJ, USA, 2006.
21. Pearson, K. On lines and planes of closest fit to systems of points in space. *Lond. Edinb. Dublin Philos. Mag. J. Sci.* **1901**, *2*, 559–572. [CrossRef]
22. Switzer, P.; Green, A.A. *Min/Max Autocorrelation Factors for Multivariate Spatial Imaging: Technical Report No. 6*; Department of Statistics, Stanford University: Stanford, CA, USA, 1984; 14p.
23. Long, C.A.; Sokal, R.R.; Sneath, P.H.A. Principles of Numerical Taxonomy. *J. Mammal.* **1965**, *46*, 111–112. [CrossRef]
24. King, B. Step-Wise Clustering Procedures. *J. Am. Stat. Assoc.* **1967**, *62*, 86–101. [CrossRef]
25. Chavent, M.; Genuer, R.; Saracco, J. Combining clustering of variables and feature selection using random forests. *Commun. Stat.* **2016**, *50*, 426–445. [CrossRef]
26. Park, C.H. A Feature Selection Method Using Hierarchical Clustering. In *Mining Intelligence and Knowledge Exploration*; Lecture Notes in Computer, Science; Prasath, R., Kathirvalavakumar, T., Eds.; Springer: Cham, Switzerland, 2014; Volume 8284.
27. Derksen, S.; Keselman, H.J. Backward, forward and stepwise automated subset selection algorithms: Frequency of obtaining authentic and noise variables. *Br. J. Math. Stat. Psychol.* **1992**, *45*, 265–282. [CrossRef]
28. Guyon, I.; Weston, J.; Barnhill, S.; Vapnik, V. Gene Selection for Cancer Classification using Support Vector Machines. *Mach. Learn.* **2002**, *46*, 389–422. [CrossRef]
29. Hastie, H.; Tibshirani, R.; Friedman, J. *The Elements of Statistical Learning. Data Mining, Inference, and Prediction*; Springer Series in Statistics Springer New York Inc.: New York, NY, USA, 2009.
30. Both, C.; Dimitrakopoulos, R. Applied Machine Learning for Geometallurgical Throughput Prediction—A Case Study Using Production Data at the Tropicana Gold Mining Complex. *Minerals* **2021**, *11*, 1257. [CrossRef]
31. Gholami, A.; Asgari, K.; Khoshdast, H.; Hassanzadeh, A. A hybrid geometallurgical study using coupled Historical Data (HD) and Deep Learning (DL) techniques on a copper ore mine. *Physicochem. Probl. Miner. Process.* **2022**, *58*, 147841. [CrossRef]
32. Jorjani, E.; Bagherieh, A.H.; Mesroghli, S.; Chehreh Chelgani, S.; Jorjani, H. Prediction of yttrium, lanthanum, cerium and neodymium leaching recovery from apatite concentrate using artificial neural networks. In Proceedings of the MetPlant 2008, Perth, Australia, 18–19 August 2008; pp. 433–446.
33. Nakhaei, F.; Mosavi, M.R.; Sam, A.; Vaghei, Y. Recovery and grade accurate prediction of pilot plant flotation column concentrate: Neural network and statistical techniques. *Int. J. Miner. Process.* **2012**, *110–111*, 140–154. [CrossRef]
34. Srivastava, V.; Bandopadhyay, S.; Akdogan, G.; Ghosh, T. Dynamic neural network based model for SAG mill power prediction. In Proceedings of the 38th International Symposium on the Application of Computers and Operations Research in the Mineral Industry, Golden, CO, USA; 2017; Volume 38, pp. 227–233.
35. Panda, L.; Tripathy, S.K. Performance prediction of gravity concentrator by using artificial neural network-a case study. *Int. J. Min. Sci. Technol.* **2014**, *24*, 461–465. [CrossRef]
36. Ford, M.A. Simulation of Ore dressing Plants. Ph.D. Thesis, University of the Witwatersrand, Johannesburg, South Africa, 1979.
37. King, R.P. Flotation. In *Modeling and Simulation of Mineral Processing Systems*, 2nd ed.; Schneider, C.L., King, E.A., Eds.; SME: Englewood, CO, USA, 2012; pp. 341–415.
38. Sutherland, K.L. Physical chemistry of flotation XI. Kinetics of the flotation process. *J. Phys. Chem.* **1948**, *52*, 394–425. [CrossRef]
39. Flint, L.R.; Howarth, W.J. The collision efficiency of small particles with spherical air bubbles. *Chem. Eng. Sci.* **1971**, *26*, 115–168. [CrossRef]
40. Reay, D.; Ratcliffe, G.A. Removal of fine particles from water by dispersed air flotation: Effect of bubble size and particle size on collision efficiency. *Can. J. Chem. Eng.* **1973**, *51*, 178–185. [CrossRef]
41. Yoon, R.H.; Luttrell, G.H. The effect of bubble size on fine particle flotation. *Miner. Process. Extr. Metall. Rev.* **1989**, *5*, 101–122. [CrossRef]
42. Clift, R.; Grace, J.R.; Weber, M.E. *Bubbles, Drops and Particles*; Academic Press: Cambridge, MA, USA, 1978.
43. Schulze, H.J. *Physico-Chemical Elementary Processes in Flotation*; Elsevier: Amsterdam, The Netherlands, 1993.
44. Karamanev, D.G.; Nikolov, L.N. Free rising spheres do not obey Newton's Law for free settling. *AIChE J.* **1992**, *38*, 1843–1846. [CrossRef]
45. Tomlinson, H.S.; Fleming, M.G. *Flotation Rate Studies. Proc. 6th International Mineral Processing Congress*; Roberts, A., Ed.; Pergamon Press: Oxford, UK, 1965.
46. King, R.P.; Hatton, T.A.; Hulbert, D.G. Bubble loading during flotation. *Trans. Instn. Min. Metall.* **1996**, *83*, C112–C115.
47. Bradshaw, D.J.; O'Connor, C.T. Measurement of the sub-process of bubble loading in flotation. *Miner. Eng.* **1996**, *9*, 443–448. [CrossRef]
48. Trahar, W.J.; Warren, L.J. The floatability of very fine particles—A review. *Int. J. Miner. Process.* **1976**, *3*, 103–131. [CrossRef]
49. Drzymala, J. Characterization of materials by Hallimond tube flotation. Part 2: Maximum size of floating particles and contact angle. *Int. J. Miner. Process.* **1994**, *42*, 153–168. [CrossRef]
50. Murphy, D.G.; Zimmerman, W.; Woodburn, E.T. Kinematic model of bubble motion in a flotation froth. *Powder Technol.* **1996**, *87*, 3–12. [CrossRef]

51. Neethling, S.J.; Cilliers, J.J. A visual kinematic model of flowing foams incorporating coalescence. *Powder Technol.* **1999**, *101*, 249–256. [CrossRef]
52. Woolley, A.R. *Alkaline Rocks and Carbonatites of the World. Part 1: North and South America*; British Museum: London, UK, 1987; 216p.
53. Brod, J.A. Petrology and Geochemistry of the Tapira Alkaline Complex, Minas Gerais State, Brazil. Ph.D. Thesis, Durham University, Durham, UK, 1999.
54. Ribeiro, C.C. Geologia, Geometalurgia, Controles e Gênese dos Depósitos de Fósforo, Terras Raras e Titânio do Complexo Carbonatítico Catalão I, GO. Ph.D. Thesis, Universidade de Brasília, Brasília, Brazil, 2008.
55. Grasso, C.B. Geologia e Geometalurgia dos Minérios de Fosfato na Base do Manto de Intemperismo da Mina do Barreiro, Araxá. Ph.D. Thesis, Universidade de Brasília, Brasília, Brazil, 2005.
56. Gaudin, A.M. Quantifying Concentrating Operations. In *Principles of Mineral Dressing*, 1st ed.; McGraw-Hill Book Company, Inc.: New York, NY, USA; London, UK, 1939.
57. Tarján, G. Fundamentals and genarl principles of Separation. In *Mineral Processing Volume 1*, 1st ed.; Akadémiai Kiadó: Magyarország, Budapest, 1981.
58. Misra, P.; Yadav, A.S. Improving the Classification Accuracy using Recursive Feature Elimination with Cross-Validation. *Int. J. Emerg. Technol.* **2020**, *11*, 659–665.
59. Kavitha, S.; Varuna, S.; Ramya, R. A comparative analysis on linear regression and support vector regression. In Proceedings of the 2016 Online International Conference on Green Engineering and Technologies (IC-GET), Online, 19 November 2016.
60. Breiman, L. Random forests. *Mach. Learn.* **2001**, *45*, 5–32. [CrossRef]
61. Dorogush, A.V.; Ershov, E.; Gulin, A. CatBoost: Gradient boosting with categorical features support. *arXiv* **2018**, arXiv:1810.11363.
62. Breiman, L.; Friedman, J. Estimating Optimal Transformations for Multiple Regression and Correlation. *J. Am. Stat. Assoc.* **1985**, *80*, 580–598. [CrossRef]
63. Lundberg, S.M.; Lee, S.I. A Unified Approach to Interpreting Model Predictions. In Proceedings of the 31st Conference on Neural Information Processing Systems (NIPS 2017), Long Beach, CA, USA, 4–9 December 2017.
64. Izenman, A.J. *Modern Multivariate Statistical Techniques*; Springer: New York, NY, USA, 2008.
65. Molnar, C. *Interpretable Machine Learning*, 2nd ed.; Lulu.com: Morrisville, NC, USA, 2020.
66. Bishop, C.M. *Neural Networks for Pattern Recognition*; Clarendon Press: Oxford, UK, 1995.

minerals

Article

Predicting Flowability at Disposal of Spent Heap Leach by Applying Artificial Neural Networks Based on Operational Variables

Nelson Herrera [1,2,*], María Sinche Gonzalez [1], Jarkko Okkonen [3] and Raul Mollehuara Canales [1]

1 Oulu Mining School, University of Oulu, P.O. Box 3000, 90570 Oulu, Finland; maria.sinchegonzalez@oulu.fi (M.S.G.); raul.mollehuaracanales@oulu.fi (R.M.C.)
2 Departamento de Ingeniería en Metalurgia y Minas, Universidad Católica del Norte, Antofagasta 1270709, Chile
3 Geological Survey of Finland GTK, Vuorimiehentie 5, P.O. Box 96, 02151 Espoo, Finland; jarkko.okkonen@gtk.fi
* Correspondence: nelson.herreranunez@oulu.fi; Tel.: +56-552651010

Abstract: The mining sector actively seeks to improve operational processes and manage residual materials, especially in areas used for heap leaching disposal. The flowability of residues following deposition can have an impact on storage capacity, productivity, and workers' safety. In this study, an artificial neural network (ANN) approach is applied to evaluate the accuracy of three models in predicting the flowability of spent heap leach when it is discharged into the dump, considering three or five segregation categories. The models with five categories exhibited the highest level of accuracy, with learning responses ranging from 72% to 78% and predictions from 88% to 96%. These indicate that ANN models have the potential to be a decision-making tool for the discharge strategy in the dump. Modules containing lithologies such as clays and phyllosilicates exhibited increased susceptibility to separation due to their water retention capacity, which negatively impacted their permeability and conductivity. The decomposition of iron oxide, along with clays and low hardness, led to the formation of fines, limited permeability, and inadequate solution drainage. Rock competence and low formation of fines provide good permeability, and better drainage conditions for the solution, and help maintain the stability of the spent heap leach in the dump.

Keywords: mineral extraction; deep learning; process control; prediction accuracy; artificial neural networks; mineral waste disposal; heap leaching piles

1. Introduction

The mining industry is continuously adjusting to the challenges posed by low-grade deposits and the need to process larger quantities of material for the extraction of the metal of interest, consequently increasing the amount of waste that includes sterile material (without economic value) and unwanted by-products [1]. For this reason, more attention is placed on the study of minerals that are difficult to treat both chemically and physically [2] and the implementation of new sustainable solutions to improve waste management practices [3]. The process of benefiting copper minerals is carried out according to the characteristics of the ore (e.g., oxides or mixed with secondary sulphides) generally treated by hydrometallurgy processes [4,5] where a selective dissolution of the mineral occurs through leaching, following solvent extraction and electro-winning to be recovered as a cathode [6]. In the leaching stage, a mass transfer of the element of interest is dependent on the hydraulic flow regime of the fluid and the physicochemical reactions present in the system [7].

Currently, there is a strong presence of hydrometallurgical processes in Chile estimated in 60% of the mining plants [8] with the heap leaching process being part of refined copper production representing around 19.6% of the world's copper production [9]. Due to the decreasing availability of high-grade ores, research is mostly focused on the processing of challenging low-grade ores [10]. Among the several technologies, the heap leaching method is widely employed due to its straightforwardness, minimal environmental impact, and cost-effectiveness. However, further research is necessary to optimise it, as the effects of ore heterogeneities are observed to change over time, in relation to their physical, chemical, and mineralogical features [11,12].

The operation in the leaching modules directly affects the characteristics of the leaching waste debris or spent heap leach. As described by Ghasemzadeh et al. [13], variables such as the height of the heap, particle size distribution, surface roughness, bulk density, acid concentration, and irrigation rate can influence the behaviour and structure of the spent heap leach due to physicochemical interactions modifying the fluid acidity, plasticity, and permeability. This last effect can change the percentage of fine granules because of chemical and mechanical wear and tear or blockage, causing migration and segregation due to a lack of porosity. Elements like calcium, magnesium, manganese, and iron are often to blame [12]. It can also cause material compaction due to its own weight and liquefaction phenomenon during the transport of the spent heap leach on conveyor belts to final deposition. The higher number of mineralogical alteration zones leads to a higher clay content, which changes the drainage solution [14]. This changes the material's permeability, pore sizes, and hydraulic conductivity, which in turn changes the granulometric sensitivity. Another relevant factor corresponds to the mineralogical content and clay, affecting permeability, rock competition, and acid consumption. A study by Liu et al. [15] used near-infrared spectroscopy (NIR) to find units with low permeability and high rock competition. These units included plagioclase, chlorite, and calcite. Sericite, kaolinite, quartz, and szomolnokite were also present.

Numerous authors have demonstrated the importance of geotechnical characterization in maintaining process stability. For example, Watanabe et al. [16] focused on analysing the chemical comminution and identifying the modification of the permeability and the static liquefaction, where it was demonstrated that a high content of fines decreases the drainage capacity of the minerals by decreasing the permeability, altering the solid-leaching solution contact, and causing chemical crushing that increases the presence of fine particles and supports solid segregation. According to Bard [17], when spent heap leach is discharged, static liquefaction happens. This causes a loss of shear strength when going from a drained load condition to an undrained one. This leads to an increase in pore pressure and the product of its own weight in materials with fine granulometry (20% less than 75 μm).

The disposal of spent heap leach requires specific engineering designs that depend directly on the operating and location conditions [14]. The material can be stacked at levels that reach an average of 4 m in height and that can generate dumps with a height of over 70 m [18,19]. It is common to obtain, as a result of the stacking process, that spent heap leach presents different segregation behaviours, making it difficult to discharge into the dump and making it impossible to configure the original design. An area is considered in the original dump design where the stacking must meet important requirements, such as extensive land breadth and a slope between 0.7% and 5%, to ensure pile stability and allow for the collection of drained flow [20]. In many cases, the design cannot be applied, requiring a request for the extension of the dump to the adjoining land to meet the demand for the available volume for the spent heap leach disposal and comply with the environmental and easement limits. The segregation and limited drainage of the spent heap leach have also generated problems associated with equipment failures, such as the misalignment of conveyor belts due to the instability generated by the material at different moisture concentrations, or in the spreader equipment or tripper, which, transported over the deposited spent heap leach, can produce geotechnical problems, generating subsidence of this equipment, which has resulted in stoppages and delays in the removal and unloading of the material, which can lead to the stacking bridge approaching the removal bridge, causing that the stacker reaches the safety distance to the bucket wheel (generally 150 m, preventing the formation of new modules to be leached. Nowadays, the mining industry seeks more complex ores and looks for improvements in both the operation-process line and the disposal of the rejected materials.

In recent years, researchers have used numerical and CFD (computational fluid dynamics) simulation alongside traditional analysis, yielding interesting results. However, these methods present limitations due to uncertainties caused by computing values and their dependence on the approximations made [21]. Researchers have applied deep learning methods like artificial neural networks (ANN) to tasks such as predicting and regressing surface quality in flotation and heap leaching processes. For instance, Zadeh [22] used fuzzy logic, Bergh et al. [23], Correa et al. [24], Umucu et al. [25], and Leiva et al. [26] did research on mineral extraction that compares how well Bayesian networks and ANNs work. In the case of applying computational techniques in the copper industry, Haghighi et al. [27] analysed the low-grade mineral identification and reduction of production costs, and in 2015, they improved the copper recovery obtained through leaching by ANN prediction [28]. Salmani et al. [29] use an ANN to analyse the copper flotation process to predict copper float rates under different operating conditions, considering different doses of chemical reagents, feed rate, and particle size, obtaining a quality prediction in the test process of 93%.

When spent heap leaches are removed and transported to their final destination, their flowability (or segregation) characteristics can influence the material stacking stability, which affects the original storage capacity, production, and worker safety. This study aims to develop accurate models that can predict the segregation behaviour of the spent heap leach. These models will assist in making strategic decisions regarding the discharge and final disposal of this material. Additionally, they will help optimise the use of the dump area, ensuring it is efficiently adapted to daily operations. We created three artificial neural network (ANN) models to predict the segregation categories that may occur when the spent heap leach is discharged in the dump. These models were developed using seven input variables related to mineralogy, lithology, granulometry, agglomeration conditions, drainage time, cover conditions, and moisture.

2. Methodology

2.1. ANN Application Background

Artificial neural networks (ANN) is a deep learning tool inspired by the functioning of the human nervous system, using interconnected artificial neurons as a base [30,31]. A classic structure of a neural network is made up of an input layer, a hidden layer, and an output layer where the organisation of the present layers is specified [32]. This study follows the workflow of a Gaussian naïve Bayes classification method. The dataset comprises continuous variables assumed to have a Gaussian distribution. The classification procedure excludes dependency among the input variables. The input variables (X_i) to the neurons are responsible for delivering the initial information to the network, where i goes from 1 to n, and considering a weighting (W_{ij}) where j ranges from 1 to n depending on the input variables. This weight is a probabilistic value based on the Gaussian basis function W (Equation (1)) that indicates the intensity with which the variable affects the neuron and the possible response of the process [24]. In the input layer, the values are first standardised by subtracting the sample mean of the n training cases and by dividing over the sample standard deviation. The procedure continues in the pattern layer, where the contribution of X_i to the probability density function W in their correspondent group j is calculated. For this, it uses the activation function g_{ij} (Equation (2)), which quantifies the contribution of the ith value of X to estimate the density function W for the group j to which it belongs (Equation (1)) [30,33,34]. Artificial neurons in the hidden layer determine their activation state using an activation or transfer function. This function operates within ranges of -1 to 1 or 0 to 1, where a value of 1 represents a fully active neuron, while -1 or 0 indicates complete inactivity [32]. The information about all the features in each training group is represented by Equation (3), which is further processed in the summary layer [33,35,36].

$$W = exp\left(\frac{\parallel X - \mu \parallel^2}{\sigma^2}\right) \tag{1}$$

$$g_{ij} = W\left(\frac{X - X_i}{\varphi}\right) \tag{2}$$

$$g_j(X) = \frac{1}{n_j}\sum_{i=1}^{n} g_{ij} \tag{3}$$

where:

i: Corresponds to input values.

j: Corresponds to output groups.

X_i: input variable in activation function.

n_j: Number of observations belonging to the output group j.

W: Weight function (Gaussian basis or probability density function).

φ: Scaling parameter.

Only the scaling parameter (φ) can be varied, which affects how quickly the influence of an observation on the density at point X decays as its distance to this point increases, starting from a value of 1, but varying in search of maximising the percentage of correctly classified observations [33,37]. The misclassifying cost, c_j, refers to the misclassifying effect of a predictive response, and h_j is the probability that an observation belongs to a response group without considering the input variables, representing a relative proportion of the sample. This is because, in some cases, it is more detrimental to classify an observation incorrectly in one group than in another. The purpose of this layer is to assign a score to each output group that will later be directed to the output layer. This is done by multiplying the estimated density function by the prior probability and the cost of classifying incorrectly, as described in Equation (4). For this study, we use a non-differential misclassification cost where the probability of misclassification is the same for all the study groups [33,36].

$$Score_j = h_j c_j g_j(X) \tag{4}$$

where:

h_j: Prior probability.

c_j: Misclassification cost.

$g_j(X)$: Probability density function.

Jackknifing re-sampling was used for cross-validation, removing one point at a time from training groups and determining the frequency of the correct classification. The jackknifing procedure seeks to reduce biases based on logarithmic dispersion methods by randomly eliminating part of the data to make valid interferences for those values that are not distributed linearly [38,39]. As a result, in the output layer, the answer to the problem is provided using binary neurons, which activate if the score j > score k when k is different from j; otherwise, they are inactive. The number of responses in the output layer is defined by the number of neurons being used, similar to the characteristics of the input layer [40]. Figure 1 represents a conceptual map of the network operation.

ANN can be used to solve regression problems by calculating a continuous function of the input variables or classification problems by producing discrete function values that correspond to a certain class using supervised and unsupervised learning. In these circumstances, one tries to adapt the input variable weights such that the procedure produces the desired response for the model's training and verification [41]. The neural network in supervised learning determines the difference between the information provided by the input signals and the output by introducing a set of patterns that contain system inputs and process responses. In unsupervised learning, the neural network simply considers the input data, ignoring the signal it has to provide. It then compares the data it receives to find common features, correlations, and patterns in order to assign synaptic weights and ultimately produce a final signal output response. The classification analysis using nearest neighbours is based on the proximity of an initial data point, assuming that similar behavioural patterns may be identified in the resulting response [33,35]. The training set values are preserved, so that if an error occurs, it is mostly observed with the nearest neighbour. The model will generate two predictions: the first prediction will be the response with the highest probability, which corresponds to the nearest neighbour, and the second prediction will be the response with the second highest probability, corresponding to the second nearest neighbour.

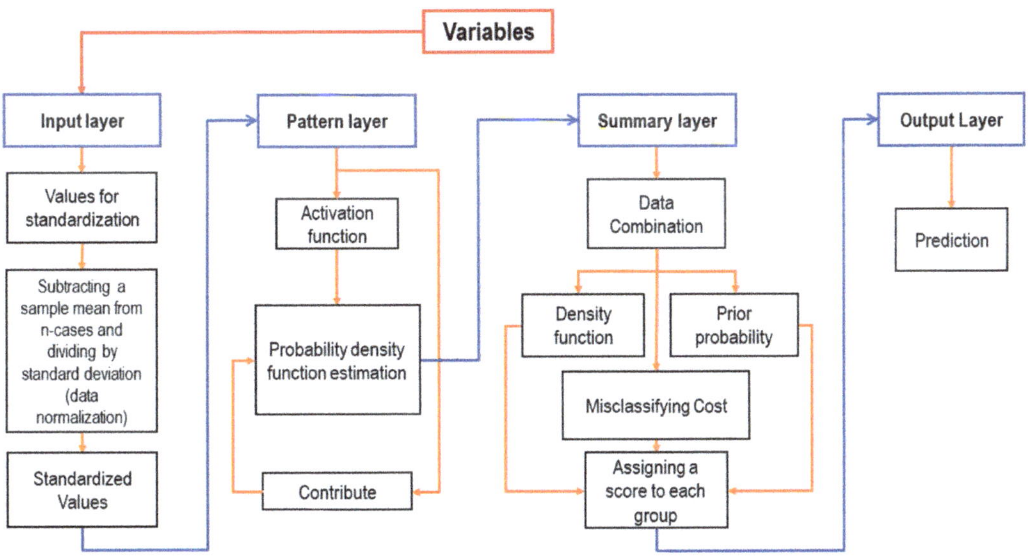

Figure 1. Artificial neural network function diagram.

2.2. Case Study Background

The study utilised both operational and laboratory datasets obtained from a copper mine in Chile. The mine employs a dynamic heap leaching method, with a height ranging from 3.8 to 4.0 m, and a leaching cycle lasting 84 days. The utilisation of sulfuric acid is affected by the particle size distribution within the range of 18% to 150 μm, with concentrations ranging from 4.5 to 8.5 g/L [20]. The drainage period begins when the irrigation of heap leaching ceases and concludes when the removal of spent heap leach begins. The intended duration for this procedure is 4 days. However, operational difficulties related to material removal can cause delays in certain modules, resulting in an extended drainage period that may exceed 13 days before proceeding to the removal stage. The spent heap leach is extracted and conveyed to a dump located 4500 m away from the site. The dump has an average moisture content of 13.5%, a surface permeability of 2×10^{-3} cm/s, and a pH ranging from 1.3 to 1.5 in the drainage/pore water. The spent heap-leach material in this study contains about 12 to 46% of clays, and hydrated phyllosilicates such as biotite, muscovite, and chlorite, with kaolinite and muscovite having a significant impact on moisture retention. The particle size is one of the most important factors of the process [2], and for this case, particles can range from 12 to 25 mm. (P_{80} pre-leach of 12.7 mm). The dump contains cuttings that deposit between 27 and 49% of the fines under 75 μm. The site operations measure the moisture level in the spent heap leach, which ranges from 8% to 17.6%. However, there were instances where the moisture content reached 22%, resulting in an average moisture content of 13%. This higher moisture content may cause the material to flow over wider areas than initially projected during unloading, as per the design. The landfill has a total area of 6.36 million square metres and has two layers for environmental protection. The initial layer consists of a 16- to 24-m-thick layer of gravel with a particle size distribution (P_{80}) ranging from 38.1 to 50.8 mm. This layer effectively manages the phreatic level and prevents excessive saturation [20]. Subsequently, a layer of impermeable andesite lava volcanic rock is applied as a protective covering. This rock contains less than 5% non-organic material and is devoid of acidic solutions. The dump has a maximum capacity of 726 million tonnes, which needs to be piled in layers at the deposition site. However, the implementation of a 5-level storage system was not possible due to the limited possibilities for stacking the segregated material along the dump. This condition prompts the increased utilisation of the basal area of the landfill, requiring an expansion of the dump surface in order to uphold its intended lifespan of 23 years. The dump incorporates a drainage system that utilises granular material sourced from the mine tailings. The granular material has a maximum size of 305 mm and contains less than 5% particles [42].

3. Experimental Design

3.1. Dataset

The approach collects available data from previous studies and operating conditions. To generate the dataset, we included operational data from 122 copper-oxide ore heap leaching modules in 6 months of operation, which also includes the characterization of their correspondent spent leach at the time of discharge in the dump. In this study, 26 variables were analysed as presented in Figure 2, which are grouped into mineralogical characteristics, lithology, drainage time, moisture, granulometry, and channels present in the cover, in addition to considering conditions used in stages prior to leaching, such as dosages used in agglomeration.

The characterization and operation data were obtained from the processing plant, and analyses were done in the laboratory, including mineralogical data, lithology, particle size (coarse and fine), solution drainage conditions from the modules, impermeable layer conditions, and moisture data from the spent heap leach.

For the creation of the dataset, we considered the following assumptions:

i. The dump slope in the discharge of the spent heap leach was first estimated and fixed as constant due to no significant variability;
ii. The speed and capacity of the conveyor belts were fixed constant, it is a standardised process without failures or stop times, and it does not vary significantly;
iii. Having a regular movement during the operation, the inclination of the discharge plume is considered constant, not a significant variable in the behaviour since it does not generate variation in the physical behaviour analysed.

For the input of the ANN model, we carefully chose the most pertinent variables that have an impact on the spent heap leach during disposal. These variables are categorised and shown in Figure 2. For training and corroboration of the models, Figure 3 describes the five spent heap leach segregation categories used in the study. The flowability observed in the material disposal, recorded hourly by the spreader, serves as the foundation for the categorization. Subsequently, data from 30 modules covering one month of operation will be utilised to validate the constructed predictive model and identify findings and recommendations.

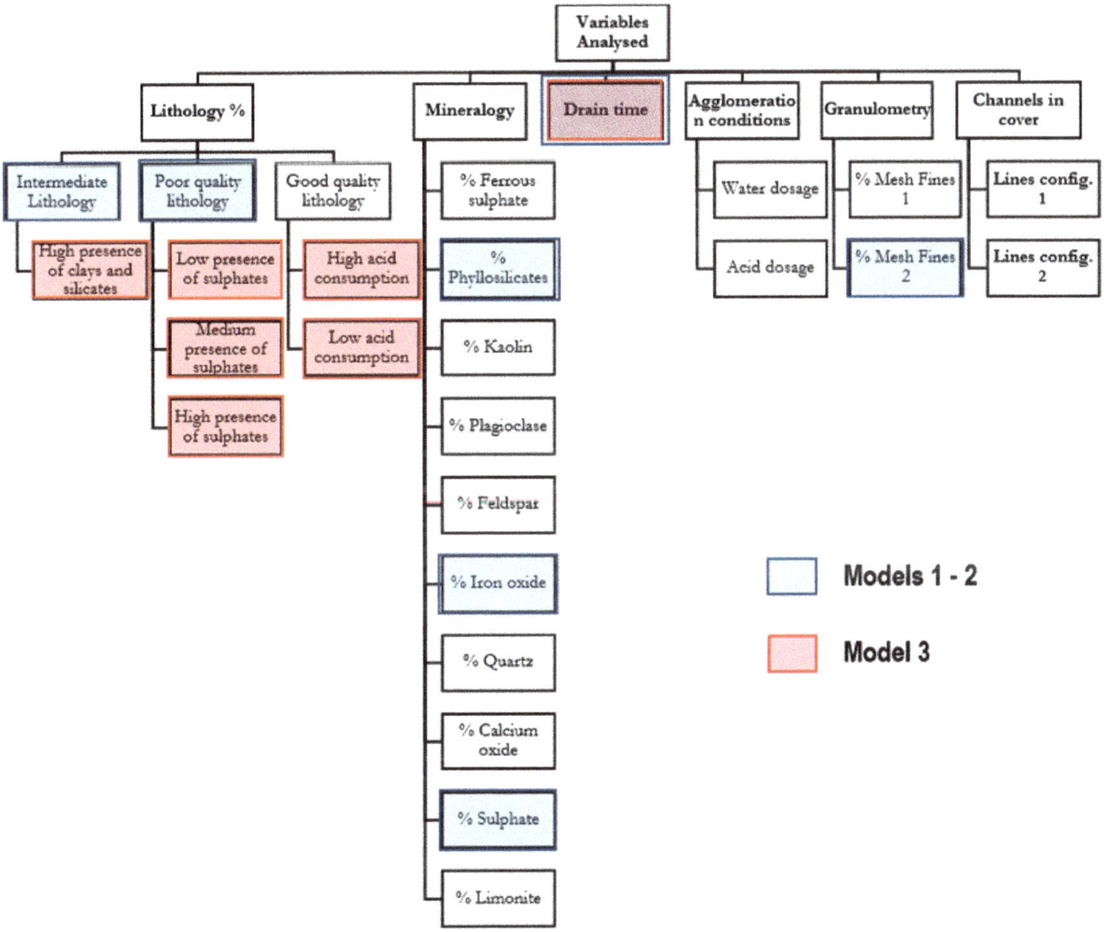

Figure 2. Predictor variables were analysed and used for model design.

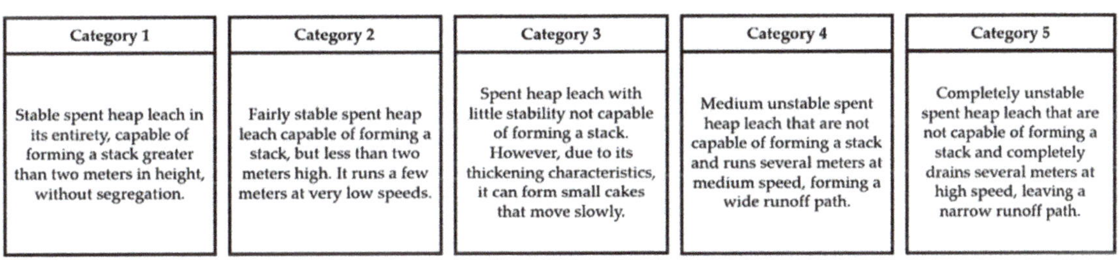

Figure 3. Spent heap leach segregation categories.

3.2. Segregation Categories

The evaluation of segregation quality in the discharged spent heap leach was performed, resulting in the identification of five categories (Figure 3) based on the behaviour of the material. Category 1 represents the driest material with the least fluidity, while category 5 corresponds to wet material with complete fluidity. The compilation of this data is generated at the time of unloading the spent heap leach from the spreader, considering the corresponding category, the height of the plume, the slope of the dump, and the transported flow.

3.3. Data Analysis

Once the metadata is generated, an analysis is conducted to establish strategies for the data acquired in the field and from the available variables from the operation plant, plus a benchmarking of the behaviour of spent heap leach that considers four mining companies close to the operation of the case studied to identify relevant variables. As shown in Figure 2, the mineralogical characteristics, granulometry, drain time, and moisture are considered critical factors that affect the flowability of the deposit of the spent heap leach. The postulates of López et al. [20], Núñez [43], and Bard [17] reinforce the idea that the fine grain size factor and distribution within a module can significantly affect permeability, which is crucial for constructing predictive models. Once the database has been generated, the predictive models designed are organised (Figure 4) following the next response variables: Model 1 with 5 response variables (solid, semi-solid, intermediate, semi-liquid, and liquid), model 2 with 3 response variables (solid, intermediate, and liquid), both considering the general behaviour of the module, and model 3 using 5 response variables but based on the behaviour by module advance. This corresponds to the monitoring of mineralogical conditions in the modules carried out as the process progresses, which corresponds to a range of 12 to 14 measurements to exceed 1500 conditions to use as a database. The comparison regarding general and advanced databases also seeks to analyse the prediction behaviour generated by different sizes of datasets, with the dataset used in models 1 and 2 being significantly smaller compared to that generated from the advance, and corroborate that by analysing nonlinear behaviours in small datasets, it is possible to aim at finding causal relationships, particularly using ANN as presented by Pasini [44], Condon [45], and Feng et al. [46]. The software RAPIDMINER Studio Version 9.10 [47] is used to construct predictive models. First, we used a supervised method to train the neural network by looking at all the possible outcomes. Then, we used an unsupervised method to let the network make its own predictions, and finally, we checked that the predictions were correct. Initially, we use supervised training to determine the process response based on seven predictive variables (Figure 2). These variables are chosen from analyses conducted with the metallurgy department of the mining company, benchmarking of other mining processes with similar characteristics, and correlation analysis associated with the segregation categories of the spent heap leach. The combination of variables is then expanded based on the generation of the highest percentage of success in network training. A statistical summary of these variables is presented in Table 1.

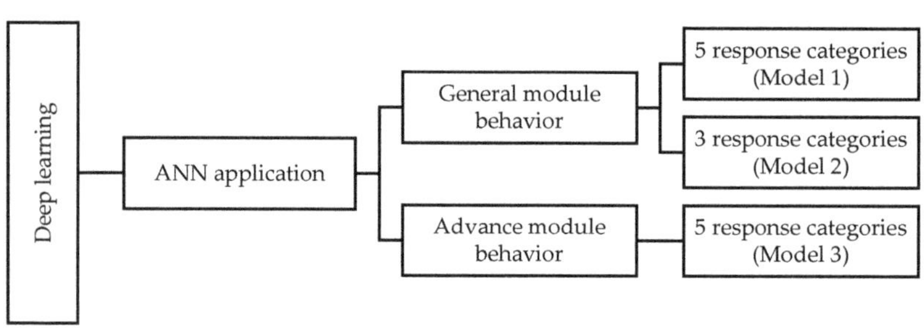

Figure 4. ANN model design strategy to predict the segregation behaviour in the dump.

Table 1. Statistical summary of the variables used in the models.

Variable ID	Variable Name	Average	Standard Deviation	Low Range	Medium Range	High Range
1	% Poor quality lithology	8.6	10.7	0.0	4.0	50.0
2	% Intermediate lithology	48.7	27.1	0.0	49.5	100.0
3	% High presence of clays and silicates	43.5	27.8	0.0	41.3	100.0
4	% Low presence of sulphates	8.2	14.9	0.0	0.0	77.6
5	% Medium presence of sulphates	1.0	3.7	0.0	0.0	30.8
6	% High presence of sulphates	0.8	2.8	0.0	0.0	23.5
7	% High acid consumption	32.7	25.9	0.0	31.3	100.0
8	% Low acid consumption	13.8	18.6	0.0	4.0	100.0
9	% Phyllosilicates	4.3	1.7	0.0	4.3	8.4
10	% Iron oxide	2.1	1.1	0.0	2.1	6.2
11	% Sulphate	4.3	7.1	0.0	0.4	46.0
12	% Mesh Fines 2 (−75 μm)	11.8	3.7	6.7	11.0	23.9
13	Drain Time (days)	7	3	3	7	16

The artificial neural network (ANN) scheme used in this study (Figure 5) consists of a pattern layer with 10 neurons, which receives input data corresponding to the variables selected for each model, assigns them a weight (W_i), and processes it through various connections. Five neurons in a summatory cell combine the output of each neuron in the pattern layer. This summatory cell computes the weighted sum of the inputs and applies an activation function. Finally, the output layer consists of 5 neurons based on the segregation categories that generate the network's response based on the computed values from the summatory cell.

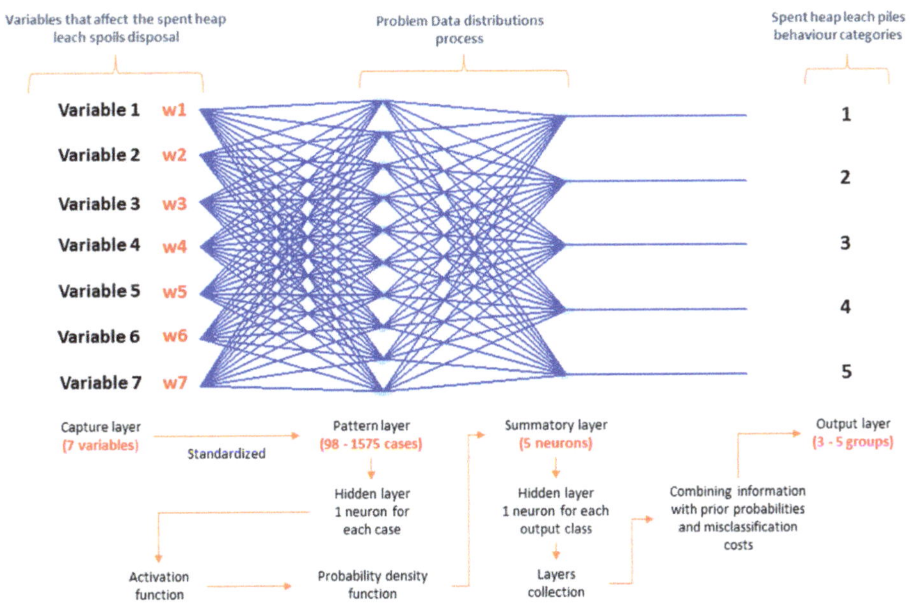

Figure 5. Artificial neural network scheme used for model generation.

In the first stage, a training phase is conducted to build models that can successfully categorise flowability. An optimised activation function is implemented using jackknifing for models 1 and 2, while for the third model, the nearest neighbour approach is used.

The second stage corresponds to evaluating the models and identifying how the chosen variables influence the prediction, considering different amounts of fluency categories.

4. Results and Discussion

Figure 6 displays a multivariate analysis that presents density graphs on the major diagonal. These graphs demonstrate the correlations among the variables utilised in the database for models 1 and 2, which are related to the overall behaviour of the modules. From a statistical analysis, the intermediate lithology and poor-quality lithology variables exhibit higher average values compared to the other parameters. The sulphate percentage and poor-quality lithology variables stand out as the most variable parameters. There is a strong correlation between intermediate lithology and variables such as sulphate percentage, iron oxide, and plagioclase, indicating their interdependence. Mesh fines 2 percentage shows a moderate positive correlation with iron oxide and plagioclase, implying that higher levels of these components lead to increased fines. Moisture exhibits a positive correlation with iron oxide and plagioclase, implying an association between these variables and increasing the moisture content in the module.

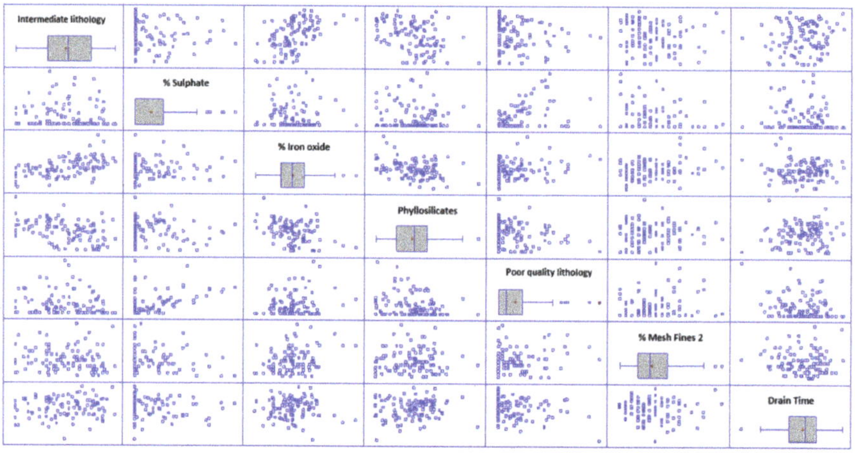

Figure 6. Correlation between the variables used in models 1 and 2 by multivariable analysis.

The data correlation implies that the most important correlation can be observed between intermediate lithology and sulphate percentage, iron oxide, and plagioclase variables. This correlation highlights the potential trade-off between poor-quality lithology and moisture content, indicating that as one variable worsens, the other may improve. It may be worth investigating the specific factors that contribute to this negative correlation to identify strategies for optimising both variables simultaneously.

Regarding the correlation present in the variables used in the database for model 3 related to the advanced behaviour of the modules (Figure 7), it was observed that intermediate lithology with a high presence of clays and silicates had a positive correlation but a relatively low impact on segregation behaviour. Sulphate presence exhibited a low correlation but high weight in the ANN models, influencing the segregation of the spent heap leach. This is explained considering that its presence occurs in a range between 1 and 30%. Iron oxide showed a positive correlation but with small weights in the models. On the other hand, phyllosilicates had a high ANN weight, signifying a significant impact on the ANN model's output. Poor-quality lithology that considers the presence of granodioritic minerals with potassium alterations and drain time as variables both had moderate correlations and ANN weights. With this, the importance of considering the composition of clays, silicates, and sulphates in understanding acid consumption and drain time is detected.

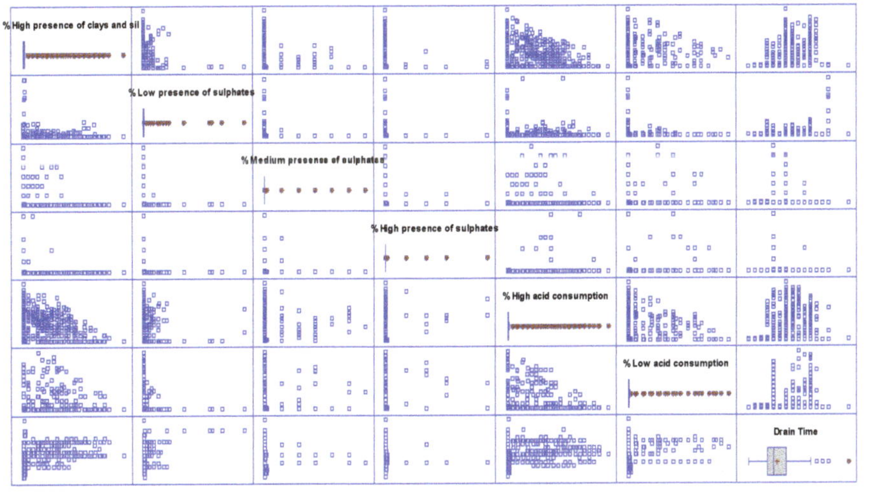

Figure 7. Correlation between the variables used in model 3 by multivariable analysis.

Figure 8 presents a comparison between the correlation and ANN weights for the variables used in the models. Positive correlation weights suggest a positive relationship and higher ANN weights indicate a stronger influence of the variable on the model's output. In the general module data, variables such as intermediate lithology, sulphate percentage, and phyllosilicates exhibited higher correlation weights, indicating a moderate to strong linear relationship with the target variable, whereas the ANN weights varied across these variables. Phyllosilicates had the highest ANN weight, suggesting an important impact on the prediction models. In the advanced module, data variables such as the high percentage presence of clays and silicates, the high percentage presence of sulphates, and the medium percentage presence of sulphates demonstrate positive correlation weights. The ANN weights for these variables varied as well, with the medium percentage presence of sulphates having the highest weight, indicating its strong influence in the ANN model.

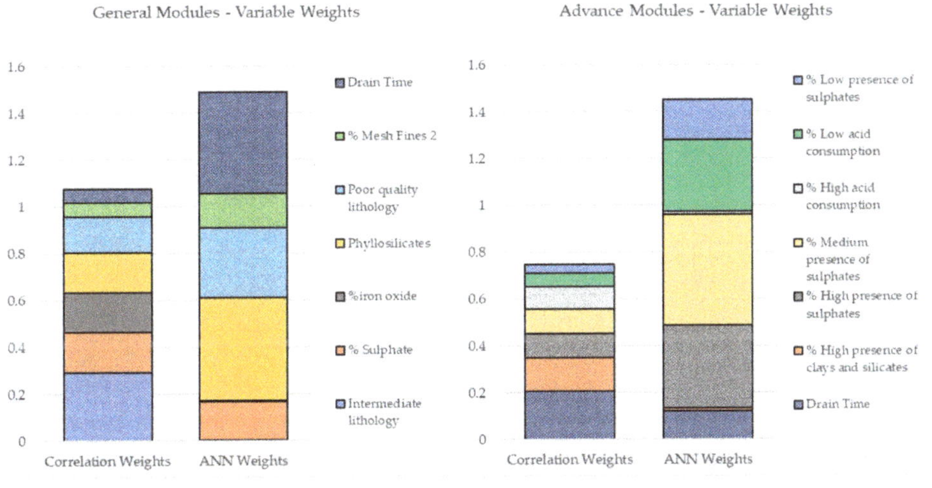

Figure 8. Correlation and ANN weight comparison for general and advanced modules variables.

4.1. ANN Model Training Stage

The training result with seven predictor variables is presented in Table 2, where model 1 obtained a 72.5% correct classification in training considering the general behaviour of the modules being applied to 98 cases, with a spacing parameter optimised by jackknifing of 0.1109. Model 2, considering the same predictor variables and cases used for the generation of model 1, with a spacing parameter optimised during training by jackknifing of 0.0429, obtains a success rate of 73.47%. The training of model 3 was incremented to 5 response categories (solid, semi-solid, intermediate semi-liquid, and liquid) and elaborated considering the percentages of stacking and removal of the modules, reaching 78.10% of correctly classified cases for a total of 1575 cases analysed and using the nearest neighbour as a spacing parameter.

Table 2. Classification training results for models 1, 2 and 3.

	Model 1		Model 2		Model 3	
Spent Heap Leach Behaviour Categories	**Cases**	**Percentage Correctly Classified (%)**	**Cases**	**Percentage Correctly Classified (%)**	**Cases**	**Percentage Correctly Classified (%)**
Solid	1	0.0	10	77.1	78	73.1
Semi-solid	9	77.8	-	-	134	59.7
Intermediate	48	83.3	48	70.0	618	85.6
Semi-liquid	32	56.3	-	-	610	77.1
Liquid	8	75.0	40	70.0	135	69.6
Total	98	72.5	98	73.5	1575	78.1

The error obtained in the model classification occurs when spent heap leach data is interpreted as a contiguous category. For model 1, of the 48 cases identified with intermediate flow and segregation, forty were correctly classified, three were classified as semi-solid, and five as semi-liquid.

For the case of semi-liquid behaviour, of the 32 cases studied, 18 were correctly classified, 13 were classified as intermediate behaviour, and 1 was liquid behaviour. This situation confirms what was expressed by Alom et al. [41], where the error can be minimised as more cases are entered into the data for training, where the errors mostly occur with neighbouring behaviours that the spent heap leach presents.

Model 2 presents a better classification due to the reduction of response variables, and the errors produced mostly occur in the closest category. Model 3 is the one that achieves a better classification of the behaviour of the unloading spent related to the cases used to train the network, as expressed by Albalasmeh et al. [48], minimising the error, covering all the feasible options that may occur, adjusting the synaptic weights of the neurons, and modifying the outputs according to the error made in each learning step until getting as close as possible to the desired output.

4.2. ANN Models Prediction and Corroboration

Once the training of the three models has been carried out, an unsupervised classification method is applied in order to analyse the level of prediction achieved. To analyse the behaviour capacity, two levels of classification response probability are applied, the first corresponding to the nearest neighbour, that is, the response with the highest probability of being obtained, and the second corresponding to the second closest neighbour, that is, the answer with the second highest probability of coming out. In the results given in Tables 3 and 4, the percentage correctly classified difference for the intermediate category is approximately 28.5%, indicating a significant improvement in model 3's classification accuracy for this category. This suggests that model 3 was preferable at classifying spent heap leach with intermediate behaviour, achieving a higher rate of accuracy. In the category of semi-liquids, the percentage difference is approximately 61.5%, indicating that model

3's correct classification percentage has increased, suggesting that model 3 performed significantly better than model 1 with 5 categories.

Table 3. Corroboration results for models 1, 2, and 3 considering the 1st highest probabilities.

	Model 1		Model 2		Model 3	
Spent Heap Leach Behaviour Categories	Cases (%)	Percentage Correctly Classified (%)	Cases (%)	Percentage Correctly Classified (%)	Cases (%)	Percentage Correctly Classified (%)
Solid	0.0	0.0	3.9	100.0	1.0	100.0
Semi-solid	3.9	100.0	-	-	1.0	100.0
Intermediate	69.2	77.8	69.2	66.7	40.1	75.3
Semi-liquid	26.9	42.9	-	-	27.1	65.4
Liquid	0.0	0.0	26.9	57.1	30.7	55.9
Total	100.0	73.6	100.0	74.6	100.0	79.3

Table 4. Corroboration results for models 1, 2, and 3 considering the 1st and 2nd highest probabilities.

	Model 1		Model 2		Model 3	
Spent Heap Leach Behaviour Categories	Cases (%)	Percentage Correctly Classified (%)	Cases (%)	Percentage Correctly Classified (%)	Cases (%)	Percentage Correctly Classified (%)
Solid	0.0	0.0	3.9	100.0	1.0	100.0
Semi-solid	3.9	100.0	-	-	1.0	100.0
Intermediate	69.2	100.0	69.2	100.0	40.1	100.0
Semi-liquid	26.9	85.7	-	-	27.1	69.2
Liquid	0.0	0.0	26.9	100.0	30.7	88.1
Total	100.0	95.2	100.0	100.0	100.0	91.5

For the category of liquids, the percentage difference is approximately 54.3%, indicating a substantial improvement in the classification accuracy of model 3. Model 1 made 90 predictions, of which it corroborated 26 cases, obtaining 73.55% certainty as the first response. This increases to 95.24% certainty for the process considering the two predictions corresponding to the first and second highest probability that a certain behaviour of the spent heap leach will occur. The prediction analysis strategy of model 2 considers, like model 1, the generation of 90 predictions obtaining 1.06% less success than that obtained in model 1 with corroboration from 26 cases considering the first probable answer, but this difference increasing by 4.76% considering the 2nd highest probability, where the range of physical qualities is smaller so there are no intermediate behaviours (semi-solid or semi-liquid), so it can be considered more drastic when making the decision. Model 2 reaches a prediction level of 100% certainty with the first and second highest probabilities, defining strategies for both categories with the highest probability of exiting.

For model 3, considering the behaviour due to module advancement, 192 corroborations were conducted, obtaining a 79.33% accuracy in the prediction, considering the first highest probability of a certain category occurring, increasing to 91.47% when considering two probabilities of occurrence. The errors produced in the models correspond to the responses given to the categories close to the true response, mostly in the intermediate behaviour and the semi-solid and semi-liquid classifications.

The error generated under the 5 categories provides an opportunity for correction during the operation since these categories are contiguous and defined with similar spent heap leach discharge strategies, thereby not affecting the space allocated to the dump. These results demonstrate that it is indeed possible with the currently monitored variables and with the application of ANN to obtain a high degree of prediction as the new data related to spent heap leach is integrated into the original data. Model 3 performed slightly better than models 1 and 2 in terms of the overall percentage correctly classified, which leads us to consider that the use of a small dataset can be a possibility to evaluate an ANN performance considering ensuring that the input data is diverse, covers patterns, and is relevant to the behaviour. There are variations in the accuracy for the segregation categories; the solid category improved significantly, semi-solid and liquid dropped in the larger dataset, and the accuracy for the intermediate segregation category remained consistently high in both models. Figure 9 presents the prediction percentage for the three models, considering the training and prediction stages.

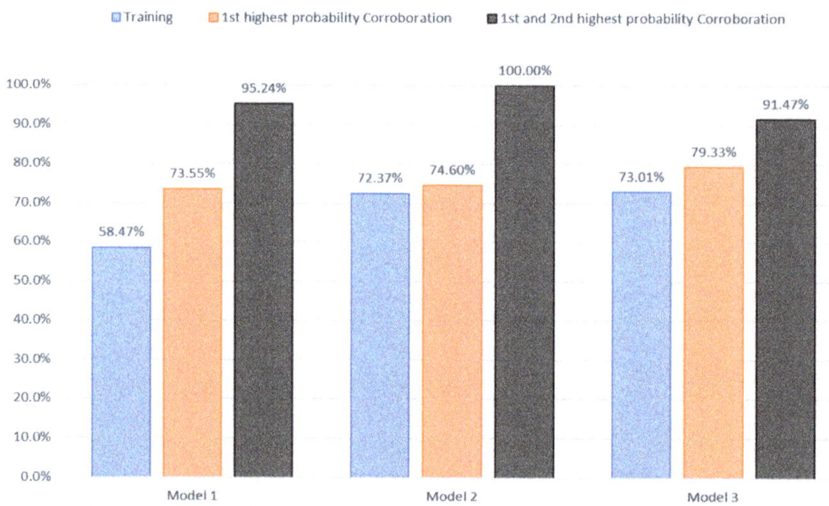

Figure 9. Models' prediction corroboration considering the most likely segregation category.

If we compared the predictive models based on evaluation metrics, the results indicate that the ANN advance module outperforms the ANN general module in terms of predictive performance but considering the differences in cases used in the database, it is possible to say that the ANN general module results can be considered "not as strong" as the predictions obtained with the advance module.

The ANN advance module demonstrates a lower root mean squared error (RMSE) of 0.819 ± 0.023 compared to the RMSE of 0.853 ± 0.278 for the ANN general module, indicating better accuracy in its predictions. The ANN advance module model also achieves better accuracy and superior performance for average absolute error (0.613 ± 0.016 vs. 0.682 ± 0.012) and relative error (16.12% ± 0.28% compared to 17.47% ± 0.31%). Additionally, the ANN advance module exhibits a higher correlation of 0.429 ± 0.083, while the ANN general module correlates 0.191 ± 0.074, indicating a stronger relationship between the predicted variables and actual segregation categories.

Analysing the weight given by the variables at the prediction level, the presence of fines in the ore is one of the most influential variables that explain the behaviour of the spent heap leach. As mentioned by López et al. [20], the particle size profile for the leaching process seeks to maximise the surface contact area in order to maximise the extraction of the element of interest, which results in changes in the physicochemical characteristics of the mineral that can affect the heap permeability. In this study, the granulometry in this process is between 11 and 18% under 75 μm (200 Tyler mesh), reaching a P_{80} of 12.7 mm, which affects the behaviour of the material in the dump. If the fines percentage increases, greater compaction is generated, decreasing the size of the pores, and reducing their hydraulic conductivity, as also established by Marchant [14], preventing good drainage in the piles.

Other variables, such as the fluctuation in ferrous sulphate concentration (low, medium, and high), have been recognised as significant factors for the model. This is due to the presence of them exceeding 30% in certain modules, resulting in a spent heap leach with minimal rock competition and high solubility in water, leading to chemical crushing. The presence of phyllosilicates as the major component leads to a rise in the fines percentage, as demonstrated by Bard et al. [17]. This drop in particle sizes adversely affects permeability, resulting in a reduced drainage capacity.

The lithology of intermediate quality is primarily composed of limonite and clays, such as muscovite and kaolinite. This lithology significantly affects the model due to its low permeability and capacity to undergo chemical crushing, facilitated by the presence of fine-grained iron oxide and low hardness (1.5–1.6 on the Mohs scale). When lithologies of good quality, such as granodiorite with potassium are present, it results in strong rock competition, preventing the generation of fine materials. This leads to excellent permeability within the material, improved hydraulic conductivity, and efficient drainage of solutions. Phyllosilicates play a crucial role in transporting water, as indicated by Ganzhorn et al. [49], highlighting their significance as a constant factor throughout the three models. In models 1 and 2, the inclusion of sulphates as a variable significantly improves the accuracy of the model. This is because sulphates generate a substantial number of fines, which have a physical and chemical impact on the behaviour of spent heap leach by affecting its permeability. Minerals with relatively low hardness, such as brochantite (3.5 to 4.0) [50] and chalcanthite (2.5), can be disintegrated on the surface [51].

4.3. Segregation Behaviour and Moisture Relation

As it was established by Pieretti et al. [52], the moisture content of the spent heap leach is one of the most influential variables in the behaviour of these materials in their operational transport and deposition. The use of this variable as input data depends on the time of availability of this characterization, and for the case studied, it is performed once the module has been downloaded, with an average waiting period of 7 days for analysis. This situation means that currently, it is not possible to use this variable as part of a previous predictive analysis if the result delivery times cannot be improved. When analysing the behaviour of the moisture percentage against the segregation behaviour as shown in Figure 10, there is a clear relationship, whereas as the moisture increases, the quality of the spent heap leach becomes more liquid, reaching category 5. When employing 5 response categories, the moisture present is modified to exhibit a logarithmic pattern, in contrast to the observed polynomial.

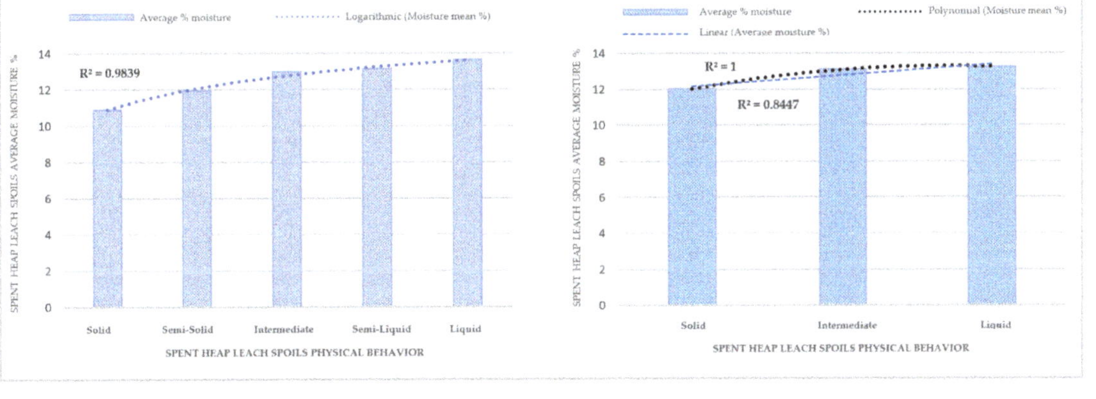

Figure 10. Average moisture percentage for physical behaviour of spent heap leach with 5 and 3 response categories.

5. Conclusions

In this study, it was possible to verify that the application of deep learning is feasible to predict the behaviour of the spent heap leach when they are dumped. The neural network method is efficient for this kind of analysis prediction, with 95.24% accuracy for model 1 when the two highest probabilities of data correlation are taken into account. From the three models generated, the models using five variables as a classification response are more reliable as they have intermediate segregation behaviours that can help minimise prediction errors by designating similar discharge strategies. The versatility of using an artificial neural network in this type of operation considers the option of using two models together using different predictor variables to know the general segregation by the spent heap leach disposal. The errors produced in the models occur with the closest neighbour to the correct category; therefore, it would not affect the discharge strategy of the spent heap leach in the dump.

However, when the moisture content is over 8%, the material flows using larger areas than originally estimated by design. Currently, considering the moisture percentage as a predictive variable in a day-by-day operation is not possible because its measurement is performed once the modules are removed, and the results of the chemical analysis are not available before the spent heap leach is discarded in the dump. However, since the objective of this work was to identify the most relevant variables to predict behaviours at the time of depositing the spent heap leach, moisture has a clear relationship with segregation and flows; therefore, it must be used as a parameter to consider for the elaboration of shifting. Evaluating a small dataset with adequate input data can be the first approach to assessing segregation behaviour. Although a larger dataset generally yields more reliable and robust results, it has been identified that carefully designing, representing, and studying a small dataset beforehand can provide valuable insights into the classification performance.

The modules with a high presence of sulphate showed poor rock competence and great ability to form fines, causing permeability problems and preventing proper drainage of the leach solution, and the spent heap leach will have less stability in the dump. The high presence of clays and silicates has an important effect on permeability, inhibiting the hydraulic conductivity of the modules, as clays have a high capacity to retain moisture. The presence of fine granulometry in the modules (higher than 18% −150 μm) is one of the variables with the greatest weight when developing the models, since a high presence in the module directly affects the permeability of the bed, decreasing the hydraulic conductivity and thus the fluidity. The flowability is better for the lithologies considered to be of good quality, formed with granodiorite and potassium alterations, providing good permeability with high rock competence and low formation of fines. With a higher presence of lithologies in the module, better drainage conditions for the solution will be obtained, helping to maintain the stability of the spent heap leach in the dump. Phyllosilicates provide fine granulometry to the modules and are considered water carriers. They cause solution drainage problems, producing a drainage deficit and causing spent heap leach with high segregation. In the presence of iron oxides presenting clays as impurities with low hardness, a scenario of easy disintegration is presented, causing fines, low permeability, and a deficit in the drainage of the solution.

6. Recommendations

Using this methodology, it is relevant to obtain more data over time to complement the metadata and provide a deeper level of characterization associated with the composites that make up the modules. To use the moisture percentage as a predictive variable, a 24–48 h sampling procedure must be established after the module irrigation ends for the characterization response time to occur before the module removal and to consider with robust information the possible relationship with the spent heap leach behaviour in the dump.

Author Contributions: N.H., M.S.G., J.O. and R.M.C. contributed to the methodology, conceived, and designed the experiments; analysed the data, and wrote the paper, N.H. All authors have read and agreed to the published version of the manuscript.

Funding: This research received no external funding.

Data Availability Statement: The data used in this study is limited to publicly available information, in compliance with rigorous confidentiality standards pertaining to the analysed mining operation to protect sensitive information. To obtain further information or resolve any uncertainties regarding the data utilised in this research, it is advisable to contact the corresponding author, who will address any questions pertaining to data accessibility and utilisation.

Acknowledgments: All authors participating in this study have given their approval and assent to the manuscript's content. They also express their heartfelt gratitude and appreciation for the help and direction offered by the Oulu Mining School at the University of Oulu and the Geological Survey of Finland GTK.

Conflicts of Interest: The authors declare no conflict of interest.

References

1. Vargas, T.; Estay, H.; Arancibia, E.; Díaz-Quezada, S. In situ recovery of copper sulfide ores: Alternative process schemes for bioleaching application. *Hydrometallurgy* **2020**, *196*, 105442. [CrossRef]
2. Ghorbani, Y.; Becker, M.; Mainza, A.; Franzidis, J.-P.; Petersen, J. Large particle effects in chemical/biochemical heap leach processes—A review. *Miner. Eng.* **2011**, *24*, 1172–1184. [CrossRef]
3. Muniruzzaman, M.; Karlsson, T.; Kauppila, P. *Prediction of the Drainage Water Quality from Mine Wastes with Reactive Transport Modelling*; Geological Survey of Finland: Espoo, Finland, 2018.
4. Lagos, J.; Ponce, R.; Valiente, P.; Giglio, S. Geology of the Antucoya Porphyry Copper Deposit: An Early Cretaceous System in the Coastal Cordillera, Antofagasta Region, Chile. In Proceedings of the 14° Congreso Geológico Chileno, La Serena, Chile, 4–8 October 2015; p. 365.
5. Trujillo, J.Y.; Cisternas, L.A.; Gálvez, E.D.; Mellado, M.E. Optimal design and planning of heap leaching process. Application to copper oxide leaching. *Chem. Eng. Res. Des.* **2014**, *92*, 308–317. [CrossRef]
6. Valenzuela-Elgueta, J.; Jimenez, Y.; Delgado, A.; Ahualli, S. Electrokinetics and stability of silica/clay mixtures at high copper concentration. Implications in the mining of copper. *Miner. Eng.* **2019**, *134*, 193–201. [CrossRef]
7. Wu, A.; Yin, S.; Yang, B.; Wang, J.; Qiu, G. Study on preferential flow in dump leaching of low-grade ores. *Hydrometallurgy* **2007**, *87*, 124–132. [CrossRef]
8. Leiva, E.; Palacios, C. *Caracterización y Distribución de Minerales oxidados de Cobre y Alteración en el Yacimiento Cuprífero Mantos de la Luna, II región de Antofagasta*; Universidad de Chile: Santiago, Chile, 2010.
9. International Copper Study Group. *The World Copper Factbook 2023*; International Copper Study Group: Lisbon, Portugal, 2023; Available online: www.icsg.org (accessed on 22 December 2023).
10. Ilankoon, I.; Tang, Y.; Ghorbani, Y.; Northey, S.; Yellishetty, M.; Deng, X.; McBride, D. The current state and future directions of percolation leaching in the Chinese mining industry: Challenges and opportunities. *Miner. Eng.* **2018**, *125*, 206–222. [CrossRef]
11. McBride, D.; Gebhardt, J.; Croft, T.; Cross, M. Modeling the hydrodynamics of heap leaching in sub-zero temperatures. *Miner. Eng.* **2016**, *90*, 77–88. [CrossRef]
12. Yin, S.; Wang, L.; Wu, A.; Free, M.L.; Kabwe, E. Enhancement of copper recovery by acid leaching of high-mud copper oxides: A case study at Yangla Copper Mine, China. *J. Clean. Prod.* **2018**, *202*, 321–331. [CrossRef]
13. Ghasemzadeh, H.; Pasand, M.S.; Shamsi, M.M. Experimental study of sulfuric acid effects on hydro-mechanical properties of oxide copper heap soils. *Miner. Eng.* **2018**, *117*, 100–107. [CrossRef]
14. Marchant, A. *Influencia de la Mineralogía de Alteración en el Comportamiento Hidráulico y Mecánico del Mineral en el Proceso de Heap Leaching Minera Antucoya*; Universidad de Concepción: Concepción, Chile, 2017.
15. Liu, M.; Wen, J.; Tan, G.; Liu, G.; Wu, B. Experimental studies and pilot plant tests for acid leaching of low-grade copper oxide ores at the Tuwu Copper Mine. *Hydrometallurgy* **2016**, *165*, 227–232. [CrossRef]
16. Watanabe, Y.; Yokoyama, S.; Shimbashi, M.; Yamamoto, Y.; Goto, T. Saturated hydraulic conductivity of compacted bentonite–sand mixtures before and after gas migration in artificial seawater. *J. Rock Mech. Geotech. Eng.* **2023**, *15*, 216–226. [CrossRef]
17. Bard, E.; Campaña, J. Aspectos Geotecnicos en el Diseño Pilas y Botaderos de Ripios Lixiviados. In Proceedings of the V Congreso Chileno de Ingeniería Geotécnica, Santiago, Chile, 28–30 November 2004.
18. Chavez, R. *Declaración de Impacto Ambiental del Proyecto "REGULARIZACION BOTADERO DE RIPIOS"*; Compañía Minera Cerro Negro S.A; Superintendencia del Medioambiente del Gobierno de Chile: Santiago, Chile, 2017.
19. Miskolczi, I.; Rykaart, M. Water Matters—Final Draindown Modelling for Mine Waste Facilities. In Proceedings of the Tailings and Mine Waste 2019, Vancouver, BC, Canada, 17–20 November 2019.

20. López, E.; Verdugo, R. *Estudio Experimental de la Permeabilidad de Materiales Depositados en Pilas de Lixiviación de Cobre*; Universidad de Chile: Santiago, Chile, 2012.
21. McBride, D.; Gebhardt, J.; Croft, N.; Cross, M. Heap Leaching: Modelling and Forecasting Using CFD Technology. *Minerals* **2018**, *8*, 9. [CrossRef]
22. Zadeh, L. Soft computing and fuzzy logic. *IEEE Softw.* **1994**, *11*, 48–56. [CrossRef]
23. Bergh, L.; Yianatos, J.; Leiva, C. Fuzzy supervisory control of flotation columns. *Miner. Eng.* **1998**, *11*, 739–748. [CrossRef]
24. Correa Deves, R. *Redes Neuronales Artificiales en Ingeniería y Física Nuclear Caracterización de Espectros PIXE*; Editorial de la Universidad de Granada: Granada, Spain, 2006; ISBN 978-84-338-4156-8.
25. Umucu, Y.; Deniz, V.; Bozkurt, V.; Çağlar, M.F. The evaluation of grinding process using artificial neural network. *Int. J. Miner. Process.* **2016**, *146*, 46–53. [CrossRef]
26. Leiva, C.; Flores, V.; Salgado, F.; Poblete, D.; Acuña, C. Applying Softcomputing for Copper Recovery in Leaching Process. *Sci. Program.* **2017**, *2017*, 6459582. [CrossRef]
27. Haghighi, H.K.; Moradkhani, D.; Salarirad, M.M. Modeling of Synergetic Effect of LIX 984N and D2EHPA on Separation of Iron and Zinc Using Artificial Neural Network. *Trans. Indian Inst. Met.* **2014**, *67*, 331–341. [CrossRef]
28. Haghighi, H.K.; Rafie, M.; Moradkhani, D.; Sedaghat, B.; Abdollahzadeh, A. Modeling on Transition of Heavy Metals from Ni–Cd Zinc Plant Residue Using Artificial Neural Network. *Trans. Indian Inst. Met.* **2015**, *68*, 741–756. [CrossRef]
29. Nuri, O.S.; Allahkarami, E.; Irannajad, M.; Abdollahzadeh, A. Estimation of selectivity index and separation efficiency of copper flotation process using ANN model. *Geosystem Eng.* **2017**, *20*, 41–50. [CrossRef]
30. He, W.; Ruhani, B.; Toghraie, D.; Izadpanahi, N.; Esfahani, N.N.; Karimipour, A.; Afrand, M. Using of Artificial Neural Networks (ANNs) to predict the thermal conductivity of Zinc Oxide–Silver (50%–50%)/Water hybrid Newtonian nanofluid. *Int. Commun. Heat Mass Transf.* **2020**, *116*, 104645. [CrossRef]
31. Sarraute, C. Aplicacion de las Redes Neuronales al Reconocimiento de Sistemas Operativos. May 2013. Available online: http://arxiv.org/abs/1305.4686 (accessed on 6 January 2023).
32. Catalán Martínez, D. *Modelización Mediante Redes Neuronales del Ensuciamiento de Membranas Cerámicas de Ultrafiltración*; Universitat Politècnica de València: Valencia, Spain, 2012.
33. Dantas, C.A.; Nunes, R.O.; Canuto, A.M.P.; Xavier-Júnior, J.C. Dynamic feature selection based on clustering algorithm and individual similarity. In *Lecture Notes in Computer Science (Including Subseries Lecture Notes in Artificial Intelligence and Lecture Notes in Bioinformatics)*; Springer: Berlin/Heidelberg, Germany, 2017; pp. 467–474. [CrossRef]
34. Silva, I.N.; Spatti, D.H.; Flauzino, R.A.; Liboni, L.; Alves, S. *Artificial Neural Networks a Practical Course*; Springer: Berlin/Heidelberg, Germany, 2017.
35. Magdon-Ismail, M.; Atiya, A. Neural Networks for Density Estimation. *Adv. Neural Inf. Process. Syst.* **1998**, *11*, 171–178.
36. Dikshit, A.; Pradhan, B.; Santosh, M. Artificial neural networks in drought prediction in the 21st century—A scientometric analysis. *Appl. Soft Comput.* **2021**, *114*, 108080. [CrossRef]
37. STATPOINT Inc. Clasificador de Redes Neurales. STATGRAPHICS—Ver. 4/25/2007. 2007. Available online: https://www.statgraphics.net/wp-content/uploads/2011/12/tutoriales/Clasificador%20de%20Redes%20Neurales.pdf (accessed on 6 January 2023).
38. Behar Gutiérrez, R.; Yepes Arango, M. *Sobre Algunas Tecnicas de Remuestreo: El Metodo de 'Jackknife'*; Departamento de Información y Sistemas, Universidad del Valle: Cali, Colombia, 2013.
39. Jara, A.; Figueroa, J. *Estudio de Mantenimiento Basado en la Confiabilidad Acorde al Sistema Asset Management Aplicado a la Flota de Bulldozer CAT D10T, Minera Anglo American Operación Mantoverde*; Universidad del Bio-Bio: Concepción, Chile, 2014.
40. Tabbussum, R.; Qayoom Dar, A. Performance evaluation of artificial intelligence paradigms-artificial neural networks, fuzzy logic, and adaptive neuro-fuzzy inference system for flood prediction. *Environ. Sci. Pollut. Res.* **2021**, *28*, 25265–25282. [CrossRef] [PubMed]
41. Alom, M.Z.; Taha, T.M.; Yakopcic, C.; Westberg, S.; Sidike, P.; Nasrin, M.S.; Hasan, M.; Van Essen, B.C.; Awwal, A.A.S.; Asari, V.K. A State-of-the-Art Survey on Deep Learning Theory and Architectures. *Electronics* **2019**, *8*, 292. [CrossRef]
42. ARCADIS. *Análisis de Estabilidad de Botadero de Ripios de lixiviación. Situación Actual y Mejoras de Largo Plazo*; ARCADIS Chile S.A.: Antofagasta, Chile, 2018.
43. Núñez, B. *Estudio del Comportamiento de Ripio en la Planta PTMP de Chuquicamata*; Universidad Católica del Norte: Antofagasta, Chile, 2016.
44. Pasini, A. Artificial neural networks for small dataset analysis. *J. Thorac. Dis.* **2015**, *7*, 953–960. [CrossRef]
45. Condon, D. *Performance of Artificial Neural Networks on Small Structured Datasets*; AMSI Vacation Research Scholarships 2018–2019; Australian Mathematical Sciences Institute: Parkville, Australia, 2019.
46. Feng, S.; Zhou, H.; Dong, H. Using deep neural network with small dataset to predict material defects. *Mater. Des.* **2018**, *162*, 300–310. [CrossRef]
47. Kotu, V.; Deshpande, B. *Predictive Analytics and Data Mining*; Morgan Kaufmann: Boston, MA, USA, 2015. [CrossRef]
48. Albalasmeh, A.; Mohawesh, O.; Gharaibeh, M.; Deb, S.; Slaughter, L.; El Hanandeh, A. Artificial neural network optimization to predict saturated hydraulic conductivity in arid and semi-arid regions. *CATENA* **2022**, *217*, 106459. [CrossRef]
49. Ganzhorn, A.-C.; Pilorgé, H.; Le Floch, S.; Montagnac, G.; Cardon, H.; Reynard, B. Deuterium-hydrogen inter-diffusion in chlorite. *Chem. Geol.* **2018**, *493*, 518–524. [CrossRef]

50. Price, M.; Walsh, K. *Pocket Nature Rocks and Minerals*; Dorling Kindersley: London, UK, 2005.
51. Rudolf, D.; Lubos, R. *Minerales*; Tikal Ediciones: Madrid, Spain, 2009; ISBN-10: 8430572163, ISBN-13: 978-8430572168.
52. Pieretti, M.; Karlsson, T.; Arvilommi, S.; Muniruzzaman, M. Challenges in predicting the reactivity of mine waste rocks based on kinetic testing: Humidity cell tests and reactive transport modeling. *J. Geochem. Explor.* **2022**, *237*, 106996. [CrossRef]

minerals

MDPI

Article

Coal Structure Prediction Based on Type-2 Fuzzy Inference System for Multi-Attribute Fusion: A Case Study in South Hengling Block, Qinshui Basin, China

Xuepeng Cui [1,†], Youcai Tang [1,†], Handong Huang [1,*], Lingqian Wang [2,*], Jianxing Wang [3], Zifan Guo [1], Chen Ma [1] and Meng Sun [4]

[1] State Key Laboratory of Petroleum Resources and Prospecting, College of Geophysics, China University of Petroleum, Beijing 102249, China
[2] College of Science, China University of Petroleum, Beijing 102249, China
[3] Unconventional Petroleum Research Institute, China University of Petroleum, Beijing 102249, China
[4] Exploration and Development Research Institute, PetroChina Huabei Oilfield Company, Renqiu 062550, China
* Correspondence: hhd@cup.edu.cn (H.H.); lingqianwang@cup.edu.cn (L.W.)
† These authors contributed equally to this work.

Abstract: The accurate prediction of coal structure is important to guide the exploration and development of coal reservoirs. Most prediction models are interpreted for a single sensitive coal seam, and the selection of sensitive parameters is correlated with the coal structure, but they ignore the interactions between different attributes. Part of it introduces the concept of the geological strength index (GSI) of coal rocks in order to achieve a multi-element macroscopic description and quantitative characterization of coal structure; however, the determination of coal structure involves some uncertainties among the properties of coal, such as lithology, gas content and tectonic fracture, due to their complex nature. Fuzzy inference systems provide a knowledge discovery process to handle uncertainty. The study shows that a type-2 fuzzy inference system (T2-FIS) with multi-attribute fusion is used to effectively fuse pre-stack and post-stack seismic inversion reservoir parameters and azimuthal seismic attribute parameters in order to produce more accurate prediction results for the Hengling block in the Shanxi area. The fuzzy set rules generated in this paper can provide a more reliable prediction of coal structure in the GSI system. The proposed system has been tested on various datasets and the results show that it is capable of providing reliable and high-quality coal structure predictions.

Keywords: coal structure prediction; GSI system; multi-attribute fusion; fuzzy inference systems

1. Introduction

Coal structure indexes the internal structural characteristics of coal reservoirs under the action of tectonic stress. The fractional anisotropy of coal structure can directly reflect the differential distribution characteristics of the coalbed methane (CBM) selection area and is an important parameter for coal reservoir quality evaluation [1]. Using continuous logging parameters to predict coal structure is a valuable tool for geologists and mining engineers. By understanding the sensitivity between logging parameters and coal structure, it is possible to make reasonable predictions about the most suitable mining areas and to enhance hazard warning capabilities in the high fracture areas of coal mines, such as gas protrusions and roof collapses [2,3].

Although an increasing number of sensitive logging reflections have been identified, experimental results from different core analyses [4] have demonstrated their limited predictive relevance. Shi et al. (2020) [5] intended to develop a quantitative coal structure identification model based on the kernel Fisher discriminant method through the idea of dimensionality reduction, which was applied to CBM in the Qinshui Basin. Mou et al. (2020) [6] developed a new method to predict index sensitivity by using tectonic coal stress concentrations. Zhang et al. (2021) [7] used a combination of logging curves and structural curvature to estimate the coal structure index of the degree of damage and predicted it by multiple linear regression methods. Some studies have calculated coal structure vulnerability indices by integrating two tools, the analytic hierarchy process and geographic information systems. These studies have identified some early warning values for water-inrush hazard areas for roughly classified coal structure results [8]. Other studies have implemented an elastic impedance inversion method for the fluid factor and brittleness coefficient, using both as effective pore-fluid bulk moduli to estimate coal structural distribution [9,10]. However, the use of multiple solutions for the inversion reduces the accuracy of the predictions, whereas they become more accurate with the addition of rich logging parameters, but the well extrapolation process still requires the inversion of seismic waves in order to solve for the relevant parameters, resulting in a certain error accumulation [11,12].

The Geological Strength Index (GSI) is a rock quality evaluation tool, formally proposed by Evert Hoek et al. (1997) [13], which has been widely used to assist in predicting coal structures. It is based on a numerical scale that assesses various parameters, such as lithology, stratigraphy, and structural complexity. However, the GSI chart relies on visual examination and if the number of core samples examined is insufficient, there may be an increase in subjective uncertainty [14]. Fuzzy inference systems (FISs) offer a more reliable method of predicting coal structures in a way that approximates human reasoning and decision-making, taking into account the uncertainty of the data. Fuzzy Logic Systems (FLSs) offer a non-traditional approach to perceptual modeling and have been successfully applied in the fields of control, information retrieval and system modeling, etc. The related singleton fuzzification works have been widely implemented to build predictive models of coal structures [15–19] and have been integrated into mainstream engineering software, such as Geolog, MATLAB and Python [20–22]. Compared to its type-1 fuzzy set (T1-FS) counterpart for rock parameters, the interval type-2 fuzzy logic system (IT2-FLS) in this study shows better performance in four areas, i.e., uncertainty simulation, rule base volume, control surface smoothness and adaptability.

In this paper, the high-resolution reservoir inversion results and fracture seismic attribute information are used to fuzzy fuse multiple attributes according to the GSI chart. The integrated attributes depend on the rule number and training results of the fuzzy model, which can effectively and accurately predict subsurface reservoir structures. In the data mining process, we inherit the methods used to obtain previous results and realize the data-driven sample acquisition based on the data, which saves the cost of human labeling. The training set is combined with the information fusion capability of the fuzzy system, which, in turn, improves the prediction accuracy of the coal body structure. Therefore, this study uses IT2-FLS combined with GSI in order to provide a theoretical basis for the accurate prediction of coal structure by multi-attribute fusion.

2. Geological Setting

The South Hengling Block is located on the eastern synclinorium of the Qinshui Basin and covers an area of 3.9 km^2 (Figure 1). The area is part of the North China Fault Block and is subject to east-west extrusive stresses in the basin tectonics, with well-developed trap columns, folds and fault structures. By the end of 2022, the Hengling Block had significant exploration potential with new proven coal-bed methane reserves of 7707 million cubic meters [23]. The district has two development strata. One is the No. 15 coal seam of the Carboniferous Taiyuan Formation located above the weathering and stripping surface of the Ordovician. The thickness of the seam ranges from 106.30 m to 158.69 m and averages 123.71 m (Figure 2b). The buried depth of the stable mineable coal seam is greater than 1500 m, with a coal thickness of 4.35 m–7.08 m and an average of 5.18 m. It contains one to three layers of interbedded gangue, mainly cleat and granular coal, locally developed primary structural coal, of which the macro coal rock type is mainly semi-bright coal, with a small amount of bright coal on top. The coal type is mainly anthracite. The other set of continuous deposits is the No. 3 coal seam in the central part of the Shanxi Formation, buried around 1410 m–1580 m deep, with a stratigraphic thickness ranging from 26.92 m to 66.80 m and averaging 49.22 m (Figure 2a). The No. 3 coal seam is thinner and partially mineable, with a thickness of 0.54 m–2.59 m and an average of 0.9 m. It contains zero to one layers of interbedded gangue, dominated by fractured-fractured-grained coal, which is semi-bright coal. The types are anthracite and lean coals. The specific assignment of the two sets of mineable coal seams in the study area is given in Table 1.

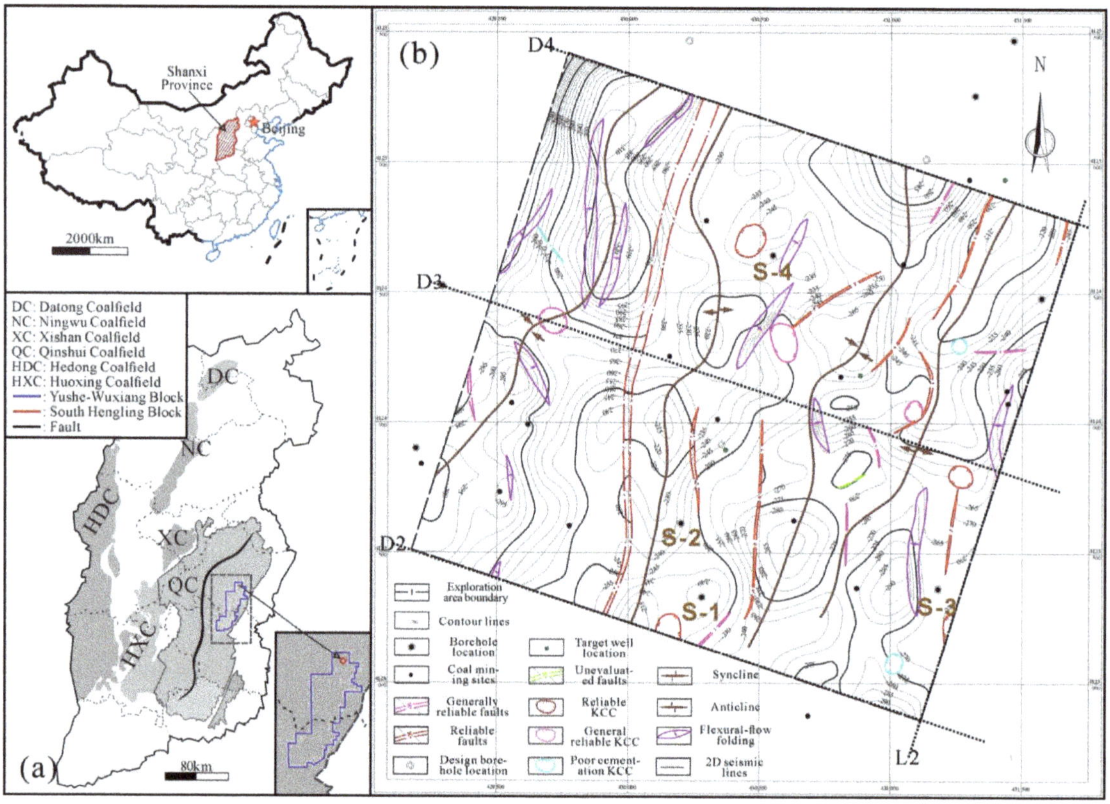

Figure 1. Location of Shanxi Province in China and the South Hengling Block of Qinshui Basin (**a**), and the elevation plan of the bottom interface of the Taiyuan Formation in the study area (**b**).

Table 1. List of mineable coal seam characteristics.

Srata	No.	Thickness (m) Min-Max / Average	Distance (m) Min-Max / Average	Structure	Mineability	Lithology		Note
						Roof	Floor	
Shanxi Formation	3	0.54–2.59 / 0.90	24.8–48.5 / 37.8	Simple	Partially mineable	mudstone, sandy mudstone, siltstone, carbonaceous mudstone	sandy mudstone, mudstone	Gangue 0–1 layers
Taiyuan Formation	15	4.35–7.08 / 5.18	86.2–126.5 / 97.8	More complex	Mineable area-wide	mudstone, siltstone, coarse-grained sandstone	mudstone, sandy mudstone	Gangue 1–3 layers

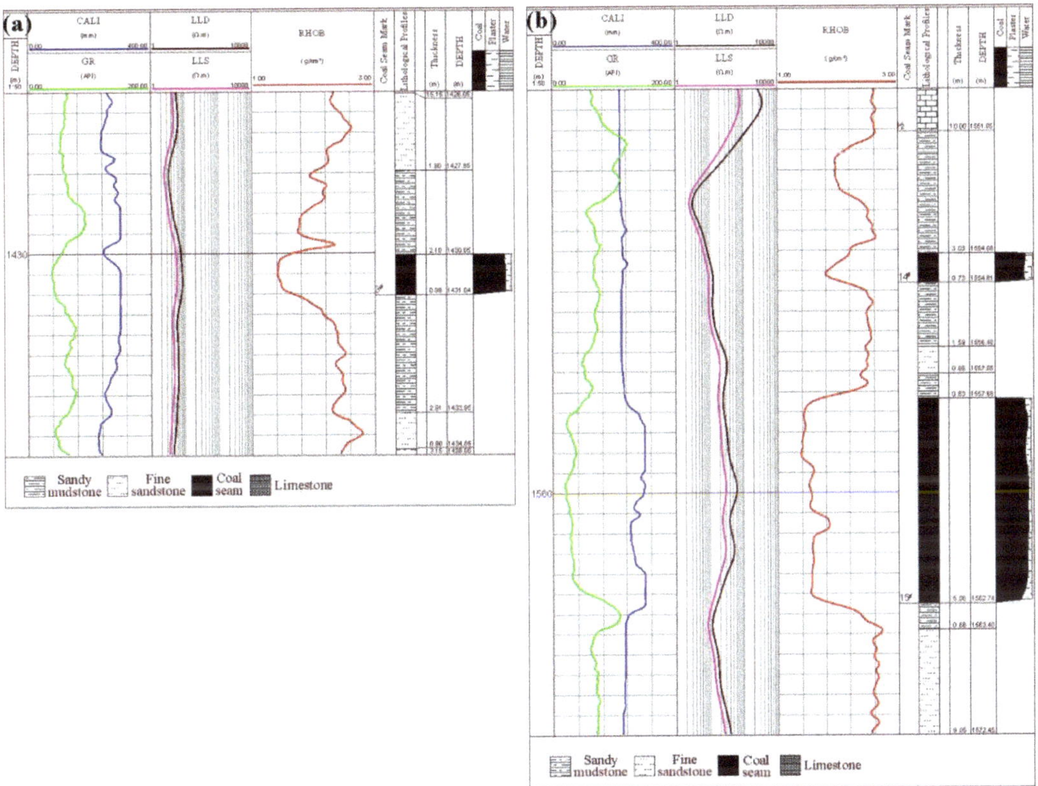

Figure 2. The 1:50 logging curves and lithologic interpretations of Coal Seam No.3 (**a**) and Coal Seam No.15 (**b**).

Figure 3 shows the typical seismic profile of a survey line in the work area. The spectrum analysis revealed that the spectrum of seismic data in the study area decreased with increasing depth, and the spectrum showed a double-peaked phenomenon. The curves with different colors in the figure represent five layer sequence interfaces, from top to bottom, K10, T3, T8, T11 and T15 interfaces, and this paper only focuses on the study of the T3 and T15 layer sequence interfaces.

The T3 section is the Shanxi Group stratum where the No. 3 coal seam was located, with a main frequency of about 45 Hz and a seismic resolution of about 18 m. There are obvious differences in wave impedance between the No. 3 coal seam and its top and bottom plate of surrounding rocks, forming a strong energy, prominent waveform and continuous traceable reflection wave. The section T15 stratum is the Taiyuan Group stratum where the No. 15 coal seam was located. The main frequency reduced to 25 Hz and the highest resolution of inversion was about 25 m. The weathering crust at the top interface of this section of coal seam was in unconformity contact with the overlying strata, forming a weak reflection wave energy and poor continuity, and most of the profiles present a composite wave pattern with poor continuity. In summary, the geological conditions in this area are relatively complex.

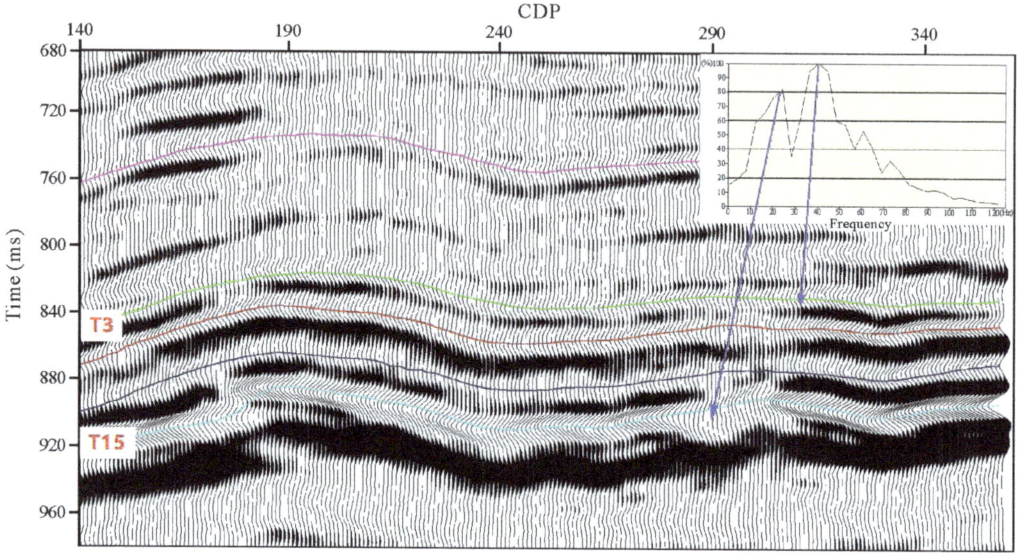

Figure 3. A typical common depth point (CDP) seismic profile of a survey line in the work area. The blue arrows represent the location of the area where the double-peaked phenomenon corresponds to the two sets of coal seams. The curves with different colors represent five layer sequence interfaces, from top to bottom, K10, T3, T8, T11 and T15 interfaces.

3. Methods

3.1. Logging the Multi-Parameter Interpretation Model

According to the description of the Chinese national standard GB/T 30050-2013 document [24], the classification of coal structure is roughly divided into four types of structures, including primary, cataclastic, fragmentation and mylonitized structures. This criterion has been combined with GSI in order to quantify the strength of the indicator factors for different structures [25]. The industrial consensus is to use wireline logs to extract the physical parameters of the coal body in order to effectively quantify the coal structure [26,27]. The method relies on multi-parameter logging datasets and lots of experience accumulated by engineers, but the non-uniqueness of seismic attribute inversion makes it difficult to apply to 3D seismic prediction work [28]. By analyzing the descriptive information of the drilled coal cores in the geological survey report for the well fields that have been surveyed, information about the relevant coal structure can be mined and the type of coal structure can be judged based on the modified quantitative GSI chart.

It can be seen from the petrophysical analysis of the two sets of coal seams in the concentrated area (Figure 4) that the density, caliper log and deep laterolog values of the coal seams are low, but due to the damage suffered by the coal structure, which makes the coal gangue entrained with some radioactive material, it may have abnormally high values of gamma and spontaneous potential. To some extent, the petrophysical quantity plates can provide insight into the statistical distribution law related to the coal seam. By referring to information from the neighboring areas [29], it was determined that the following logging multi-parameter prediction model for coal structure was established using gamma, density, deep laterolog, interval transit time and caliper curves:

$$GSI = 0.17Gr + 105rho - 1.9\ln(Rd) - 0.12AC - 0.25CAL - 61 \quad (1)$$

where GSI is the structural GSI value of the coal body, Gr is the gamma value, rho is the bulk density, Rd is the resistivity, AC is the interval transit time and CAL is the caliper curves. By establishing an empirical formulation of the regression equation shown in Equation (1), the rich multi-parameter logging data is converted into a descriptive and quantitative characterization of the coal seam. As shown in the Figure 5, the true and fitted values match well, and the multi-parameter prediction model yields highly accurate predictions and enables a better understanding of the characteristics and trends of the data. Therefore, the GSI values obtained from this model will be used as the output data of the fuzzy training system in the next subsection.

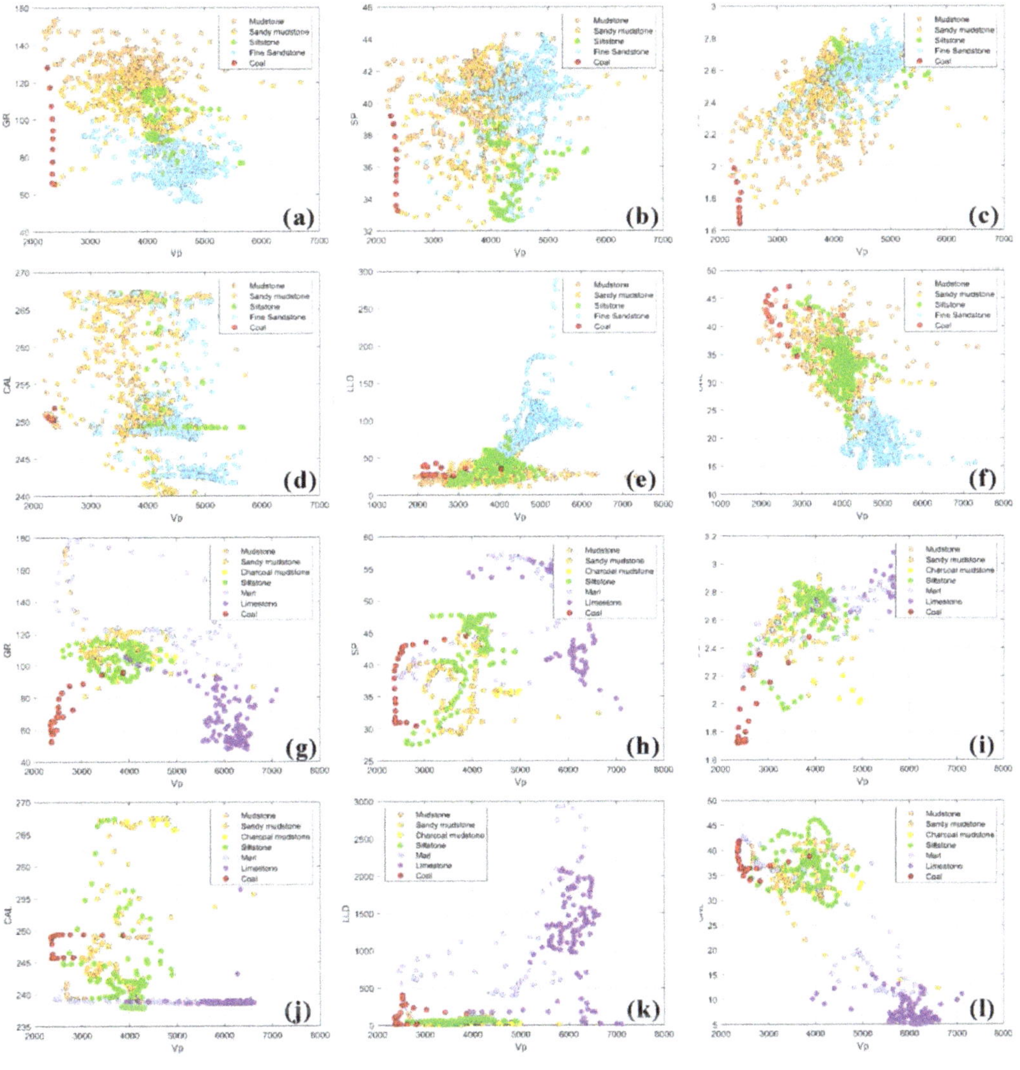

Figure 4. Cross plots of P-wave velocity with GR: gamma (**a**,**g**), SP: spontaneous potential (**b**,**h**), DEN: density (**c**,**i**), CAL: caliper log (**d**,**j**), LLD: deep laterolog (**e**,**k**) and CNL: compensated neutron log (**f**,**l**) in the petrophysical analysis, where (**a**–**f**) belong to coal seam No. 3 and (**g**–**l**) belong to coal seam No. 15.

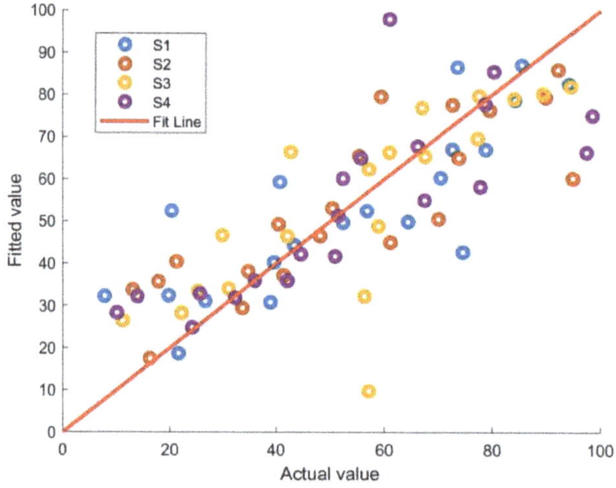

Figure 5. Comparison of the actual GSI value and the fitted GSI value.

The comprehensive evaluation curve of the coal structure on the well was obtained by fitting the GSI of the characteristic parameters from the logging data. It can be seen from Figure 6 that the GSI values of the shallow coal body structure are generally larger than 50, which belong to the primary structure and fractured structure. The GSI value of the deep-seated coal structure is generally less than 30, which belongs to the fractured structure and vesicular structure. This is related to the changes in coal by geological conditions during the formation process. With an increase in depth, the composition, structure and properties of coal will also change. Specifically, as the depth of the coal seam increases, factors such as pressure, temperature, and pore water chemical environment will change, and these changes will cause physical and chemical reactions in the coal, such as compaction, metamorphism and cracking of the coal. These reactions can cause the organic matter, minerals and pore structure of coal to change and form a more complex structure. In addition, deeper coal seams are usually subject to higher ground and tectonic stresses, which can also lead to fracture and deformation of the coal and the formation of more complex structures.

It should be noted that the establishment of the empirical formula needs to consider the influence of several factors, such as coal seam characteristics, geological conditions and logging instrument performance. The reliability and accuracy of the empirical formula are also affected by the amount of sample data, feature selection and model training, so it is necessary to focus on data quality and model tuning during the process of establishing the empirical formula. Furthermore, the GSI values obtained from this polynomial model can be used as the output training set for the fuzzy logic system.

3.2. Construction of Geological Strength Index (GSI) with Type-2 Fuzzy Logic

The popularity of the early GSI was to build a bridge of communication between the experimentally measured strength parameters of rock masses and the deformation characteristics of rock outcrops [30]. It has been shown that the GSI chart can quantify the degree of tectonic deformation as a quantitative classification indicator, which gives measurability to descriptive language that relies on visual methods to obtain words with unclear meanings, such as bulky, well-interlaced, partially disturbed, etc. Within the limits of the GSI chart, this artificial deblurring process will yield results with little variation, depending on the understanding and experience of different people. However, the artificial deblurrer is not universally applicable for seismic wavefield signals from non-outcropping rock blocks. The research and application of type-one fuzzy logic in the mathematical manipulation of perception began as early as 1965 [31]. IT2-FLS, as an extension of type-one and generalized type-two fuzzy logic, can add additional fuzzy set freedom and training performance, and is currently the most widely used system. The IT2-FLS consists roughly of five parts (Figure 7a), of which the clear input–output data pairs are:

$$(x_i; y), i = 1, 2, \ldots, N, \tag{2}$$

where $x_i = (x_1, x_2, \cdots, x_n) \in \mathbb{R}^n$ and $y \in \mathbb{R}$, N represents the number of categories at the input and n represents the number of samples in each category. Equation (2) represents the multiple-input single-output model.

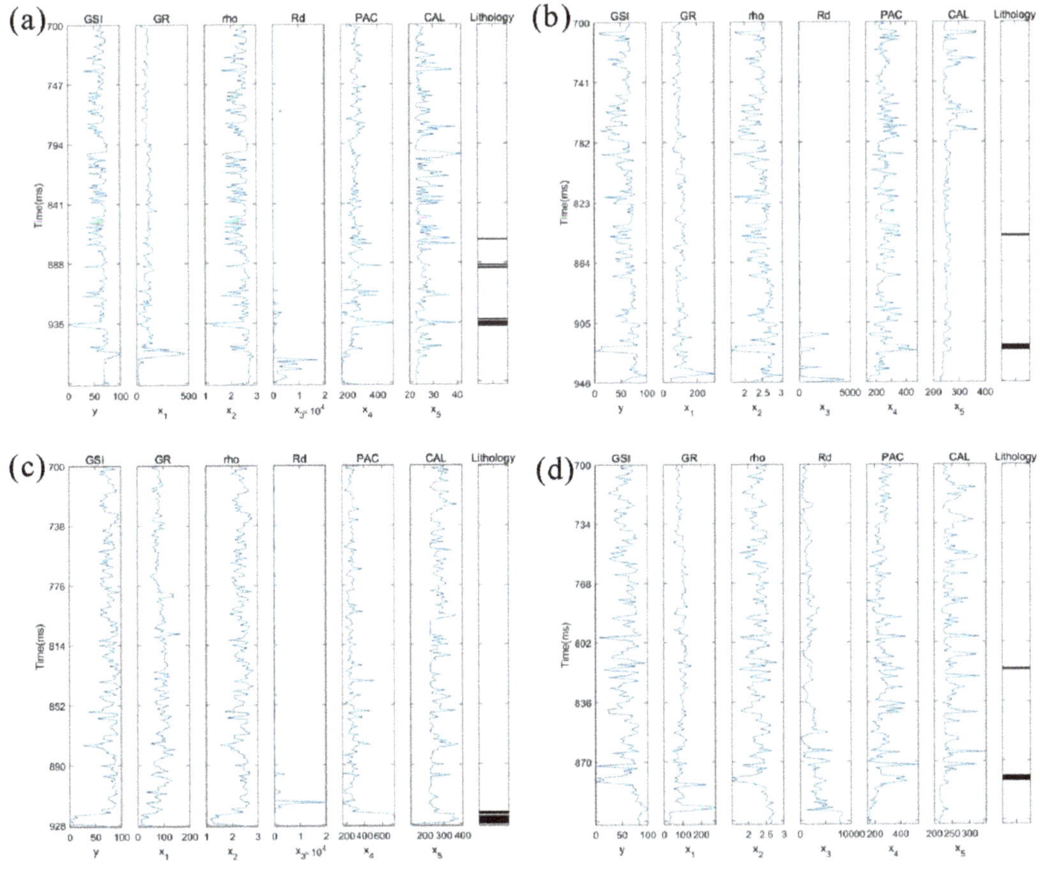

Figure 6. (**a–d**) Schematic diagram of the predicted GSI and other characteristic parameters for the logging of four wells in the work area.

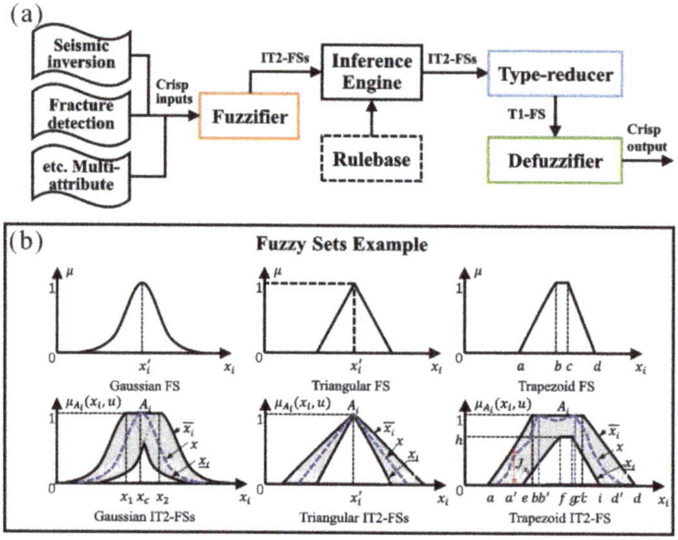

Figure 7. Structural design of IT2-FLS (**a**) with the membership functions design (**b**). Taking the Trapezoid IT2-FS as an example, nine points (a, b, c, d, e, f, g, i, h) can be used to represent an IT2 FS, the blue dashed line is the center of gravity of the IT2 FS, $\overline{x}$ represents the upper FS limit, $\underline{x}$ represents the lower FS limit, and the gray part sandwiched by the upper and lower limits is the footprint of uncertainty (FOU).

A fuzzifier is a set of membership functions (MFs) that maps the input variable $\boldsymbol{x}_i$ between 0 and 1, denoted as:

$$A_i = \int_{x_i \in D_{A_i}} \int_{u \in J_{x_i} \subseteq [0,1]} \mu_{A_i}(x_i, u) / (x_i, u), \tag{3}$$

where A_i is a fuzzy set with a membership level of $\mu_{A_i}(x_i, u)$. $u \in [0,1]$ is called a secondary membership variable with a definition domain $J_{x_i} = [\mu_{\underline{x_i}}(x_i), \mu_{\overline{x_i}}(x_i)] \subseteq [0,1]$, which is called the footprint of uncertainty (FOU) of A_i (Figure 4b). The width size of FOU is directly related to the imparted uncertainty [32], whereas $\overline{x_i}$ and $\underline{x_i}$ represent the upper and lower bounds of the data set, respectively. There are three types of MFs commonly used in IT2-FLS (Figure 7b) and the choice of MFs is not rigid and often depends on the actual application of the fuzzy system being implemented. For example, Gaussian MFs with wider coverage are often chosen for nonlinear signal response approximation simulations, whereas linear MFs are more suitable for applications such as data space quantization and pattern recognition classification.

The role of the rule base is to extract the influence of the rule between the output variable (y) and the input variable (x_i). The Mamdani rule was the first rule proposed [33], and the Takagi–Sugeno–Kang (TSK) model was transformed into multiple linear subsystem fits based on Mamdani [34]. These two IF–THEN rules are expressed as follows:

$$Mamdani: \text{ IF } x_{i1} \text{ is } A_{i1}^{(l)} \text{ and } \ldots \text{ and } x_{in} \text{ is } A_{in}^{(l)}, \text{ THEN } y \text{ is } B^{(l)} \tag{4}$$

$$TSK: \text{ IF } x_{i1} \text{ is } A_{i1}^{(l)} \text{ and } \ldots \text{ and } x_{in} \text{ is } A_{in}^{(l)}, \text{ THEN } y \text{ is } \boldsymbol{p}^{(l)^T}\boldsymbol{x} + p_0^l \tag{5}$$

where $l = 1, 2, \ldots, M$ denotes the rule index number, M represents the maximum number of rules, $A_{ij}^{(l)}$, $B^{(l)}$ and p_0^l are all interval fuzzy sets, $B^{(l)} = \left[\underline{b}^{(l)}, \overline{b}^{(l)}\right]$ denotes the centroid of

Mamdani rule, $\boldsymbol{p}^{(l)} = \left(p_1^{(l)}, p_2^{(l)}, \cdots, p_n^{(l)}\right)^T$ is the weight factor of the TSK rule and, when $\boldsymbol{p}^{(l)} = (0, 0, \cdots, 0)^T$, the TSK rule is equivalent to the Mamdani rule. For the convenience of calculation, the Mamdani rule was chosen in this paper.

The inference engine is designed to calculate the firing interval of the lth rule, divided into:

$$\boldsymbol{B}^{(l)} = \left[\mu_{\underline{x}_1^{(l)}}(x_1) \times_T \cdots \times_T \mu_{\underline{x}_n^{(l)}}(x_n), \; \mu_{\overline{x}_1^{(l)}}(x_1) \times_T \cdots \times_T \mu_{\overline{x}_n^{(l)}}(x_n)\right] \equiv \left[\underline{b}^{(l)}, \overline{b}^{(l)}\right] \quad (6)$$

where $\times_T$ is the minimum t-norm operation, or:

$$\boldsymbol{B}^{(l)} = \left[\mu_{\underline{x}_1^{(l)}}(x_1) *_T \cdots *_T \mu_{\underline{x}_n^{(l)}}(x_n), \mu_{\overline{x}_1^{(l)}}(x_1) *_T \cdots *_T \mu_{\overline{x}_n^{(l)}}(x_n)\right] \equiv \left[\underline{b}^{(l)}, \overline{b}^{(l)}\right] \quad (7)$$

where $*_T$ is the product t-norm operation. The role of $B^{(l)}$ is to center on the weighted average of the rule output data. Both of the firing intervals have a wide range of applications, and the distinction in terms of performance is not very obvious. Therefore, either of the two inference engines can be chosen.

A type reducer is a technical module used to defuzzify the fuzzy set, and in this paper, is the central-of-sets type reducer was chosen, defined as:

$$y = \bigcup_{\substack{b^{(l)} \in B^{(l)}(x,u) \\ y \in \mathbb{R}}} \frac{\sum_{l=1}^{M} B_c^{(l*)} b^{(l)}}{\sum_{l=1}^{M} b^{(l)}} = \left[y_{left}, y_{right}\right] \quad (8)$$

where $B_c^{(l*)}$ is the center of the output fuzzy set $B^{(l)}$, determined according to the following definition:

$$\prod_{j=1}^{m} \mu_{A_{ij}^{(l*)}}(x_{ij}) \geq \prod_{j=1}^{m} \mu_{A_{ij}^{(l)}}(x_{ij}) \quad (9)$$

Here, y_{left} and y_{right} are defined as:

$$y_{left} = \frac{\sum_{l=1}^{L} B_c^{(l*)} \overline{b}^{(l)} + \sum_{l=L+1}^{M} B_c^{(l*)} \underline{b}^{(l)}}{\sum_{l=1}^{L} \overline{b}^{(l)} + \sum_{l=L+1}^{M} \underline{b}^{(l)}} \quad (10)$$

$$y_{right} = \frac{\sum_{l=1}^{R} B_c^{(l*)} \underline{b}^{(l)} + \sum_{l=R+1}^{M} B_c^{(l*)} \overline{b}^{(l)}}{\sum_{l=1}^{R} \underline{b}^{(l)} + \sum_{l=R+1}^{M} \overline{b}^{(l)}} \quad (11)$$

Finally, the defuzzifier was used to obtain a clear output:

$$y = \frac{y_{left} + y_{right}}{2} \quad (12)$$

IT2-FLS can replace the textual description language in the traditional GSI system chart when considering the combinations of different MFs (Figure 8). This study benefits from the availability of a free open-source MATLAB/Simulink toolbox developed by Taskin et al. (2015) [35], which makes it possible to extend the findings of this study to practical applications.

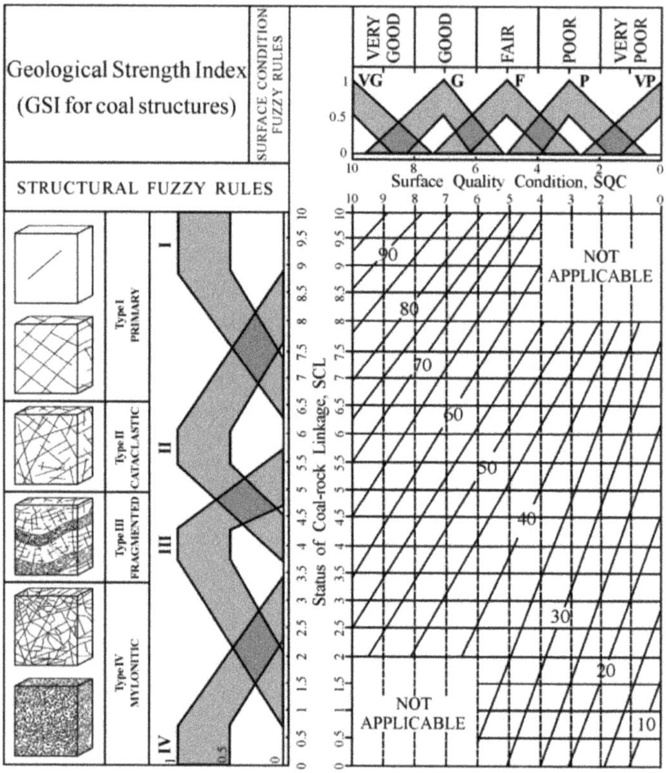

Figure 8. GSI system chart with IT2-FLS. "NOT APPLICABLE" means that it is not applicable in this range.

As shown in Figure 8, the process of embedding an aggregated fuzzy system into a GSI system first requires the identification of the parameters used to calculate the GSI, including the lithology parameters and the fracture development degree parameters. Then, each parameter should be converted into linguistic variables by fuzzification techniques in order to obtain a fuzzy set for each parameter. Next, all the fuzzy sets need to be input into the aggregated fuzzy system for processing in order to calculate the final geological strength index. This process requires the use of hierarchical analysis to determine the weights of each input parameter. The output of the aggregated fuzzy system will be mapped to the corresponding position of the GSI system in order to obtain the final GSI value. By embedding the aggregated fuzzy system into the GSI table, more accurate and reliable GSI values can be obtained, which helps to better predict the coal structure.

3.3. Multi-Attribute Data Mining Based on Fuzzy Sets

The main coal-bearing rock systems in this area are the Shanxi and Taiyuan formations, of which the Shanxi Formation is a deltaic sedimentary system, the upper part of the Taiyuan Formation is a deltaic sedimentary system and the lower part is a barrier coastal sedimentary system. The sedimentary microphases developed in the Shanxi Formation are mainly interdiversion depressions, diversion channels and peat bogs, etc. The sedimentary microphases developed in the upper part of the Taiyuan Formation are mainly submerged diversion channels and interdiversion bays, and the sedimentary microphases developed in the lower part of the Taiyuan Formation are mainly barrier islands, lagoons, sand pings, mud pings, mixed pings, open terraces and peat bogs (Figure 9), whose lithology is directly related to the preservation of CBM. However, the various logging parameters in the 3D work zone were difficult to obtain by inversion, so we needed to mine the data comprehensively within a limited number of attribute parameters.

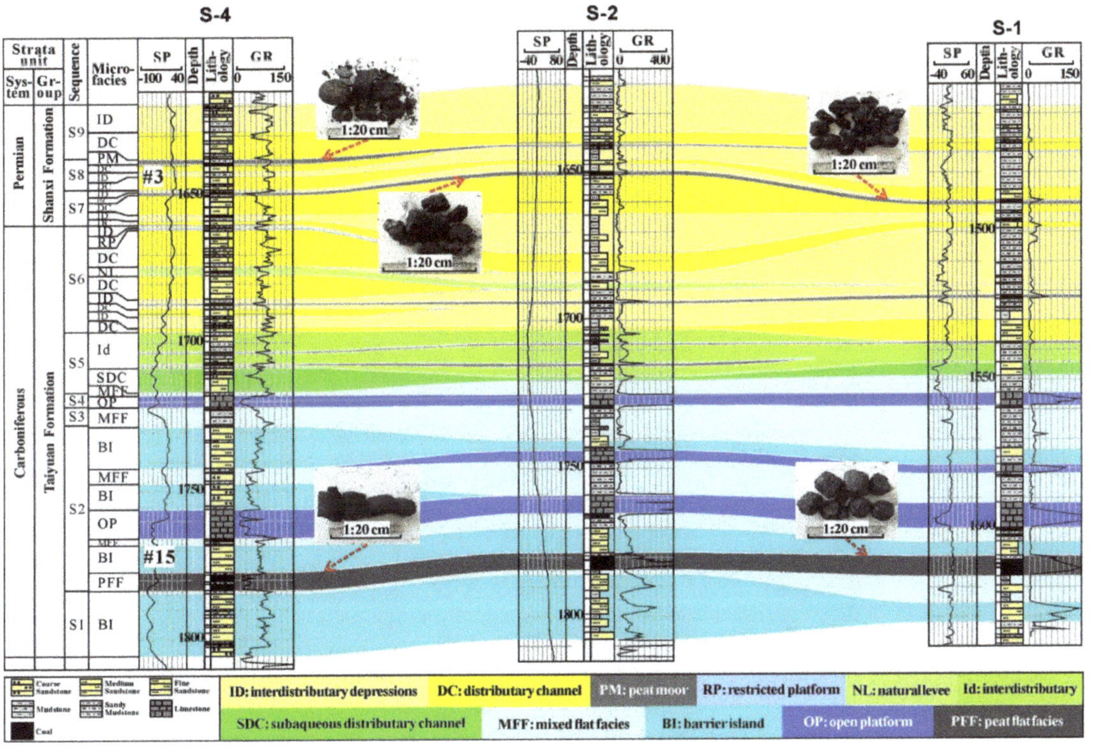

Figure 9. Profile of sedimentary connecting wells of coal samples from coalbed methane blocks in Hengling.

In this study, a seismic phase-controlled nonlinear chaos inversion technique [10] was used, which has great advantages for the identification of thin coal seams. As shown in Figure 10a, a comparison of the post-stack inversion profile of well S-2 with a logging curve shows that the reservoir characteristics of the profile were obvious, with low-velocity coal seams (dark blue), high-velocity limestone (red), medium- to high-velocity sand (yellow) and mud shale (light blue), and the inversion profile is very effective in identifying lithology. The pre-stack inversion profile (Figure 10b) on the S-2 well corresponded well to the gas-bearing properties, with high values of full hydrocarbon detection. The two 3D data volumes were entered into the IT2-FLS as rock block characterization parameters (Figure 11b,c).

We also calculated the coherent and curvature data volumes for the coal-rock linkage in the GSI chart (Figure 11d,e). In order to fully employ the wide azimuthal stacked forepath set data obtained from the Offset Vector Tile (OVT) domain processing, we applied the optimal surface voting crack detection technique of Wu et al. (2018) [36] to the azimuthal seismic information because the azimuthal information added by OVT helps in the extraction of the 3D spatial seed points. As shown in Figure 12d, we can observe the fracture growth with six different elements of orientation information, and we can combine the multi-directional fracture characteristics (Figure 12a–c) to obtain the integrated fracture parameter of the OVT domain (Figure 11f). This new fracture attribute parameter is more recognizable than the traditional fracture attribute parameter feature, so we input this parameter into the IT2-FLS as a comprehensive indicator of the current status of coal-rock linkage (SCL).

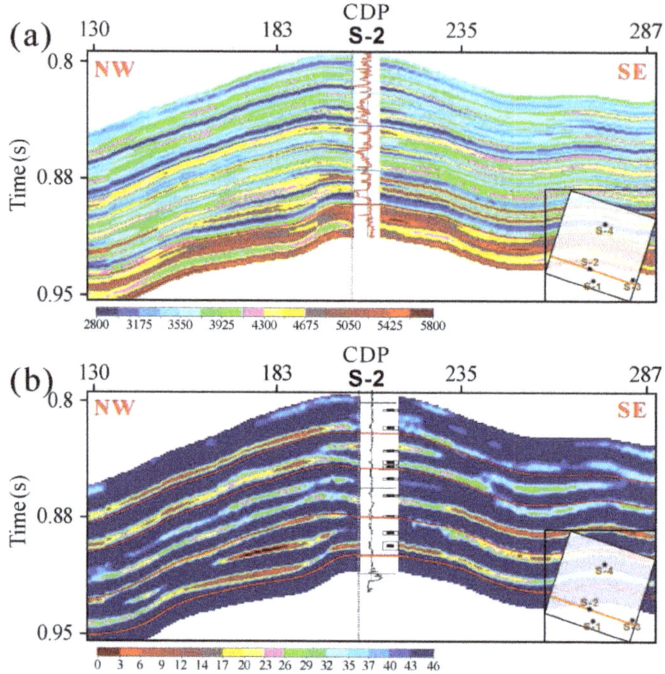

Figure 10. (**a**) Post-stack nonlinear impedance inversion profile and (**b**) Lamé coefficient multiplied density profile. The logging curves in the figures are the P-wave velocity and hydrocarbon detection curves, respectively. The graphical identification is coal rock. The locations of the post-stack and pre-stack inversion profiles are shown in the lower right corner, respectively.

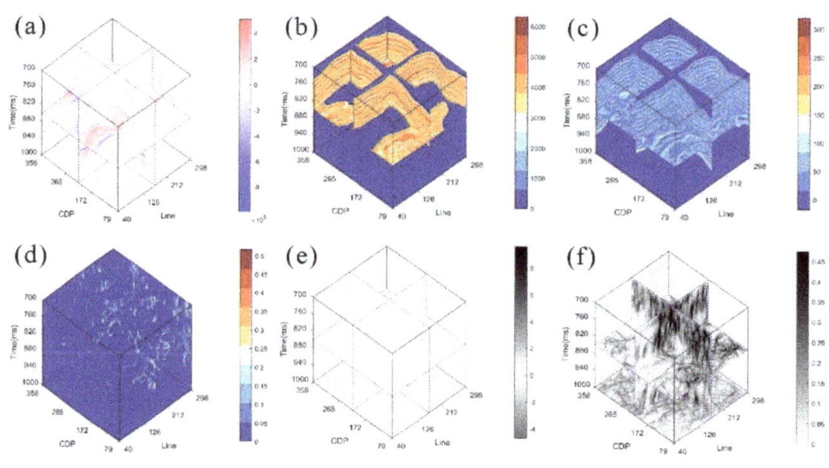

Figure 11. (**a**) Seismic pure wave volume, (**b**) post-stack impedance volume, (**c**) pre-stack Larmé coefficient times density volume, (**d**) coherence volume, (**e**) curvature volume and (**f**) optimal surface voting volume for OVT domains with integrated multi-directional information.

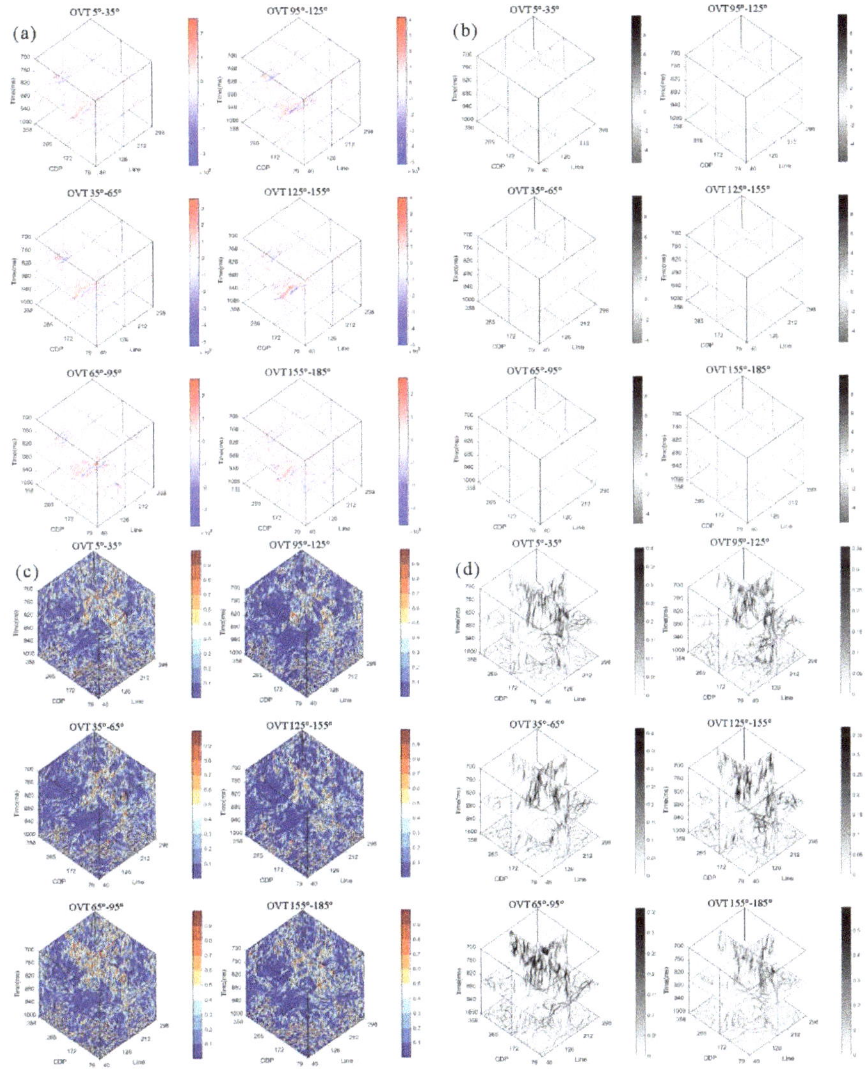

Figure 12. (**a**) Seismic pure wave volumes, (**b**) curvature volumes, (**c**) coherent volumes and (**d**) optimal surface voting volumes for six orientations in the OVT domain.

Finally, we combine the surface quality conditions (SQCs) in the GSI table of IT2-FLS with inverse physical parameters, such as p-wave velocity, density and the Lamé coefficient, and the current SCL with seismic property indices, such as the seismic coherence cube, curvature and OVT domain optimal voting volume, etc. Combined with our current published methods [37], we can achieve multi-attribute 3D data mining of target lithologies. The method is inspired by seismic data collected with drilling logs, and extracts annotation information and interpretation results along the well trajectory, which effectively solves the most time-consuming process of creating and collecting labels in a huge amount of seismic data.

4. Application of Fuzzy Logic Multi-Attribute Fusion

Considering the coal structural system, the task of this study was to investigate the structural spreading of the coal mass in the target area using a multi-attribute 3D seismic fusion method. The fuzzy output y is the multi-parameter predictive coal structure model of the logging wells calculated by regression Equation (1), and the input (x) is the factors that may affect the coal structure. We selected five attributes ($x_i = (x_1, \cdots, x_5)$), including seismic impedance, Larmé×density, Vp/Vs ratio, curvature and OVT optimal surface voting volumes, and obtained the input data pairs ($(x_i; y), i = 1, 2, \ldots, 5$) by automatically acquiring the coal rock locations of each data set by the program. In order to show the 3D structure of coal rocks, we only kept the parameters related to coal rocks for each group of attributes (Figure 10). The structure of the inversion volume related to lithology is shown in Figure 13a–c, and the development of multi-directional fractures is shown in Figure 13d–f.

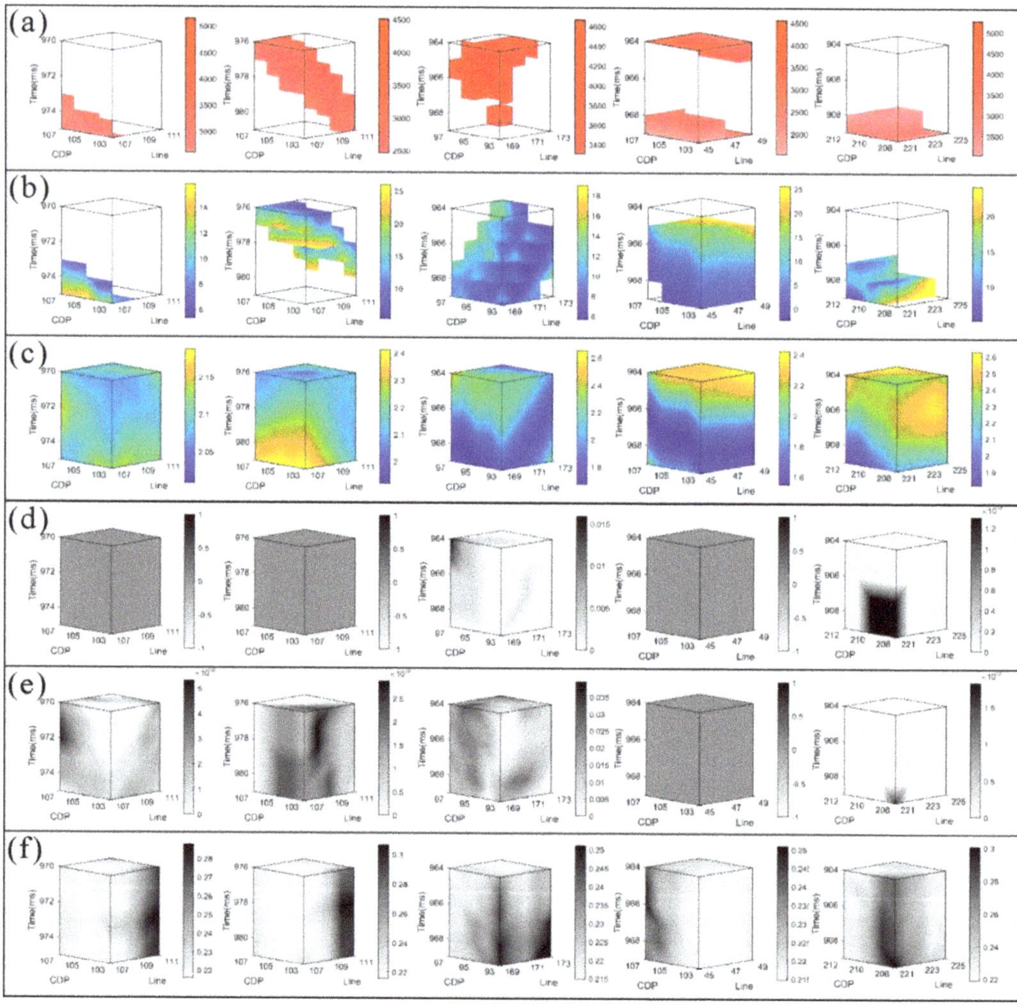

Figure 13. Multi-attribute presentation of coal structure based on data mining. These are the post-stack impedance volume (**a**), pre-stack Larmé×density volume (**b**), pre-stack Vp/Vs volume (**c**), coherent volume (**d**), curvature volume (**e**) and OVT optimal surface voting volume (**f**), respectively.

According to the IT2-FLS-GSI chart, we covered five fuzzy sets (A_i^5) in each inverse body attribute and four fuzzy sets (A_i^4) in the crack detection attribute, and used the collected data corresponding to the training set (D_{train}) to obtain the complete fuzzy rule base. The fuzzy sets ($B_c^{(l)}, l \in \{1, \cdots, K\}$) only need to be selected from the centers of the predefined fuzzy sets ($\{B^1, \ldots, B^K\}$). The defuzzified GSI prediction model (y) was obtained (Figure 14), which can be quantified to represent the GSI values of the input data pairs (x). As seen in Figure 14a, the multi-attribute fuzzy fusion GSI model and the logging multi-parameter GSI model both have the same prediction trend and smoother output, whereas the error size of both is maintained between ±20. Adding or subtracting rules or changing the MFs will have some effect on the error, and achieving accurate results requires constant fine-tuning and pruning of the model. Based on the rule base, we plotted the visual appearance of the fuzzy prediction model. Both are generally consistent in appearance, but the results from the secondary tests show that the continuous fuzzy prediction model (Figure 14b) is more accurate than the piecewise fuzzy prediction model (Figure 14c). After applying to the whole working area, Figure 15a shows the predicted structure of coal mass No.3, with a higher content of primary coal in the southeast and west. Figure 15b shows the predicted structural section of coal mass No.15. It can be seen that the central and northwestern parts of the study area have a lower GSI, more tectonic coal content and softer coal quality.

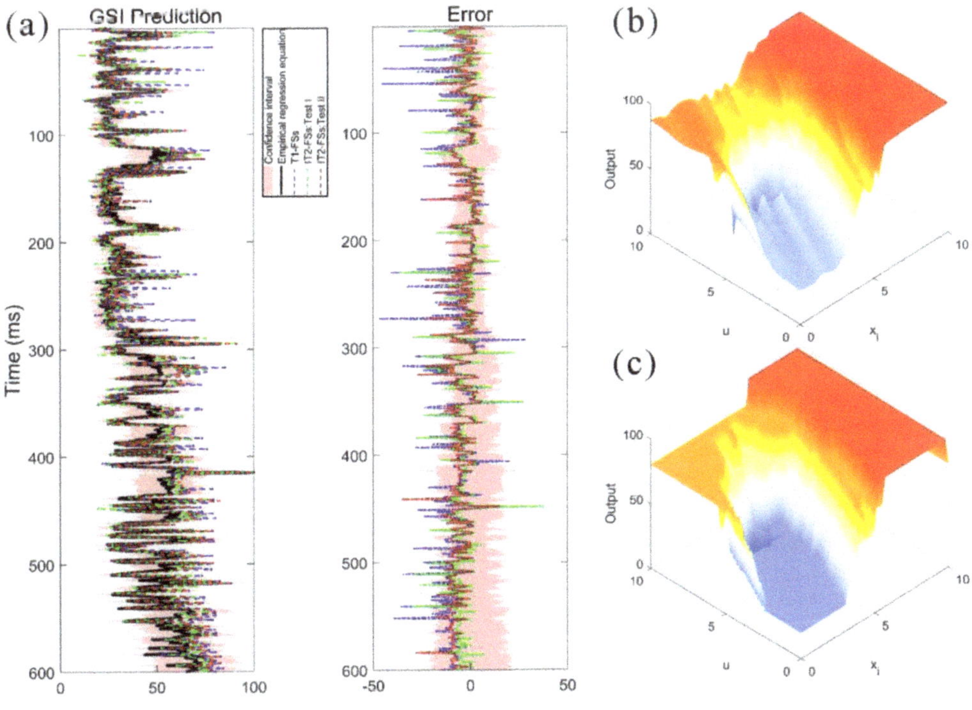

Figure 14. (**a**) GSI and error analysis of fuzzy system prediction, where the pink area represents the confidence interval, the black line is the GSI calculated by the logging multi-parameter regression equation, the black dashed line is the T1-FS prediction result and the green and red dashed lines are the results of two trials of piecewise and continuous types of IT2-FS, respectively; (**b**) continuous fuzzy prediction model; and (**c**) piecewise fuzzy prediction model.

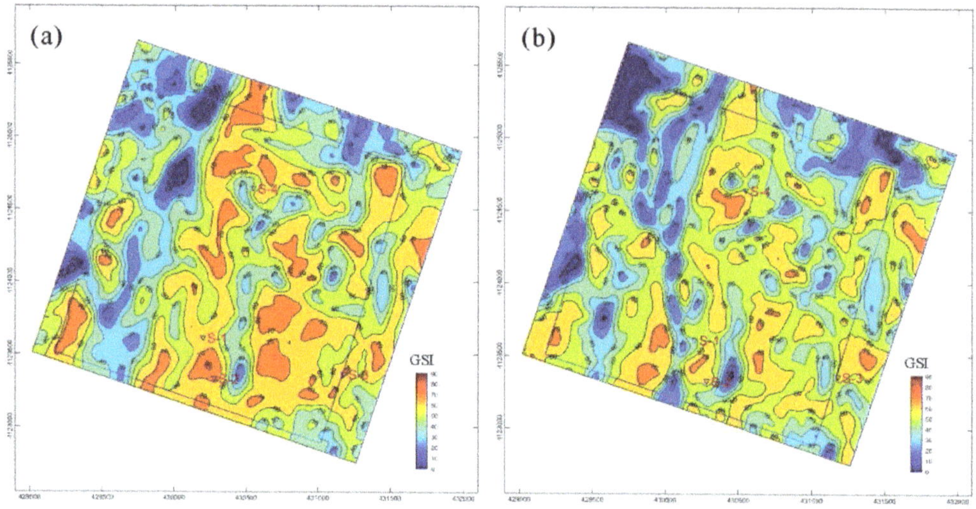

Figure 15. Predicted section of coal structure from coal seam No. 3 (**a**) and coal seam No. 15 (**b**).

5. Discussion

In this paper, a type-two fuzzy logic prediction model was proposed to compare with the coal structure interpreted using logging data. The overall coal structure prediction results of the study area were basically consistent with the core-taking description, with a matching degree of more than 75%. According to the description of the coal structure of the actual coal core, it was found that the northwestern part of coal seam No.15 in the study area had a lower GSI, which indicates more tectonic coal content and softer coal quality, whereas the eastern part had a higher GSI value, indicating more primary structural coal content and harder coal quality. Coal seam No. 3 was mainly dominated by crushed grain coal, followed by fractured coal, in which the GSI values were higher in the southeast and west, and the content of primary coal was higher. The predicted results of this study can be used as the basis for further exploratory work in this area.

From the predicted results in Figure 15, in connection with the geological structure of the work area in Figure 1b, it can be seen that there was a certain correspondence between the location of large faults in the geological structure and the high value of GSI. Generally speaking, the activity of the fault will lead to fragmentation and deformation of the surrounding rocks, and a certain stress field will also be formed. All of these factors will have an impact on the mechanical properties of the surrounding rocks. Therefore, when the area with high GSI values happens to be located near a large fault, this may indicate that there is a high concentration of stress in the area, which leads to an increase in the mechanical strength index of the rock. Therefore, the influence of geological formations on high GSI values should be fully considered when planning and designing projects in order to reduce the risk of possible geological hazards.

Different types of coal structures affect the development of CBM through pore structure differences. Xu et al. (2019) [38] concluded that fractured-fragmented and primary-fragmented coals are favorable for gas adsorption percolation, but the former has more developed adsorption pores, whereas fractured-fragmented coals are subjected to strong tectonic stress, which leads to the destruction of pore structure and are thus unfavorable for CBM storage and transportation. The results of the analysis of the coal structure of coal seam No. 3 and coal seam No. 15 in this area show that fractured coal and fractured grain coal are mainly developed in this area, which is presumed to be favorable for the production of coalbed methane.

However, this study is based on limited sample data and does not take into account possible spatial heterogeneity. Therefore, future research should use more data and more samples in order to verify these predicted results, and should consider the impact of spatial heterogeneity.

6. Conclusions

This paper introduces a multi-attribute fusion coal structure prediction method based on a type-two fuzzy logic inference system, and combines the quantification capability of GSI, which can be useful for predicting semantic uncertainty. The multi-attribute data source in the fuzzy logic system inherits the content of the previous related results, which improves the accuracy of the data source and the convenience of the data mining process. We also make full use of the multi-directional nature of the seismic data in the OVT domain in order to achieve the fine portrayal of fracture detection, and these datasets provide strong support for fuzzy logic-based multi-attribute fusion. Through practical applications in the study area, we focused on the effective use of data, such as 3D inversion data volumes and seismic attribute volumes, thus solving the problem that conventional prediction methods cannot be carried out in well-free areas and achieving quantitative discrimination of coal structure. This study is beneficial for further work on the reservoir evaluation of CBM. For example, it is important to further investigate the relationship between coal structure and permeability, gas content and formation pressure.

Author Contributions: Conceptualization, X.C. and Y.T.; methodology, X.C. and Y.T.; software, X.C. and H.H.; validation, X.C., Y.T., H.H. and L.W.; formal analysis, X.C., H.H. and L.W.; investigation, X.C., J.W., Z.G. and C.M.; resources, X.C. and H.H.; data curation, X.C., J.W., Z.G. and C.M.; writing—original draft preparation, X.C.; writing—review and editing, X.C., Y.T., H.H. and L.W.; visualization, X.C.; supervision, M.S.; project administration, Y.T., H.H. and L.W.; funding acquisition, H.H. and M.S. All authors have read and agreed to the published version of the manuscript.

Funding: The authors were grateful to receive support from the National Natural Science Foundation of China (No. 41974124).

Data Availability Statement: The original contributions presented in this study are included in the article. Further inquiries can be directed to the corresponding author.

Acknowledgments: We sincerely thank Ahmet Taskin and Tufan Kumbasar for providing access to the IT2-FLS MATLAB/SIMULINK toolbox.

Conflicts of Interest: The authors declare that the research was conducted in the absence of any commercial or financial relationships that could be construed as a potential conflict of interest.

References

1. Li, L.; Liu, D.; Cai, Y.; Wang, Y.; Jia, Q. Coal structure and its implications for coalbed methane exploitation: A review. *Energy Fuels* **2020**, *35*, 86–110. [CrossRef]
2. Huang, B.; Qin, Y.; Zhang, W.; Wang, G. Identification of the coal structure and prediction of the fracturability in the No. 8 coal reservoir, Gujiao block, China. *Energy Explor. Exploit.* **2018**, *36*, 204–229. [CrossRef]
3. Lv, F.; Gao, W. A prediction model of coal structure based on logging parameters in Liupanshui Coalfield, Guizhou, China. *Arab. J. Geosci.* **2021**, *14*, 1–18. [CrossRef]
4. Wang, Y.Z.; Mao, C.; Liu, J.L. Pore Characteristics of Coal Reservoir in Jixi Basin. In Proceedings of the International Field Exploration and Development Conference, Chengdu, China,, 16–18 October 2019; Springer: Singapore, 2020; pp. 842–851. [CrossRef]
5. Shi, J.; Zeng, L.; Dong, S.; Wang, J.; Zhang, Y. Identification of coal structures using geophysical logging data in Qinshui Basin, China: Investigation by kernel Fisher discriminant analysis. *Int. J. Coal Geol.* **2020**, *217*, 103314. [CrossRef]
6. Mou, J.; Liu, H.; Zou, Y.; Li, Q. A new method to determine the sensitivity of coal and gas outburst prediction index. *Arab. J. Geosci.* **2020**, *13*, 465. [CrossRef]
7. Zhang, J.; Zhang, Y.; Wang, J.; Wen, X. Prediction of gas pressure in thin coal seams in the Qinglong Coal Mine in Guizhou Province, China. *J. Pet. Explor. Prod. Technol.* **2021**, *11*, 4075–4086. [CrossRef]
8. Qiao, W.; Li, W.; Zhang, X.; Niu, Y.; Chen, Y.; Wang, Y.; Xing, T. Prediction of floor water disasters based on fractal analysis of geologic structure and vulnerability index method for deep coal mining in the Yanzhou mining area. *Geomat. Nat. Hazards Risk* **2019**, *10*, 1306–1326. [CrossRef]

9. Zhang, S.; Huang, H.; Dong, Y.; Yang, X.; Wang, C.; Luo, Y. Direct estimation of the fluid properties and brittleness via elastic impedance inversion for predicting sweet spots and the fracturing area in the unconventional reservoir. *J. Nat. Gas Sci. Eng.* **2017**, *45*, 415–427. [CrossRef]
10. Huang, H.; Yuan, S.; Zhang, Y.; Zeng, J.; Mu, W. Use of nonlinear chaos inversion in predicting deep thin lithologic hydrocarbon reservoirs: A case study from the Tazhong oil field of the Tarim Basin, China. *Geophysics* **2016**, *81*, B221–B234. [CrossRef]
11. Yang, Z.; Li, Z.; Zhu, J.; Yi, H.; Hu, J.; Feng, G. Deriving dynamic subsidence of coal mining areas using InSAR and logistic model. *Remote Sens.* **2017**, *9*, 125. [CrossRef]
12. Ma, J.; Yang, J.; Zhu, Z.; Cao, H.; Li, S.; Du, X. Decision-making fusion of InSAR technology and offset tracking to study the deformation of large gradients in mining areas-Xuemiaotan mine as an example. *Front. Earth Sci.* **2022**, *10*, 962362. [CrossRef]
13. Hoek, E.; Brown, E.T. Practical estimates of rock mass strength. *Int. J. Rock Mech. Min. Sci.* **1997**, *34*, 1165–1186. [CrossRef]
14. Marinos, P.; Hoek, E. GSI: A geologically friendly tool for rock mass strength estimation. In Proceedings of the ISRM International Symposium, Melbourne, VIC, Australia, 19–24 November 2000; OnePetro, 2000.
15. Zhang, Y.J.; Cao, W.G.; Zhao, M.H. Application of fuzzy sets to geological strength index (GSI) system used in rock slope. Soils and Rock Instrumentation, Behavior, and Modeling, GeoHunan International Conference, Changsha, China, 3–6 August 2009; American Society of Civil Engineers: Reston, VA, USA, 3–6 August; pp. 30–35. [CrossRef]
16. Bashari, A.; Beiki, M.; Talebinejad, A. Estimation of deformation modulus of rock masses by using fuzzy clustering-based modeling. *Int. J. Rock Mech. Min. Sci.* **2011**, *48*, 1224–1234. [CrossRef]
17. Sonmez, H.; Gokceoglu, C.; Ulusay, R. An application of fuzzy sets to the geological strength index (GSI) system used in rock engineering. *Eng. Appl. Artif. Intell.* **2003**, *16*, 251–269. [CrossRef]
18. Liu, H.; Sang, S.; Wang, G.G.; Li, Y.; Li, M.; Liu, S. Evaluation of the synergetic gas-enrichment and higher-permeability regions for coalbed methane recovery with a fuzzy model. *Energy* **2012**, *39*, 426–439. [CrossRef]
19. Lei, Z.; Jienan, P.; Xiaomin, Z. Fuzzy comprehensive evaluation of mining geological condition in the No. 9 coal seam, Linhuan coal mine, Huaibei Coalfield, China. *Procedia Environ. Sci.* **2012**, *12*, 9–16. [CrossRef]
20. Moosavi, N.; Bagheri, M.; Nabi-Bidhendi, M.; Heidari, R. Porosity prediction using Fuzzy SVR and FCM SVR from well logs of an oil field in south of Iran. *Acta Geophys.* **2022**, *71*, 769–782. [CrossRef]
21. Woo, J.; Rhee, C.W.; Jeoung, T.J.; Jang, S. Application of Multi-Resolution Graph-based Clustering for electrofacies prediction: A case study from the Horn River Shale, British Columbia, Canada. *Geosci. J.* **2020**, *24*, 507–518. [CrossRef]
22. Fuchs, C.; Spolaor, S.; Nobile, M.S.; Kaymak, U. pyFUME: A Python package for fuzzy model estimation. In Proceedings of the IEEE International Conference on Fuzzy Systems (FUZZ-IEEE), Glasgow, UK, 19–24 July 2020; IEEE: New York, NY, USA; pp. 1–8. [CrossRef]
23. Jiang, W.; Zhang, P.; Li, D.; Li, Z.; Wang, J.; Duan, Y.; Wu, J.; Liu, N. Reservoir characteristics and gas production potential of deep coalbed methane: Insights from the no. 15 coal seam in Shouyang block, Qinshui Basin, China. *Unconv. Resour.* **2022**, *2*, 12–20. [CrossRef]
24. *GB/T 30050-2013*; Classification of Coal-Body Structure. China National Coal Association: Beijing, China, 2013.
25. Dong, Z.; Zhang, X.; Yang, W.; Lei, M.; Zhang, C.; Wan, J.; Han, L. Automatic Cutting Speed Control System of Boom-Type Roadheader Based on Geological Strength Index. *Minerals* **2022**, *12*, 1582. [CrossRef]
26. Hampson, D.P.; Schuelke, J.S.; Quirein, J.A. Use of multiattribute transforms to predict log properties from seismic data. *Geophysics* **2001**, *66*, 220–236. [CrossRef]
27. Zhang, Z.; Qin, Y.; Wang, G.; Sun, H.; You, Z.; Jin, J.; Yang, Z. Evaluation of coal body structures and their distributions by geophysical logging methods: Case study in the Laochang block, eastern Yunnan, China. *Nat. Resour. Res.* **2021**, *30*, 2225–2239. [CrossRef]
28. Wang, L.; Zhou, H.; Dai, H.; Yu, B.; Liu, W.; Wang, N. Poststack seismic inversion using a patch-based Gaussian mixture modelPatch-based GMM for poststack inversion. *Geophysics* **2021**, *86*, R685–R699. [CrossRef]
29. Liu, D.; Wang, Y.; Ni, X.; Tao, C.; Fan, J.; Wu, X.; Zhao, S. Classification of coal structure combinations and their influence on hydraulic fracturing: A case study from the Qinshui Basin, China. *Energies* **2020**, *13*, 4559. [CrossRef]
30. Hussian, S.; Mohammad, N.; Ur Rehman, Z.; Khan, N.M.; Shahzada, K.; Ali, S.; Tahir, M.; Raza, S.; Sherin, S. Review of the geological strength index (GSI) as an empirical classification and rock mass property estimation tool: Origination, modifications, applications, and limitations. *Adv. Civ. Eng.* **2020**, *2020*, 6471837. [CrossRef]
31. Zadeh, L.A. Fuzzy sets. *Inf. Control.* **1965**, *8*, 338–353. [CrossRef]
32. Mendel, J.M.; John, R.B. Type-2 fuzzy sets made simple. *IEEE Trans. Fuzzy Syst.* **2002**, *10*, 117–127. [CrossRef]
33. Coupland, S.; John, R. Geometric type-1 and type-2 fuzzy logic systems. *IEEE Trans. Fuzzy Syst.* **2007**, *15*, 3–15. [CrossRef]
34. Takagi, T.; Sugeno, M. Fuzzy identification of systems and its applications to modeling and control. *IEEE Trans. Syst. Man Cybern.* **1985**, *SMC-15*, 116–132. [CrossRef]
35. Taskin, A.; Kumbasar, T. An open source Matlab/Simulink toolbox for interval type-2 fuzzy logic systems. In Proceedings of the IEEE Symposium Series on Computational Intelligence, Cape Town, South Africa, 7–10 December 2015; IEEE: New York, NY, USA; pp. 1561–1568. [CrossRef]
36. Wu, X.; Fomel, S. Automatic fault interpretation with optimal surface voting. *Geophysics* **2018**, *83*, O67–O82. [CrossRef]

37. Cui, X.; Huang, H.; Hao, Y.; Li, L.; Luo, Y.; Zeng, J.; Tian, Z.; Shen, Y.; Hu, Y. Supervised post-stack 3D seismic data classification via multi-scale and multi-label consistent PCA-LDA. *Geophysics* **2022**, *88*, 1–98. [CrossRef]
38. Xu, X.; Meng, Z.; Wang, Y. Experimental comparisons of multiscale pore structures between primary and disturbed coals and their effects on adsorption and seepage of coalbed methane. *J. Pet. Sci. Eng.* **2019**, *174*, 704–715. [CrossRef]

minerals

MDPI

Review

Soft Computing Application in Mining, Mineral Processing and Metallurgy with an Approach to Using It in Mineral Waste Disposal

Nelson Herrera [1,2,*], María Sinche Gonzalez [1], Jarkko Okkonen [3] and Raul Mollehuara [1]

[1] Oulu Mining School, University of Oulu, P.O. Box 3000, 90570 Oulu, Finland; maria.sinchegonzalez@oulu.fi (M.S.G.); raul.mollehuaracanales@oulu.fi (R.M.)
[2] Departamento de Ingeniería en Metalurgia y Minas, Universidad Católica del Norte, Antofagasta 1270709, Chile
[3] Geological Survey of Finland GTK, Vuorimiehentie 5, P.O. Box 96, FI-02151 Espoo, Finland; jarkko.okkonen@gtk.fi
* Correspondence: nelson.herreranunez@oulu.fi; Tel.: +56-552651010

Abstract: In the past two decades, the mining sector has increasingly embraced simulation and modelling techniques for decision-making processes. This adoption has facilitated enhanced process control and optimisation, enabling access to valuable data such as precise granulometry measurements, improved recovery rates, and the ability to forecast outcomes. Soft computing techniques, such as artificial neural networks and fuzzy algorithms, have emerged as viable alternatives to traditional statistical approaches, where the complex and non-linear nature of the mineral processing stages requires careful selection. This research examines the up-to-date use of soft computing techniques within the mining sector, with a specific emphasis on comminution, flotation, and pyrometallurgical and hydrometallurgical processes, and the selection of soft computing techniques and strategies for identifying key variables. From this, a soft computing approach is presented to enhance the monitoring and prediction accuracy for mineral waste disposal, specifically focusing on tailings and spent heap leaching spoils database treatment. However, the accessibility and quality of data are crucial for the long-term application of soft computing technology in the mining industry. Further research is needed to explore the full potential of soft computing techniques and to address specific challenges in mining and mineral processing.

Keywords: mineral extraction; soft computing; process control; prediction accuracy; artificial neural networks; expert systems; fuzzy algorithms; mineral waste disposal; tailings; heap leaching piles

1. Introduction

The mining industry contributes to approximately 10% of global economic activities, of which industry payments for services and direct support comprise another 10%, making it a critical part of multiple production chains [1]. Over time, this industry has been a precursor for technological developments. According to the European Parliament [2], during the past 12 years, a quarter of the mining industry has doubled its investments in technology, reaching 93% implementation with successful results. Over 90% of mining companies believe that complementing their operations with technology translates into added value and helps to revolutionise their business. The advancement of technology has led to the emergence of untapped prospects in the field of big data capture systems, which have not yet been fully explored or utilised in industrial settings. For instance, these systems can be employed for routine inspections of operational equipment or for the creation of daily production records [3]. This volume of data is expected to grow exponentially over time, reaching an amount of one hundred and twenty zettabytes in 2023, corresponding to a 675-fold increase since 2005 [4–6].

The mining industry and metallurgical processes are familiar with the concept of capturing large amounts of data. However, the analysis and interpretation of this data present novel problems for operators and decision-makers who aim to enhance productivity and sustainability in their operations. The European Union has set four key goals for the year 2030, as outlined by Usman et al. [7], which include prioritising energy efficiency, reducing CO_2 emissions, and promoting the adoption of clean energy sources. To achieve these objectives, the EU emphasises the importance of sustainable raw material production through the utilisation of digital tools, as well as advancements in safety, productivity, and profit margins [8]. While the World Economic Forum has predicted that from 2017 to 2025, $425 trillion will be invested globally in the application of artificial intelligence (AI) for the productive sector [9], in Chile, the creation of national policies and initiatives such as the Roadmap Mining 4.0 [10] policy seeks to implement digital technologies in the mining industry. There is a correlation between artificial intelligence (AI), Mining 4.0, and machine learning technologies that can potentially drive transformative advancements in mining processes, enhance productivity, and enable data-driven decision-making in an increasingly interconnected and digital world. The objective of this paper is to reveal the potential in question.

2. Background and Finding the Gaps

Since 2015, the fields of data science and soft computing have significantly contributed to the development of more precise predictive models [11]. The effectiveness of these models is contingent upon the specific characteristics of the problem at hand and the amount of available data. Furthermore, the advancements in computing capabilities have facilitated the widespread adoption and refinement of these tools in various industries and processes, encompassing a wide range of specialties, through the integration of artificial intelligence and statistical techniques. Soft computing is a subfield of artificial intelligence that encompasses various paradigms and techniques designed to handle incomplete and imprecise information in crucial processes. Its primary objective is to enable companies to derive valuable solutions for tasks such as prediction, information discovery, and knowledge acquisition [12,13]. This research aims to examine the application of soft computing in mineral processing, providing a perspective on the operational impact of relevant stages present in the processing lines of valuable minerals and considering the possibility of supporting operational decisions. Based on this review, an application approach to mine waste disposal operations is introduced.

2.1. Automation in the Mining Industry and Opportunities for AI

The mining industry can benefit significantly from the application of AI to enhance process control and improve productivity. The development of sensors, transmitters, and controllers has given way to centralisation of the control and supervision of variables using evolving technologies in control systems (Figure 1) and the development of control loop applications [14]. These systems offer prompt information and rapid responses to external and internal disturbances, guaranteeing a safe and stable operation within the process. The choice of the functions oriented to the analysis of information and the application of superior decisions are fundamental functions for the control and monitoring of the process [15]. The focus of their work lies in the examination of information, and the effectiveness of their decision-making process is contingent upon the execution and necessity of decisions and problem-solving tasks that lack algorithmic resolutions.

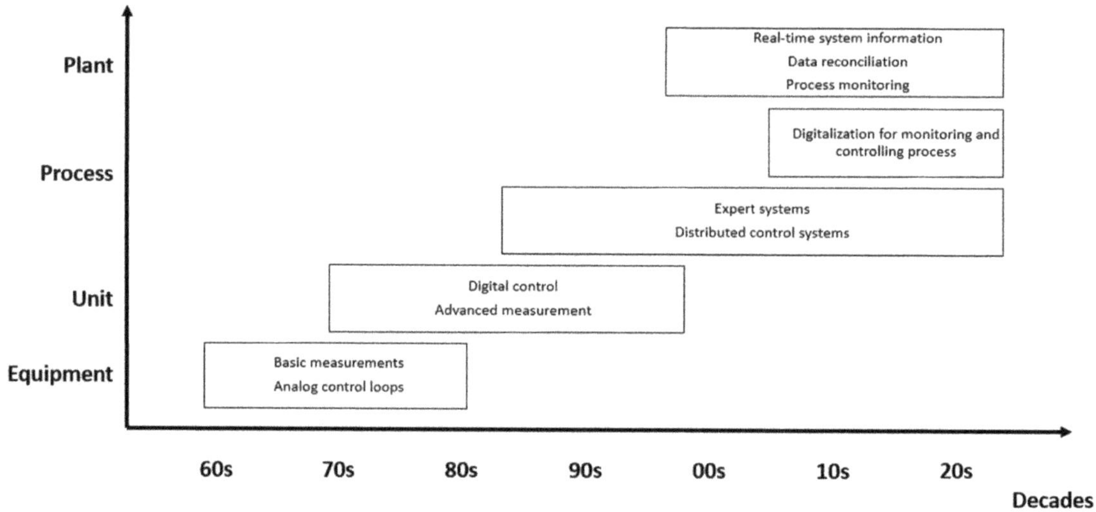

Figure 1. Automation system evolution in mineral processing based on Sbárbaro et al. [15] and Nad et al. [16].

The evolution of artificial intelligence enables complicated and interpretative solutions in contrast to empirical model approximations, offering complex representation learning, non-linearity, feature extraction, flexibility, and scalability with continuous and end-to-end learning. Artificial intelligence can handle diverse inputs, ambiguity, and uncertainty, making it suitable for different processing languages and computer vision. However, to identify solutions, we must provide explicit, known, and well-defined conditions [17]. Continuous improvement in the mining industry has made it common to find an interconnected control hierarchy such as the one presented in Figure 2, where instrumentation plays a fundamental role in ensuring the proper functioning of the plant. The second sector is dedicated to maintaining control over key variables, as evidenced in flotation processes where variables such as pulp level, foam level, air injection, and pump speed are regulated using conventional proportional, integral, and derivative (PID) control methods [18].

Reaching the highest levels of the pyramid, where we can find advanced flotation control (AFC) and optimised flotation control (OFC), will directly depend on a satisfactory control of the lower levels, where the PID controls are insufficient for the total management of the processes. The AFC structure must possess the capability to effectively mitigate disturbances arising from various sources, such as feed flow or other external factors, in addition to considering dynamic situations such as the accumulation of matter and delay times for the calculation of performance parameters such as recovery and the grade of the mineral. OFC will always seek to maximise the economic performance of the process, associating it with the recovery and grade of the concentrates obtained [19].

Metallurgical facilities frequently use PID control techniques due to their versatility and ability to be adjusted to the optimal operating parameters, achieving a response that suits the process [20]. Nevertheless, the intricate nature of multivariate processes precludes the application of conventional control techniques. In contrast, advanced control endeavours to establish a mathematical model that accurately characterises the operational and phenomenological aspects of the process under control. This model is subsequently validated through simulation. Expert systems, fuzzy logic, and predictive control models, both linear and non-linear, are viable alternatives in the industrial sector. These approaches can be effectively blended into a cohesive solution known as the predictive control of the fuzzy model [21]. Ai et al. [22] propose a novel approach that aims to overcome these limitations by using deep learning features and offline conservative double Q-learning control as an option to minimise the PID limitations in dealing with complex and nonlinear systems. During the 1990s, statistical features were used to classify froth structures in flotation cells during mineral processes [23]. This method was further developed by Hadler et al. [24] and applied by Polaris. Mines enhanced productivity and plant performance in lead zinc concentration standardised operational practices and adjusted reagent dosages based on the accumulated error over time. Their approach involved utilising set-point adaptive optimisation, incorporating a feedforward neural network (FNN) soft sensor for online feed grade estimation and a long short-term memory network (LSTM) to track fixed dosage set-points [25]. The implementation of automation in various processes has significant benefits, including the mitigation of staff exposure to hazardous environments, prevention of equipment damage, optimisation of operational expenses, and preservation of product quality [26,27].

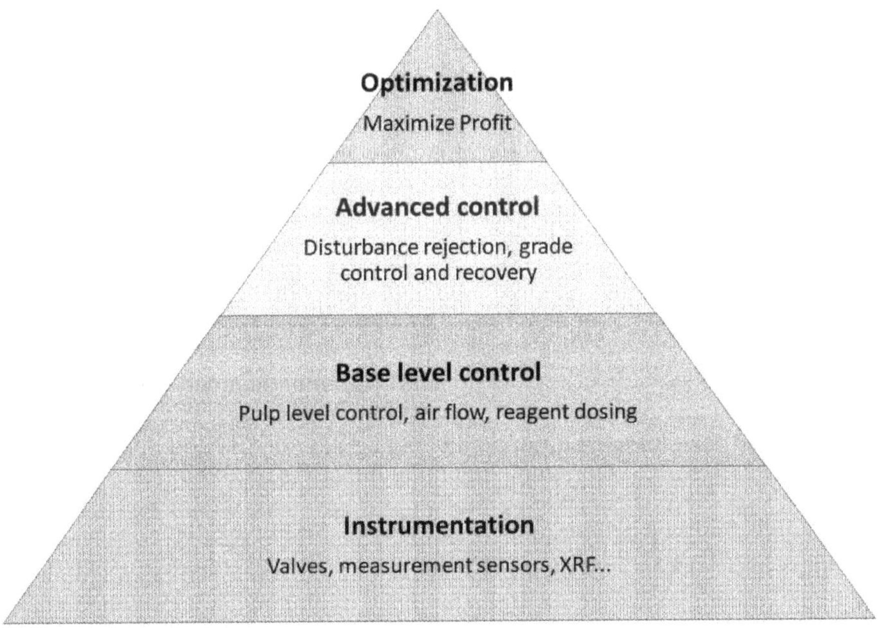

Figure 2. Level hierarchy for a flotation processes control system based on Shean et al. [18].

2.2. *Application of Soft Computing in Mineral Extraction and Processing*

The successful use of AI in the mining industry requires a collaborative effort among domain experts, data scientists, and technology vendors. To ensure the most appropriate AI solutions, each mining operation's individual demands and challenges must be carefully assessed. Concerns such as data security, staff retraining, and ethical considerations should also be addressed as part of the deployment approach. Through the analysis of mineral beneficiation and metal extraction processes, it is possible to identify some examples of the difficulty in controlling these processes optimally due to the instability of the variables involved. The composition of the slags impacts the recovery of the element of interest in pyrometallurgical processes such as the fusion process and is thus a variable that must be considered to obtain optimal recovery. However, because there are a variety of chemical reactions driving the process, it is difficult to describe a relationship between the compositions of the slags and the contents of components of interest that are present [28]. In the mineral processing field, Estrada et al. [29] investigated problems related to the flotation process such as the control of mineral granulometric sensitivity from the milling stage and milling global energy optimisation, as well as tailings transportation and deposition.

Martin et al. [30] analysed possible scenarios in which deposit destabilisation can occur, causing serious environmental and safety damage.

The study conducted by Fu et al. [31] examines the influence of several factors, such as mineralogy and viscosity, on thermodynamic changes and the manipulation of variables within kinetic models. Additionally, the application of soft computing has had a significant influence both in the areas of classification and concentration of minerals and in pyrometallurgy. Most of the applications are in artificial vision, probably due to the impulse to find methods to measure parameters such as particle sizes or chemical composition using image-based methods as a cheaper and faster alternative to techniques based on sample analysis [32]. The heap leaching process is an important part of the hydrometallurgical production line because it makes it easier to obtain the desired mineral. This process involves several control variables, including the properties of the leaching agent and the characteristics of the ore to be leached, such as its porosity, mineralogy, grade, and impurities, among others [33].

In this investigation, we have been able to analyse how the use of different soft computing tools has been incorporated in different stages related to mineral extraction and how they have been applied in situations related to both production and security, establishing the contribution and support needed to optimise mineral processes. A comprehensive analysis has been conducted on a total of 46 research publications that investigate the utilisation of soft computing techniques in various domains, including deposit operations, comminution, flotation, hydrometallurgical processes, and pyrometallurgical processes.

Table 1 presents a classification of the methods applied for each revised publication according to the applied area, and in Figure 3, the percentage of each methodology applied to the areas of interest is established, where for metallurgical processes, the artificial neural network (ANN) and model predictive control (MPC) are the most widely implemented. The relevance of identifying the variables with the greatest impact according to the associated operation has also been evidenced. The results obtained will be analysed in addition to the variables used to establish a comparison between the different methods and how their implementation has been approached. As shown in Figure 4, the areas of flotation and pyrometallurgical processes are those where the most studies have been developed, especially in the search for operational improvements for process control and prediction. This situation is directly explained by the possibility of accessing data available from existing monitoring in industrial processes, as will be seen later, and by the fact that these stages are the ones that present the greatest diversity of variable monitoring.

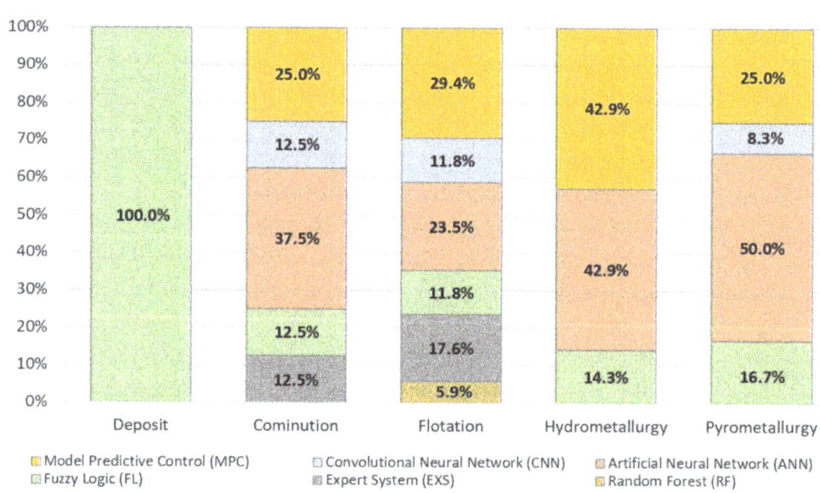

Figure 3. Comparison of soft computing methods applied in mineral extraction and processing.

Table 1. Summary of representative studies of soft computing application in mineral extraction and processing.

Author	Operation	Soft Computing Application					
		RF	EXS	FL	ANN	CNN	MPC
Sun et al. [34]	Mining Stage			•			
Danish et al. [35]				•			
Li et al. [36]				•			
Stange et al. [37]	Comminution		•		•		
Tessier et al. [38]			•				
Olivier et al. [39]							•
Estrada et al. [29]							•
Hamzeloo et al. [40]					•		
Umucu et al. [41]					•		
Cai et al. [42]				•			
Olivier et al. [43]						•	
Aldrich et al. [23]	Flotation					•	
Moolman et al. [44]					•		
Ramasamy et al. [45]							•
Chen et al. [46]							•
Cortes et al. [47]							•
Aldrich et al. [48]						•	
Riquelme et al. [49]							•
Brooks et al. [21]							•
Ali et al. [50]		•		•	•		
Hoseinian et al. [51]					•		
Shean et al. [52]			•				
Zhang et al. [53]			•				
Ai et al. [25]				•			
Fu, Y. et al. [31]			•				
Ai et al. [22]					•		
Zhang et al. [54]					•		
Komulainen et al. [55]	Hydrometallurgy						•
Moreno et al. [55]							•
Pang et al. [56]				•			
Azizi et al. [57]					•		
Hoseinian et al. [33]					•		
Gao et al. [58]							•
Xu et al. [59]					•		
Gui et al. [60]	Pyrometallurgy			•			
Deng et al. [61]				•			
D. Liu et al. [28]					•		
J. Liu et al. [62]							•
Savic et al. [63]					•		
Ghea Puspita et al. [64]					•		
Cardoso et al. [65]					•		
Qian et al. [66]					•		
Cardoso et al. [67]							•
Wang et al. [68]							•
Yang et al. [69]						•	
Zhao et al. [70]					•		

RF: Random forest, EXS: expert system, FL: fuzzy logic, ANN: artificial neural network, CNN: convolutional neural network, MPC: model predictive control.

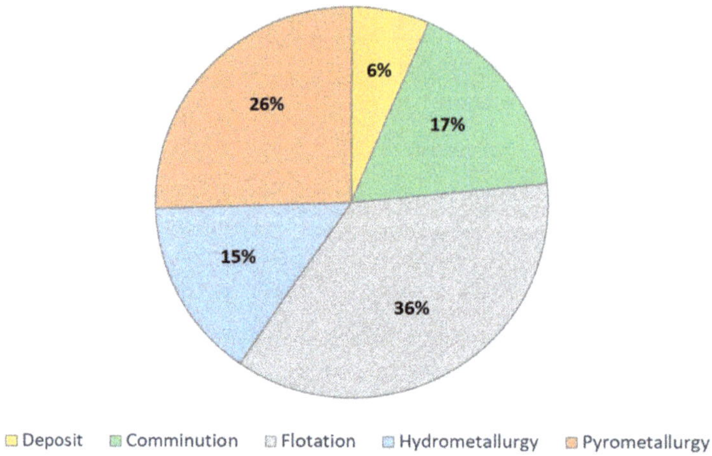

Figure 4. Soft computing application by mineral extraction and processing.

2.3. *Applications of Soft Computing in the Mining Stage*

Ongoing research and development efforts are being conducted within the mining industry to enhance safety measures and decision-making processes, employing techniques such as fuzzy logic and other methodologies (Table 2). Danish et al. [35] used fuzzy logic to predict mine fires in underground coal mines, while Li et al. [36] proposed a method to minimise the risk of gas explosion accidents in coal mines using fuzzy comprehensive evaluation. Similarly, Sun et al. [34] presented a fuzzy logic-based approach for predicting the risk of rock burst accidents in coal mines. It is possible to identify CO, O_2, N_2, and temperature as common variables in the application of fuzzy logic to the prediction and control of security mine conditions linked to variables that can be monitored in the field. With this understanding of which variables are most important for a given application, they can focus their efforts on collecting relevant data and refining their models accordingly. These studies provide evidence of the possible advantages that can be gained by employing soft computing techniques to enhance safety and decision-making procedures in ore deposit operations. From the studies applied to the mining stage, the most relevant variables that have been identified include:

(a) Density: For ore deposits, density is an important variable that provides information about the composition and mineralogy of the deposit.
(b) Water transmitting ability: Refers to the capacity of a rock or mineral to allow the flow of water through it. It is an important variable in understanding the material's hydrogeological characteristics.
(c) Fracture development degree: Corresponds to the extent and intensity of fractures or cracks within the rock or mineral. It can affect the permeability and fluid flow within the deposit.
(d) Confined water pressure: Refers to the pressure exerted by water within confined spaces or pores in the deposit. It can influence the stability and behaviour of the deposit.
(e) Watery property of the floor aquifer: This variable refers to the characteristics of the water present, such as its chemical composition, pH, and mineral content. It can impact the interaction between the aquifer and the ore deposit.
(f) Aquifuge thickness and strength: Refers to the thickness of impermeable or low-permeability layers that prevent the flow of water. Strength refers to the resistance of these layers to deformation or failure. These variables can affect the hydrogeological conditions and water movement within the deposit.
(g) Mining thickness and depth: This variable refers to the thickness of the ore body being extracted. Depth refers to the vertical distance from the surface to the ore

body. These variables are important in determining the feasibility and logistics of mining operations.

(h) Inclined productivity: Refers to the efficiency and productivity of mining operations in inclined or sloping deposits. It considers factors such as the angle of the deposit and the methods used for extraction.

Table 2. Characteristics of studies applied in the mining stage.

Paper	Problematic	Method	Data Used	Specific Variables	Results
Sun et al. [34]	Risk assessment of floor water inrush in deep mining	Improved fuzzy comprehensive evaluation, Delphi method, and analytic hierarchy process	Hydrogeological data from six industrial mining faces	Density, water-transmitting ability, fracture development degree, confined water pressure, watery property of the floor aquifer, aquifuge thickness and strength, mining thickness and depth, inclined length	The approach provides a tool for the risk assessment of floor water inrush in deep mining, where the results are consistent with the field-observed results
Danish et al. [35]	Predicting mine fires in underground coal mines	Fuzzy logic model with Mamdani inference system	Data from 10 gas monitoring stations collected from sensors in an underground coal mine	Input variables: CO, O_2, N_2, temperature; Output variable: Fire intensity	The fuzzy logic system is reliable for decision making regarding fire intensity and assessing fire intensity with variables at the same time (validated using Graham's index), and identified suspected areas for spontaneous combustion.
Li et al. [36]	Quantitative risk assessment of gas explosions in underground coal mines	Combination of fuzzy analytic hierarchy process (FAHP) and Bayesian network (BN)	EXS application of risk factors related to gas explosions in underground coal mines	Flow rate, pressure, pipe diameter, pipe roughness coefficient, pump efficiency, energy consumption, cost	Inference to predict the probability of gas explosion risks with the determination of accident causes. The identification of the weight variables helps determine the optimal combination for flow rate, pressures, pipe diameter for each pipe segment, pump efficiency, and pipe roughness coefficient.

We can say that the soft computing application for the analysed studies is established, considering the following characteristics: 1. Risk factors: This is a key issue in assessing the risk of accidents in mines. The specific risk factors may vary depending on the study but generally include factors such as gas levels, ventilation failure, and water inrush. 2. Specific methods: Fuzzy logic and AHP: The application of fuzzy logic helps to handle uncertainty and imprecision in data under this variable selection. The analytical hierarchy process (AHP) determines the weights of different factors in a decision-making process, including the risk factors selected to be analysed in this quantitative risk assessment of gas explosions in underground mines.

2.4. Applications of Soft Computing in the Comminution Stage

The primary aim of comminution plants is to achieve the optimal particle size for efficient extraction of the desired mineral while also ensuring cost-effectiveness. However, to dissociate the low-grade minerals, the material must be finely ground, which implies high energy consumption in the milling area [54]. The grinding circuits are difficult to control due to poor plant models, external disturbances, and process variables that are not easy to measure [39]. In addition, the variables and properties of the extracted mineral, such as size, composition, and hardness, affect mill performance [38]. The disturbances in the operational conditions in milling directly affect the subsequent stages, such as leaching and flotation, that depend on the product obtained in this first stage, affecting the performance of a mining plant in general.

The strategies for reducing energy consumption in the mining industry have been extensively considered in recent years due to the increase in energy prices [71,72]. Therefore, control strategies in mining processes are one of the many ways to optimise this consumption, especially if the strategies consider global optimisation and ensure the stability of the system. The milling of minerals represents up to 50% of the energy consumption in a mineral concentrator plant. For this reason, Estrada et al. [29] developed a centralised hybrid model predictive control scheme (HMPC) in the grinding process that seeks to minimise specific energy consumption of the equipment and stabilise the feeding to the plant by ensuring an output particle size of 230 μm (65 mesh). With this hybrid model, the implementation of conventional strategies is a low-cost opportunity based, for example, on expert systems that handle variables and discrete events, and applying HMPC strategies allows the inclusion of discrete events in both the model and the controller. Stange [37] examines the application of ANN to the control of grinding circuits, specifically autogenous milling. Tessier et al. [38] describe a machine vision strategy for online rock mixture composition estimation, obtaining an overall accuracy of 92.5% for dry ore combinations.

In a run-of-mine ore milling circuit, Olivier et al. [39] propose using disturbance observers, specifically a fractional order disturbance observer (FO-DOB) and a Bode ideal cut-off disturbance observer (BICO-DOB), in conjunction with a standard PI controller to improve control performance in the presence of strong external disturbances and severe model–plant mismatches. Estrada et al. [29] use an industrial data-tuned grinding simulator to provide a hybrid model predictive control technique for mineral grinding circuits. Hamzeloo et al. [40] investigate the use of image analysis and neural networks to estimate particle size distribution on an industrial conveyor belt in a copper concentrator, attaining an overall RMSE of 6.11%. Umucu et al. [41] examine the application of ANN in modelling a calcite grinding system in mineral processing. Cai et al. [42] offer an approach for underground coal mining rock burst predictions utilising micro-seismic monitoring and a fuzzy comprehensive evaluation model. Olivier et al. [43] discuss the use of deep CNN to classify feed ore images into one of four categories based on size distribution, with an overall accuracy of 96.4%. Table 3 presents a comprehensive summary of the findings derived from each individual investigation. Based on the nature of the applications conducted in the studies, it is possible to categorise the studies conducted by Stange [37] and Tessier et al. [38] as utilising machine vision and image analysis techniques for diverse applications, such as estimating rock mixture composition, monitoring flotation froth, and analysing fragmentation.

Stange [37], Umucu et al. [41], and Hamzeloo et al. [40] applied ANNs for the control and modelling of mineral processing systems, as well as for classifying feed ore images based on size distribution. Olivier et al. [43] study the possibility of obtaining a convolutional neural network (CNN) model for characterising the size distribution of feed ore in run-of-mine ore milling circuits. Cai et al. [42] propose a methodology for rock burst forecasting in underground coal mining using microseismic monitoring and a fuzzy comprehensive evaluation model.

Table 3. Characteristics of studies applied in mineral comminution processes.

Paper	Problematic	Method	Data Used	Specific Variables	Numeric Results
Stange et al. [37]	Control of grinding circuits, specifically autogenous milling	ANN for control strategies and exploration of various control approaches by EXS	No specific data mentioned; theoretical discussion of ANN use in grinding circuit control	No specific variables mentioned; exploration of various control approaches using ANNs	Proposes two control strategies using ANNs for the control of autogenous grinding circuits. ANNs have significant potential in developing a model of the hydrocyclone classifier.
Tessier et al. [38]	Online estimation of rock composition for nickel mineral treatment	Machine vision approach for feature extraction, dimensionality reduction, and class boundary establishment using support vector machines	Digital images of five different mineral types and mixtures of them	Composition of rock mixtures: colour and textural features extracted from sub-images	Good estimation for mixture compositions for dry ore but some inaccuracies for wet ore mixtures due to light reflection. The proposed approach can be used for real-time monitoring of variations in run-of-mine ore composition.
Olivier et al. [39]	Improving control performance in a milling circuit	MPC controller with a fractional order disturbance observer (FO-DOB) and a Bode ideal cut-off disturbance observer (BICO-DOB)	Simulation data generated from a non-linear MIMO plant model	Controlled variables: product particle size, fraction of the mill volume filled with material, slurry volume in the sump. Mill-manipulated variables: solid feed-rate, water feed-rate, steel balls feed-rate, water flowrate into the sump and slurry flowrate into the cyclone	The FO-DOB and BICO-DOB are useful tools for ROM ore milling circuit control. This addition to the normal PI controller gives better results than the PI controller alone because of the decrease in the ISE values. The BICO-DOB has poorer disturbance rejection performance than the other two DOB varieties but gives the best set-point tracking performance.
Estrada et al. [29]	Develop an HMPC strategy for grinding circuits	Hybrid MPC controller identification procedure for two controlled variables	Data from industrial data-tuned grinding simulator	Conveyor feed rate, water feeding sump, SAG mill speed, product hardness, specific energy consumption, product particle size; Activation/deactivation of secondary grinding circuits and product granulometric distribution	A hybrid identification procedure for two controlled variables is correctly performed, with energy consumption minimization and the maintenance of particle size output.

Table 3. *Cont.*

Paper	Problematic	Method	Data Used	Specific Variables	Numeric Results
Hamzeloo et al. [40]	Estimate particle size distribution on an industrial conveyor belt in a copper concentrator	Image analysis and ANN	Images collected from an industrial conveyor belt in the crushing circuit of a copper concentrator	Particle size distribution, image pixel values, scaling factors, size features, eigenvectors and eigenvalues, cumulative passing %, volume of particles, metal ball diameter	Model estimations of particle size distribution achieve an overall RMSE of 6.11%. For area-based size estimations, the model obtains an RMSE of 4.45%. It obtains an RMSE of 18.54% for weight-based size estimations. Other size measures had RMSE values ranging from 5.45% to 37.11% for area-based size estimations and from 18.54% to 37.11% for weight-based size.
Umucu et al. [41]	Grinding system modelling of calcite in mineral processing	ANN–MLPNN and RBFNN	Experimental data collected from laboratory conditions	Input variables: cumulative percentages of ball mill feed, ball mill conditions, and grinding time. Output variables: ball mill product cumulative percentages	The RBFNN model performs better than the MLPNN model, highlighting the importance of analysing data and using capable systems for fast decision-making. The study used different powder filling levels for the calcite sample to evaluate the statistical data obtained from the ANN models.
Cai et al. [42]	Rock burst forecasting in underground coal mining	Fuzzy comprehensive evaluation model	Microseismic monitoring data from a coal mine and laboratory acoustic emission measurements of coal samples	Fault total area, space-time diffusivity, equivalent energy magnitude, seismicity degree, time information entropy, source concentration degree, seismic diffusivity	Importantly, the proposed methodology was successfully applied to a coal mine using a combination of indices for more accurate forecasting. Microseismic monitoring is a powerful tool for rock burst forecasting
Olivier et al. [43]	Characterization of the feed ore size distribution in a milling circuit	Deep CNN application for classification	Feed ore images captured from an industrial conveyor belt with a vertically mounted camera	223 images captured and categorised in four groups	The CNN achieved an overall accuracy of 96.4% in classifying into one of four categories based on size distribution, with an overall F1-score metric of 0.97.

Finally, Olivier et al. [39] and Estrada et al. [29] propose new control strategies for mineral grinding circuits, using techniques such as disturbance observers and model predictive control.

From the studies applied to mineral comminution processes, the most relevant variables that have been identified include:

(a) Particle size: A critical variable in comminution processes, as it affects the efficiency of subsequent mineral extraction stages. Achieving the optimal particle size is essential for efficient extraction of the desired mineral.
(b) Composition and hardness: The composition and hardness of the mineral being processed can significantly impact mill performance. Different minerals may require different grinding conditions to achieve the desired particle size.
(c) Operational conditions: Factors such as the mill speed, feed rate, and grinding media size can influence the efficiency and effectiveness of comminution processes.
(d) External disturbances: Changes in ore feed characteristics or variations in power supply can affect the stability and performance of comminution circuits.
(e) Product size setpoint: This variable is the target size for the final product and is used as a control parameter in grinding circuits.
(f) Rock types: These refer to different types of rocks or minerals present in the ore mixture. They are used as labels or classes for classification purposes.

2.5. Applications of Soft Computing in the Flotation Stage

The mineral flotation process is a complex dynamic process to control that cannot be optimised using constant control strategies because each mineral exhibits different behaviours before this process [73] and the mineral grade from the deposit is in constant decline, for example, reaching 0.4% in the case of copper [74]. This means that more controlled processes are required to maximise the separation of the species of interest from those without economic value that contaminate the final product. Keeping mineral flotation processes under control is a complex task since their stability depends largely on various variables such as the processed tonnage, mineralogy, percentage of solids, reagent dosage, and granulometry, among others (Table 4).

A change in these generates disturbances in the flotation process, which are reflected in the recovery results [18]. Historically, the optimisation of recovery results was achieved with experienced and qualified operators, but as Laurila et al. [75] explain, the flotation process is currently undergoing a paradigm shift towards automated control, ushering in a new era where:

- The design of flotation circuits is being simplified, which facilitates the regulation and control of the processes [76].
- The cell size has increased over time, going from 50 m^3 to designs close to 600 m^3 [77].
- The development of new technologies for online image analysis has made it possible to provide information on the status of the equipment involved in the process, such as stirring motors, valves, sensors, and pumps, and the quality of variable measurements such as air supply, pH, and bubble size [78].

The evolution of the soft computing application in processes linked to flotation processes has undergone evolutionary development, encompassing diverse techniques and technologies for monitoring and control, highlighting the importance of accurate data and control strategies in achieving optimal performance. When analysing the studies according to the applied methodology, Chen et al. [46], Cortés et al. [47], and Brooks et al. [21] analysed the implementation of advanced control strategies in mineral processing, specifically in ball mill grinding circuits and rougher flotation circuits, and the implementation of advanced technologies, including image processing, X-ray fluorescence, and diffused reflective spectroscopy, to optimise a copper roughing circuit. The publications cover topics such as MPC, hybrid ANN models, and multivariable predictive control systems, where the results show improved process stability, increased copper recovery, and significant performance improvements compared to previous control methods.

Table 4. Characteristics of studies applied in mineral flotation processes.

Paper	Problematic	Method	Data Used	Variables	Results
Aldrich et al. [23]	Classification of different froth structures in flotation cells	Decision tree and CNN algorithms for constructing knowledge-based systems	Surface froth images of two industrial flotation cells. Training data set consisting of 400 exemplars randomly sampled from the four classes	Froth image characteristics (e.g., statistical features)	The CNN system correctly classified unknown froth forms. The system improves profitability and reduces operating instability. The net classified froth structures with 96% accuracy, but limited training data sets reduced classification performance to 68%.
Moolman et al. [44]	Grinding efficiency of dry ball mills	Machine vision application to the feed with a ANN model from a laboratory-scale ball mill	Grinding aids and stage efficiency	Grinding data as particle size distribution, specific surface area, and grinding efficiency	SGLDM and NGLDM analysis of digitised froth images shows feature extraction potential. Neural networks classify foam shapes well. Dry ball mills can grind 25% better with grinding aids. SGLDM outperformed NGLDM at 66.7% with a 90% classification rate. However, NGLDM classified froth conditions better than SGLDM features at 78.9% compared to 52.4%.
Ramasamy et al. [45]	Comparison of predictive control schemes with detuned multi-loop PI controllers for controlling ball mill grinding circuits	MPC	Experiments conducted under five different operating conditions. Process model parameters such as the breakage rate function and hydrocyclone model parameters based on steady-state data collected from the circuit	Variables controlled: cyclone overflow fraction passing 104 μm and mill throughput. Fresh feed rate (0.375 to 0.5 kg/min), sump water addition rate, mill solids (67% to 74%), slurry pumping rate, hydrocyclone model parameters, breakage rate function, and sump level	Detuned multi-loop PI controllers oscillated and could not eliminate control loop interactions. The MPC system reached setpoints without overshoot or offset and decoupled well. Constrained MPC suppresses big input moves and is more resistant to operational conditions than PI controllers. PI controllers modify variables more than MPC. The mill throughput–sump water loop with PI controls is slower.

Table 4. *Cont.*

Paper	Problematic	Method	Data Used	Variables	Results
Chen et al. [46]	Implementing a model predictive control in a ball mill grinding circuit	MPC	Ball mill grinding circuit in an iron ore concentrator plant process variable	Particle size, mill solids concentration, sump level, circulating load, fresh feed rate, mill feed water flow rate, dilution water flow rate, vibratory conveyor speed, mill feed water control valve's opening, dilution water control valve's opening, pump speed	The study shows that MPC improves grinding circuit performance, ensures operational stability, and greatly reduces overload situation warnings. MPC lowered overload condition warnings by 31.9% over PID controllers. In addition, MPC lowered overload condition warnings by 85.8%. The MPC technique showed its efficacy in the invested time period, reducing alarms and ensuring operational continuity.
Cortes et al. [47]	Stabilise rougher flotation circuit operation	MPC application based in Honeywell's Profit Controller	Rougher flotation circuit data from Concentrator A-1 at División Codelco Norte	Airflow rate set points, level control, pH, and tonnage variation	Profit FLOT enhanced control and improved process stability and copper recovery. Profit FLOT on daily shifts increased the A-1 concentrator recovery by 1.5%. Higher-profit FLOT use, practise enhancements, strategy revisions, and operator training can still capture marginal benefits. Performance and recovery rates may improve further with further optimisation.
Aldrich et al. [48]	Control in froth flotation processes	Implementation of automated systems based on CNN using machine vision	Froth images obtained from laboratory and industrial flotation cells, operational data such as the air flow rate, pulp level, and concentrate grade	Bubble size and shape, air flow rate, pulp level, concentrate grade	Machine vision and image analysis techniques can be used to monitor froth stability, but fully automated control is not possible.
Riquelme et al. [49]	Identify and measure bubble size distribution in flotation	Image processing and a parametric method with a circular Hough transform (CHT) and log-normal distribution for MPC	Bubble images from flotation columns obtained through a camera system conducted with different frother concentrations and superficial air velocities	Frother concentration, superficial air velocity, bubble size distribution parameters, bubble Sauter mean diameter, flotation recovery	CHT detects clustered bubbles better than other approaches. Log-normal distribution estimates BSD well. BSD parameter dynamics are explained well by the non-linear Wiener model. After experiments, static models were estimated using a non-linear least squares technique with R2 = 93.3% and 98.5%.

Table 4. *Cont.*

Paper	Problematic	Method	Data Used	Variables	Results
Brooks et al. [21]	Optimization of a copper roughing circuit to improve recovery in an oxide rougher circuit	Application of sophisticated technologies such as image processing, X-ray fluorescence, diffused reflective spectroscopy, and cascaded MPC	Manipulated, disturbance and controlled variables in flotation cells	Feed flows, densities, air flows, pulp levels, feed, concentrate and tail Cu grades, and froth velocity	Successful MPC installation improves float and Cu recovery. Innovative measurement technology improved the accuracy and reliability of critical control parameters such as the Cu grade, froth velocity, and concentrate and tail grades.
Ali et al. [50]	Predict the flotation behaviour of fine high-ash coal in the presence of a hybrid ash depressant	Random forest, ANN, fuzzy logic, and adaptive neuro-fuzzy inference system	Flotation experiments on fine high-ash coal with a training data set (80% of total data) and test dataset (20% of total data) containing five inputs and two outputs	Inputs: Al-PAM polymer dosage, pH, polymer conditioning time, dispersant dosage, and impeller speed. Outputs: combustible recovery and froth ash content	The models predict the performance of the coal flotation process. The fuzzy logic model had the best prediction performance in coal flotation. Overall accuracy: FL > ANN > ANFIS > RF > HyFIS.
Hoseinian et al. [51]	Develop a model to predict SAG mill power	Hybrid ANN algorithm model application	SAG mill operation dataset from Aq Darreh gold processing plant (GA population size: 100, max generation: 450)	Feed moisture, mass flowrate, mill load cell mass, SAG mill solid percentage, inlet water flow rate, outlet water flow rate, work index, and mill power	Correlation coefficient (R) of the GANN model: 0.9127 in testing compared to ANN alone with an R of 0.7947. Obtained relationship input parameters for the work index, inlet and oulet water to the SAG mill, mill load cell mass, SAG mill solid percentage, mass flowrate and feed moisture. Mean squared error (MSE) of the GANN model: 0.0451 in training, 0.0430 in testing. MSE of the ANN model: 0.1549 in training, 0.4054 in testing
Shean et al. [52]	Predicting changes in pulp height in froth flotation	Development of a dynamic model from a flotation laboratory test	Froth flotation mass balance and calibration using experimental data	Bubble size distribution, air flow rate, and pulp height under different conditions	The dynamic model predicts steady-state froth flotation pulp height variations. The experimental system responds slower to the reagent than the model because the model assumes the system is well mixed, while the experimental results show plug flow. The model can be adjusted to reflect industrial flotation control's dynamic nature.

Table 4. *Cont.*

Paper	Problematic	Method	Data Used	Variables	Results
Zhang et al. [53]	Simulating the relationship between the reagent dosage and froth surface appearance in a lead–zinc flotation plant	Hammerstein–Wiener-based model with the illumination modelling-based marker watershed method for EXS application	Reagent data and froth surface images collected from a lead–zinc flotation plant (149 pairs used for developing the Hammerstein–Wiener model and 60 pairs used for testing and validating the model)	Reagent variables (frother, activator, and collector dosage), froth surface variables (bubble size distribution, froth surface image, and highlight spot marker), model variables (Hammerstein–Wiener model)	The log-normal distribution can describe lead–zinc flotation plant bubble size distribution (BSD) in the quiet zone below the interface. The Hammerstein–Wiener model beats the Wiener model, LS-SVM model, and neural network model in fitting accuracy and performance with an RMSE of 0.0420 and an R-squared of 0.9721. The proposed approach can guide reagent dosage changes to manage mineral processing froth flotation and segment zinc froth image bubbles with 95.6% accuracy.
Ai et al. [25]	Flotation reagent control	Reduction in extracted deep learning features using a stacked autoencoder, fuzzy association setpoint calculation, and offline Q-learning-based reagent control	Feed characteristics, froth grade, flotation reagents, and froth videos. RL benchmark environment data with 1,000,000 samples using a soft actor–critic (SAC) controller	Froth images to extract four features: bubble size, bubble shape, froth velocity, and froth color.	The method outperformed other existing methods in terms of the MAE and qualified ratio of the concentrate grade. It was effective and promising for practical flotation reagent control.
Fu et al. [31]	Effect of particle size and process time on magnesite flotation using machine learning to predict flotation performance	EXS application and mathematical modelling	Flotation experiments performed on magnesite with seven size fractions and six flotation times	Feed particle size, flotation time, pulp pH, collector dosage, recovery rate of MgO and SiO_2	Optimal particle size range identified for magnesite flotation of 30 to 48 μm. The EXS method performs better than other models in predicting the MgO and SiO_2 recovery, which increased from 54.18% to 95.12% and from 50 to 400 mg/L, respectively. The SC model is an effective tool for predicting the effects of flotation parameters.

Table 4. *Cont.*

Paper	Problematic	Method	Data Used	Variables	Results
Ai et al. [22]	Set-point adaptive optimization and control strategy for antimony flotation process	Fuzzy logic functions, machine vision, and FAR mining to extract information and generate optimal set-points	Feed grade, reagent dosages, froth image, and concentrate grade. A total of 1000 groups of data in the desired concentrate grade range, with 950 groups to generate FARs and the rest for validation	Feed grade, reagent dosages, froth image, concentrate grade, and image features: froth height, froth colour, froth velocity, and bubble size	Better performance compared to manual manipulation and other automatic control methods with an FNN-based prediction accuracy for feed grade of RRMSE: 2.94% and MRE: 8.73%. Improved control performance in concentrate grade.
Zhang et al. [54]	Develop an adaptive modelling method for froth flotation reagent control	An adaptive ANN auto-regressive model (A-NNARX) for dynamic froth flotation control. In a non-linear model, the model predicts the flotation reagent control technique	Flotation industry data. The data are categorised by feed grade: low zinc, normal, high zinc, and high lead	Concentrate quality, zinc and lead feed grade, froth videos, flotation reagents, hand-crafted image characteristics, bubble size distribution. Colour, texture, and foam velocity. PCA components (control inputs, process outputs), quadratic cost function, Euclidean distance (ED) between the target foam image features and control results, histogram bins, test samples, and training data	The A-NNARX model improved the qualified ratio by 0.1666 compared to manual control, achieved better performance in terms of Euclidean distance, and increased the qualified ratio under expert control from 0.7500 to 0.8194 while decreasing the MAE error by 0.1384. A weight-level regularization method improved the capacity for deformation evaluation.

Ai et al. [25] and Aldrich et al. [23] discuss various aspects of froth flotation in the mining industry, including control strategies, machine vision through image analysis, and machine learning techniques for classifying froth structures. Moolman et al. [44] and Zhang et al. [53] include the use of digital image processing techniques to extract features from froths, the development of models to simulate the relationship between the reagent dosage and froth surface appearance, and the use of machine vision and predictive modelling to control the process. The papers provide numeric results and highlight the potential for improved process control and optimisation in flotation plants. Ali et al. [50], Fu et al. [31], and Ai et al. [22] discuss the use of artificial intelligence models to predict the flotation behaviour of fine high-ash coal in the presence of a hybrid ash depressant, the effect of particle size and flotation time on magnesite flotation, a set-point adaptive optimisation for the antimony flotation process applying an ANN soft sensor to estimate the feed grade online, and a FAR-based set-point generator to adjust the set-points of the image features. Zhang et al. [54] research an adaptive modelling method for industrial processes specifically in the context of froth flotation reagent control. The method involves an incremental learning approach to update the process model and preserve performance on old patterns while accommodating new process "excitation" patterns. The adaptive process model is then used in a nonlinear predictive control strategy for flotation reagent control. Ai et al. [25] provide an approach that involves fuzzy association setpoint calculation and offline Q-learning-based reagent control in froth flotation processes, which can improve the efficiency and effectiveness of industrial processes. Finally, Shean et al. [52] present a dynamic model for predicting changes in pulp height in aerated slurry tanks in froth flotation specifically. The model is developed and validated through the use of experimental data, which includes measurements of bubble size distributions and air flow.

The findings can be categorised into two groups when evaluated in relation to the response variable: flotation process optimisation and bubble size distribution. Chen et al. [46], Ramasamy et al. [45], Cortés et al. [47], and Ali et al. [50] discuss the use of artificial intelligence and model predictive control to optimise and improve the efficiency of mineral processing and flotation processes. These studies demonstrate the potential of these technologies to improve process control and optimisation, leading to increased profitability. Riquelme et al. [49] present a new technique for identifying and measuring bubble size distribution (BSD) in flotation columns using image processing and a parametric method. The circular Hough transform (CHT) is used to detect bubbles in the images, and a log-normal distribution is used to estimate the BSD. A non-linear Wiener model is developed to explain the dynamic behaviour of the BSD parameters. The results show that the CHT method is superior to other methods in detecting clustered bubbles, and the estimated number of bubbles is similar to what is obtained with a visual inspection. Another study, similar to the one developed by Moolman et al. [44], seeks to establish patterns from digital image processing, specifically using the spatial grey level dependence matrix (SGLDM) and neighbouring grey level dependence matrix (NGLDM) methods, to extract features from digitised images of froths in a copper flotation plant. The extracted features can be used to identify various process phenomena in the plant and develop control strategies for flotation plants. The paper also discusses the significance of froth appearance in flotation processes and the potential for better control of a plant through a more accurate and systematic interpretation of the physical features of the froth phase. A neural network was used to classify different froth structures based on the extracted features, and the study demonstrated the potential of neural networks for pattern recognition in complex processes. If an analysis of the variables included in these investigations is conducted, it is feasible to establish certain considerations:

(a) Flotation performance variables: These include parameters such as float recovery, total copper recovery, acid-insoluble copper recovery, and concentrate and tail grades. These variables are used to evaluate the effectiveness of different control strategies and technologies in improving flotation performance.
(b) Speed rate: An important parameter in flotation circuits, as it affects the residence time of particles in the circuit and can impact flotation performance. Operator control and MPC velocity control show the impact of different control strategies.
(c) Bubble size and gas holdup: These variables are important in understanding the behaviour of froth flotation systems, as they impact froth stability, mass pull, and flotation performance. The relationship between the bubble size and air rate can be used to predict changes in the pulp height.
(d) Reagent dosage: Reagents are used in flotation circuits to promote particle–bubble attachment and improve flotation performance. A relationship exists between the reagent dosage and froth surface appearance, and machine vision and predictive modelling can be used to control reagent dosage.
(e) Advanced measurement technologies: The use of advanced measurement technologies, such as image processing, X-ray fluorescence, and diffused reflective spectroscopy, improves the measurement and control of key variables in mineral processing operations. These technologies enable the more accurate and reliable measurement of parameters such as the Cu grade, froth velocity, and concentrate and tail grades.

From the analysis, artificial intelligence models have shown promising results in predicting flotation behaviour and classifying froth structures. For example, Chen et al. [46] show that decision tree methods and artificial neural networks can distinguish between froth features that are difficult to discern with the naked eye, with an overall classification accuracy of approximately 92.5% for copper flotation froths and 96% for platinum flotation froths. Similarly, Ramasamy et al. [45] show that Mamdani fuzzy logic (MFL) models performed the best among all the models tested for predicting the performance of the coal flotation process, with R-squared values of 0.9483 and 0.9243 for the training and testing phases, respectively, when predicting the froth ash content. Advanced measurement technologies, such as image processing, X-ray fluorescence, and diffused reflective spectroscopy, have shown promise in improving the measurement and control of key variables in mineral processing operations. For example, Brooks et al. [21] discuss the implementation of advanced technologies, including image processing, X-ray fluorescence, diffused reflective spectroscopy, and cascaded model predictive control (MPC), to optimise a copper roughing circuit. The results show improved process stability and increased copper recovery compared to previous control methods.

Model predictive control (MPC) has shown promise in optimising mineral processing operations and improving plant performance. Zhang et al. [54] provide an adaptive modelling method for industrial processes with a focus on froth flotation reagent control. The method employs an incremental learning strategy to update the process model and maintain performance in existing patterns while tolerating new process "excitation" patterns. The adaptive process model is then applied to a nonlinear model of predictive control. Experiments based on historical data and in a real-world lead–zinc froth flotation plant show promising results for practical application. The reagents used in the flotation process aim to change the surface properties of the mineral with a certain degree of release [15], and the hydrophobic properties cause them to separate from each other; then, they are contained in the pulp and selectively adhere to bubbles [79]. The relationship between the foam structure that develops on the pulp surface and the efficacy of the flotation process, namely in terms of the mineral grade and recovery, has been established in previous research [48]. By analysing commonalities between the studies, it is possible to identify that the use of advanced technologies and control strategies can lead to significant improvements in plant performance, including increased total and insoluble Cu recovery. Advanced measurement technologies such as image processing, X-ray fluorescence, and diffused reflective spectroscopy can enable more accurate and reliable measurements of key control parameters, such as the Cu grade, froth velocity, and concentrate and tail grades. The application of machine learning and artificial intelligence models can be used to predict flotation behaviour and classify froth structures, leading to improved control and optimisation of flotation processes.

Model predictive control (MPC) approaches have the potential to enhance the precision of control in the roughing circuit, enhance the stability of the float bank, and increase the recovery and grade. Froth image analysis has the potential to significantly improve process control and optimisation in flotation plants, but further research is needed to assess the predictability of froth grade from froth image features. These antecedents indicate that the approach developed in the soft computing application focuses more on the optimisation and control of the variables of the processes or on the process itself, but not on the control of the foam of a flotation process, since its structure can reflect the floating behaviour, indicating both the grade and the recovery [48]. In the MPC, the control signal is minimised in such a way that its squared error is reduced at each instant of time. What makes this model attractive are its algorithms since they have the natural ability to consider imposed technological and process constants as input and output values [80]. An indispensable part of this model is the use of algorithms from a dynamic model to predict the output values of a system based on past and current values, as well as future control signals. The development of this principle is complex to address in industrial systems, requires a broad knowledge of the physicochemical phenomena involved, and includes the execution of information from experiments that is exhaustive and sufficient to identify the parameters that are key in the process [58]. On the other hand, an MPC can be constructed using historical data, making this type of model more convenient than kinetic models [81]. For example, an artificial neural network of two phases was established by Sun et al. [82], in which one phase is used to derive and create a nominal model of a flotation process based on categories that do not follow an intrinsic classification and a second phase is designated to prevail against discordance with the nominal model. To achieve finite time control for a class of non-linear systems in this second phase, a radial basis function was implemented to handle the saturation of input variables that could occur and thereby generate constant error traceability.

Authors such as Kallioinen et al. [83], Harbort et al. [84], and Carr et al. [85] identified that the sizes of the flotation cells would increase their volume, with the main benefits being a reduction in capital expenditure and operating costs, lower energy consumption per cubic metre, a lower amount of instrumentation, and facilitation of the control of each cell. The use of advanced control at this stage is of great interest due to its high performance in terms of predictability and data analysis in the representation of complex processes. With this, an advanced control strategy can be implemented to achieve adequate reagent control where the model can be updated as the base data acquires new monitoring data using a model predictive control strategy [86]. This will significantly impair the prediction performance when a dynamic model is used [87]. The adaptive model technique has the potential to solve this problem by maintaining the performance of old data and keeping the dynamic model updated with the new "excitation" patterns based on the new measured data [88]. Rosenfeld et al. [89] employed a methodology consisting of four steps: training, data tuning (finetuning), retraining, and incremental learning. This strategy involved the integration of data training and self-learning approaches.

Therefore, an incremental learning method is an interesting alternative which seeks to update industrial process models based on neural networks. Unlike traditional incremental learning, where the discrepancy or restriction between an original and updated model is measured by the "deformation" of all hidden networks, the deformation of the last hidden layer of the neural network is concentrated. Furthermore, a weight regularisation method is designed for the last hidden layer of the neural network to show the performance capacity through strain evaluations. Ultimately, the foam flotation process, which exhibits significant inconsistencies, is controlled by the implementation of predictive control procedures utilising an empirically derived model that is continuously updated in an incremental manner.

2.6. Applications of Soft Computing in the Hydrometallurgy Stage

In the context of hydrometallurgical processes, various techniques have been employed to enhance the efficiency and control of the process. Methods such as artificial neural networks and multiple linear regression, data-driven model predictive control, dynamic process simulators, and fuzzy parameter self-tuning PID control algorithms have been implemented in the process, which includes leaching, column leaching, and solvent extraction (Table 5). Azizi et al. [57] and Komulainen et al. [90] use mechanistic models to explain how copper moves from the watery phase to the organic phase during the extraction process. The first study places emphasis on modelling the settler hydrodynamics, while the second one presents a dynamic process simulator for copper solvent extraction plants using mixer–settlers. Both studies report good agreement between simulations and measurements, indicating that the mechanistic models accurately describe the process trends. Komulainen et al. [90] emphasise the importance of parameter adaptation for modelling accuracy, especially for efficiency parameters. Gao et al. [58] use ANN to predict gold recovery and thiourea (TU) consumption during the leaching process. The results show a high prediction accuracy for the trained model, with a minimum absolute error of gold recovery varying from 1.46% to 3.45% and a prediction accuracy between 94.46% and 98.06%. The absolute errors of TU consumption varied from 0.079 to 0.428 kg/t, and the accuracy of the predictions was between 95.15% and 99.20%. As previously stated, the variables selected for model generation are part of the regular monitoring that is carried out in operation, which includes observation of the copper concentration [57,58,90], temperature [57,58], leaching time [58,90], and pH [58,90].

Table 5. Characteristics of studies applied in mineral hydrometallurgical processes.

Paper	Problematic	Method	Data Used	Specific Variables	Results
Komulainen et al. [90]	Developing a dynamic process simulator for copper solvent extraction plants	Mechanistic models for MPC	One month of operating data from an industrial copper solvent extraction process	Input variables: PLS concentration, lean electrolyte concentration and rate, and flow rates. Output variables: Loaded and barren organic concentrations and rich electrolyte and raffinate concentrations	Mechanistic models accurately describe the SX process trends. The mean residual is well below 2% for organics and rich electrolyte and around 6% for raffinates, which is considered very good considering the poor measurement accuracy of these streams
Moreno et al. [55]	Developing a dynamic model for mixer–settler units used in the solvent extraction (SX) process of copper plants	Dynamic modelling for MPC	Copper concentration, pH, SX operation variables, equilibrium isotherm calculations, and mass transfer expressions	Input variables: Aqueous inlet flowrate, pH, Cu^{+2} aqueous and organic inlet, organic inlet flowrate, mixer volume, settler volume, and free acidity in electrolyte. Output variables: Volumes of phases in the mixer, flowrates at the mixer exit, Cu^{+2} in aqueous phase at the mixer exit, and Cu^{+2} in organic phase at the mixer exit	By incorporating time delay and flexible model fitting parameters, a better fitting settler model was found, accurately reproducing changes in SX. The relative mean squared errors for outlet copper concentrations in the extraction unit were 0.03% (aqueous phase) and 6.76% (organic phase), while for the stripping unit they were 0.07% (aqueous phase) and 2.89% (organic phase).
Pang et al. [56]	Improving control performance in leaching rare earths using a self-tuning PID control algorithm	Fuzzy parameter self-tuning PID control algorithm	Simulation data	pH and element contents in solution. Proportional, integration, and differential adjustment factors	Simulation results show that the fuzzy parameter self-tuning PID control algorithm outperforms traditional PID control algorithms in terms of control performance, response time, and accuracy. A relationship between the pH value and the amount of the initial solution is established

Table 5. *Cont.*

Paper	Problematic	Method	Data Used	Specific Variables	Results
Azizi et al. [57]	Predicting gold recovery in the cyanidation process	ANN and MLR	Cyanide leaching circuit of gold mine	Input includes pH, solid percentage, NaCN concentration, particle size and leaching time. Output: Au recovery	ANN provides efficient and cost-effective, with highly accurate predictions of 0.556 for the training and 0.67 compared to MLR. Leaching time and particle size are the most important factors affecting gold extraction.
Hoseinian et al. [33]	Optimization of the copper oxide column leaching process	ANN and GANN	A database of 120 sets of copper leaching column tests; 96 sets were used to train the network and 24 sets were used to test the model	Particle size, column height, leaching time, and acid flow rates	Copper recovery has an inverse relation with the column height and particle size and a direct relation with the leaching time and the acid flow rate. The GANN model is more efficient than the ANN model for Cu recovery prediction with reasonable accuracy. The algorithm can be incorporated in the training phase of a network to improve prediction.
Gao et al. [58]	Developing a data-driven model predictive control approach for dynamic systems	Data-driven MPC approach that combines modified partial least squares (PLS) and MPC	Multiple input and output variables from laboratory tests	Input variables: Input flows and heat to the tank. Output variables: Cooling water temperature, atmospheric temperature, steady state temperature, and steady state level	The proposed MPC approach has high prediction precision and an ability to cope with dynamics in the process, outperforming traditional MPC and MPC in a traditional PLS framework. This was tested on a continuous stirred tank heater system
Xu et al. [59]	Using thiourea as an alternative to cyanide for gold leaching from refractory ores	Grey relational analysis and artificial neural network models	Results of leaching experiments conducted on a high-arsenic gold concentrate using A. ferrooxidans and TU as a leaching agent	Leaching time, initial pH, temperature, TU dosage, stirring speed, and ferric iron concentration	GRA and ANN models can efficiently reflect practice and provide effective suggestions for controlling optimum parameters in the leaching process. The absolute errors of gold recovery varied by 2.5% and the accuracy of the predictions was around 96%. The accuracy of the other predictions was around 97%.

There are specific variables according to the reality of the process and operating conditions, including aqueous and organic phases [57,90], ferric iron concentrations [58], and settler hydrodynamics [57]. Based on existing studies, there is an efficient and cost-effective method [55] with high prediction accuracy and improved control performance using artificial neural networks (ANN) and genetic algorithm neural networks (GANN), as presented by Xu et al. [59]. For example, Hoseinian et al. [33] report that the GANN model is more efficient than the ANN model for predicting copper recovery in column leaching processes. Gao et al. [58] establish an ANN model to predict gold recovery and TU consumption with high prediction accuracy, and Pang et al. [56] demonstrate superior control performance, reaction time, and accuracy compared to the classic PID control algorithm.

From the studies applied to mineral hydrometallurgical processes, the most relevant variables that have been identified include:

(a) pH: This variable plays a significant role in controlling the behaviour of chemical reactions. It is an important variable in the leaching process as it can affect the rate of element recovery.
(b) Particle size: Refers to the size of the particles of the element of interest in the ore. It is an important factor that can affect the efficiency of the leaching process.
(c) Temperature: A key variable that affects the efficiency and kinetics of chemical reactions, directly affecting the rate of element recovery
(d) Time: Refers to the duration of processes such as leaching. It is a critical variable as it determines the amount of time available for the element of interest to dissolve into the leaching solution.
(e) Element of interest's grade: The grade or initial quantity of the element to recover is a relevant base variable considered in predicting models for copper and gold recovery. This variable can be complemented by the change in element concentration in solution during the process.
(f) Reactive consumption: Reagents are used to increase the kinetics of chemical reactions. Here, the consumption of thiourea or agents such as ferric ions can impact the recovery efficiency.
(g) Solid percentage: Corresponds to the proportion of solid material in a solution. It is a relevant parameter in solid–liquid processes such as stirring leaching that can influence the efficiency.
(h) Stirring speed: Refers to the speed at which the leaching solution is agitated. It is a parameter that can influence the contact between the particles of the element of interest and the leaching solution, affecting the recovery rate.

2.7. Applications of Soft Computing in the Pyrometallurgy Stages

In high-temperature metallurgical processes such as copper losses, slag monitoring, silicon content prediction, and matte grade optimisation, soft computing methods such as MPC, ANN, CNN, FL, dynamic modelling and simulation, hybrid intelligent models (combining fuzzy logic, neural networks, and other techniques), feature selection methods based on mutual information, genetic algorithms, and other techniques have been applied, improving the efficiency, productivity, and cost-effectiveness of metallurgical processes (Table 6). As previously mentioned, the slag compositions have a notable influence on the recovery of the element of interest, such as gold or copper. These methods are used to develop and validate models for predicting matte grade in the copper flash smelting process or hot metal quality in a blast furnace. The base variables identified in this type of process include temperature, pressure, chemical composition, flow rate, time, and pH. Some specific variables are used depending on the pyrometallurgical process being studied, such as the grade, element content, and slag basicity and volume. From the studies, it was identified that soft computing applications improve accuracy and efficiency in processes such as the prediction of steel quality or the optimisation of blast furnace ironmaking. Additionally, the importance of feature selection is highlighted in developing accurate models and the ways ensemble learning methods can better cope with complex variations in industrial processes compared to traditional machine learning methods.

cating their versatility but also their inconsistency. The artificial neural network (ANN) method appears to be a common method that can be applied in all stages, performing well in the criteria for all stages except for hydrometallurgical processes, where the CNN method performs slightly better in terms of accuracy and reliability and FL stands out as a versatile choice.

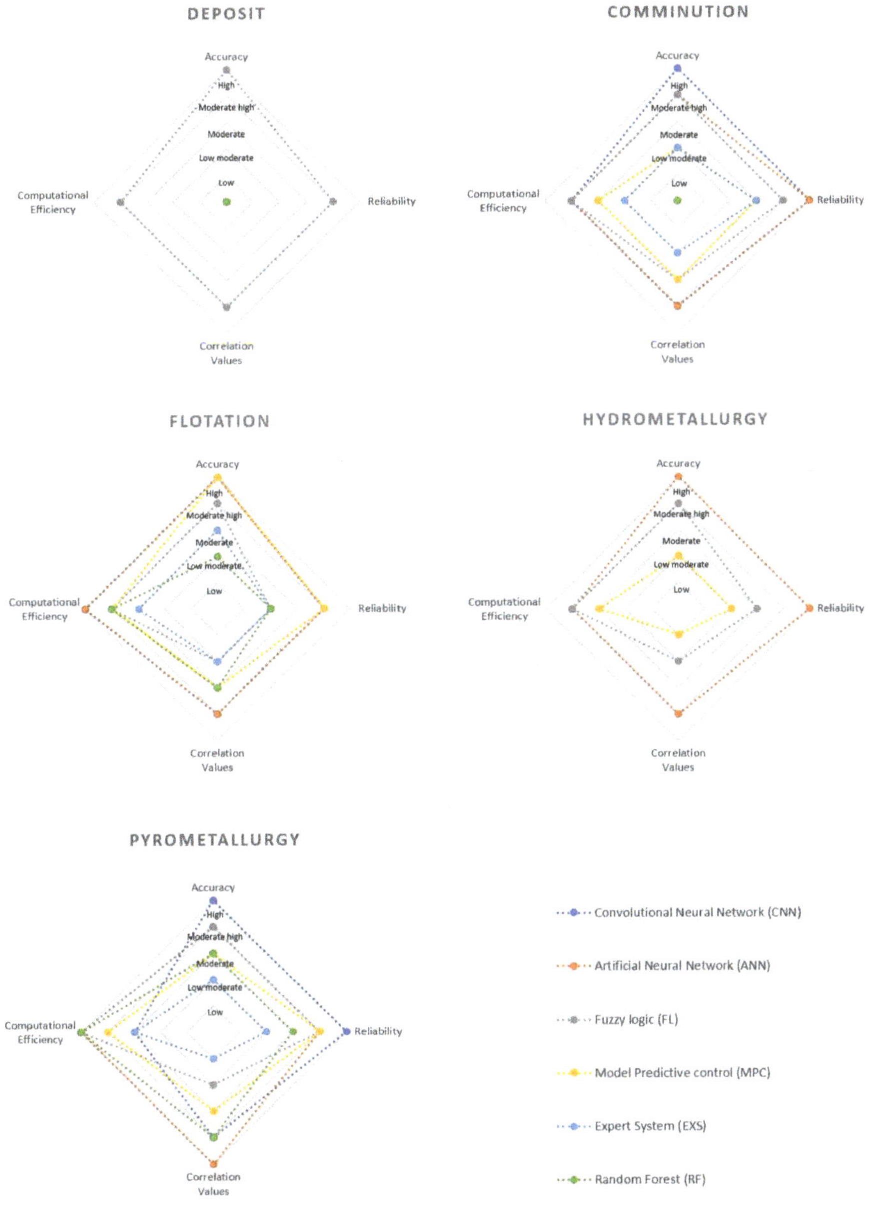

Figure 5. Qualitative comparison of soft computing methods applied in mineral processing.

4. Proposed Approach for the Application of Soft Computing in Waste Disposal in Mineral Extraction and Processing

One critical challenge to ensuring medium- and long-term operation is to achieve the adequate disposal and management of mining tailings (Residues from sulphide mineral flotation composed of material without the mineral of interest and water that are disposed in specific deposits known as TSF) and spent heap leaching spoils (Material that remains as a residue from the heap leaching process once all of the mineral of interest has been recovered). This has been declared as critical within the sustainability and responsible operation programs by the European Union to ensure proper management to avoid damaging the environment [74,91]. The incorporation of better technology for tailing storage facilities (TSFs) and spent heap leaching spoils dumps (leaching waste deposit, LWD) and the development of mechanisms for the measurement of parameters and variables are the final objective in controlling physical and chemical stability monitoring systems and establishing a robust understanding of the makeup and water balance of the process. In Chile, copper production as a concentrate will reach 89.9% in the year 2027 due to the depletion of oxidised copper minerals [74], while the use of continental water consumption exceeds 0.36 m^3/ton of ore, with the consumption expected to rise further [92]. The application of soft computing to waste disposal may provide quality, reliable, and real-time information on the waste ore and tailings according to the operational needs. In addition, authorities and communities may benefit from the accessibility of this information, to improve, make transparent, and strengthen the ties between the industry and the communities, taking an important step to comply with national legislation and to improve the sustainability culture.

The proper management and ultimate disposal of these wastes is critical to the continuity of the profit chain and are linked to a successful storage operation, adequate use of land, security, and, in a relevant way, to the environmental commitment of the mining operation. In countries with a mining tradition, such as Chile [93], Finland [94], and Canada [95], regulations and guides provide information to design, install, and manage waste rock dumps and TSFs focused on the beginning of the deposit creation, leaving the subsequent monitoring and control of these deposits to their own criteria against structural behaviour, physicochemical stabilization, and mitigation of possible failures. Industrial practices and the changes that take place in a dynamic operational environment mean that on many occasions the original design is strongly modified for safety or even does not reach the useful life that was originally planned. The stability of a tailings storage facility (TSF) can be influenced by [96,97]:

- Operating factors (input material, deposition rate, geometrical and geotechnical controls such as humidity and compaction).
- Deposit location (climate and geological factors that include the seismicity, ground foundation slope, and confinement of the land degree)
- Deposit type selected (type of the TSF, geometric configuration including height, volume, and slope angle)

Establishing the use of soft computing to advance the "mining digitalization" of processes such as the monitoring of mining tailings and spent heap leaching spoils transport and deposition is a solution that could be non-invasive and consistently obtain a higher quality control and constant evolutionary knowledge by identifying the most influential variables in these processes, minimising possible prediction errors. A dependency on the tailings and spent heap leaching spoils disposal fluency behaviour can be established based on the dynamic mineralogy of the ore and rheological and permeability characteristics. In the case of tailings, the presence of specific clays and changes in the solid concentration in the thickener discharge cause changes in the TSF behaviour, modifying the established area and volume of the disposal, changing the functionality and possibly creating environmental impacts. The evaluation of these parameters with soft computing enables correlation and identification of a behavioural pattern. This implementation can also serve as an update for operational decisions related to environmental demands in countries with a mining tradition, such as Chile and Finland. These countries have recognised the significance of valuing existing data and conducting thorough analysis in order to inform operational decision-making processes. In this approach, we establish the relevance of pulp characterization as a part of a predictive model of fluency behaviour for tailings and leaching waste ore generated from variables that are currently measured at mining sites.

A prior analysis must be generated to help identify the impact generated by the conditions used in the operation and that will translate into the tailings and spent heap leaching spoils behaviour. An alternative used to identify these variables corresponds to the application of a correlation, and thereby establishes the combination of variables to be used with soft computing tools such as ANN or CNN. Then, the parameters of the new empirical model are adjusted in a very precise way, using only the new measurement data from the new model [98]. However, a low level of learning is achieved using this method, since "catastrophic forgetting" can occur, which means that while fine tuning is established in the new model with new data, the historical data performance is drastically impaired due to the discrepancies between the old and new data sets [99]. Retraining also uses the weights from the existing empirical model and uses it to create a new empirical model; unlike data fitting, the parameters of the new empirical model are precisely adjusted using historical data and current data from the new model from a simultaneously [100]. This scope can solve the "catastrophic forgetting"; however, this method needs important periods of time in order to update the data continuously [61]. This analysis also raises the question of whether, if a robust database could be established with data from different mining operations, it could find common ground between the studied cases that could be correlated between the different operations considering critical mineralogical factors such as the presence of clays and changes in rheological behaviour due to the medium used, such as fresh water or sea water. Naturally, this presents the potential to manage different perspectives to find the right path in the soft computing application. For example, the application of an incremental method aims for the "excitation" patterns of a new process to be accommodated without compromising the performance of the patterns of a process with historical data. In general, this type of method adapts new patterns by designating new constraints or constant rules to modify the adjustable parameters of the updated model. It is more efficient than a retraining method because the empirical model does not need to be trained on all historical data [101].

The use of incremental learning as a part of an artificial neural network can be a real option considering the flexibility in the quantity and interaction of the input nodes that can be applied, but also considering the time and application capacity given the available data and the operation monitoring frequency used at an industrial level. An opportunity to use ANN for operational decision-making lies in the existing monitoring capacity of the variables related to the mining waste operation, access to this information, and how robust this data is. As part of the analysis, Figure 6 illustrates a proposed database generation to analyse the most appropriate application of soft computing in the deposition of the tailings and spent heap leaching spoils. This database will serve as the foundation for conducting a preliminary stage of weight identification for each variable, followed by subsequent analysis.

Figure 7 presents an analysis framework that explores potential connections using various soft computing techniques, including neural networks. This framework incorporates training, prediction, and confirmation stages.

As is shown, the proposal emphasises the importance of standardising the database for macroscopic data collection, considering the structure, properties, processes (hydrogeology and geochemistry), and operation (flows, Cp, disposal) as possible options for input data modelling. This approach ensures a comprehensive understanding of the parameters involved. A second aspect explores alternatives for deep learning applications in advanced control, emphasising the results obtained from the analysis presented in this publication and including neural networks, SIP model applications, SVR (support vector regression), stochastic mathematical programming formulation, and principal component analysis (PCA). These techniques offer different methodologies and approaches to effectively analysing and controlling the parameters. The third aspect focuses on dependent modelling generation and highlights critical variables for analysis, evaluation, and prediction. Hydrogeology, operational monitoring and rheology, and geochemistry are identified as potential options. By taking into account these variables, one can acquire valuable insights into the interconnectedness among them and make accurate predictions about future behaviour.

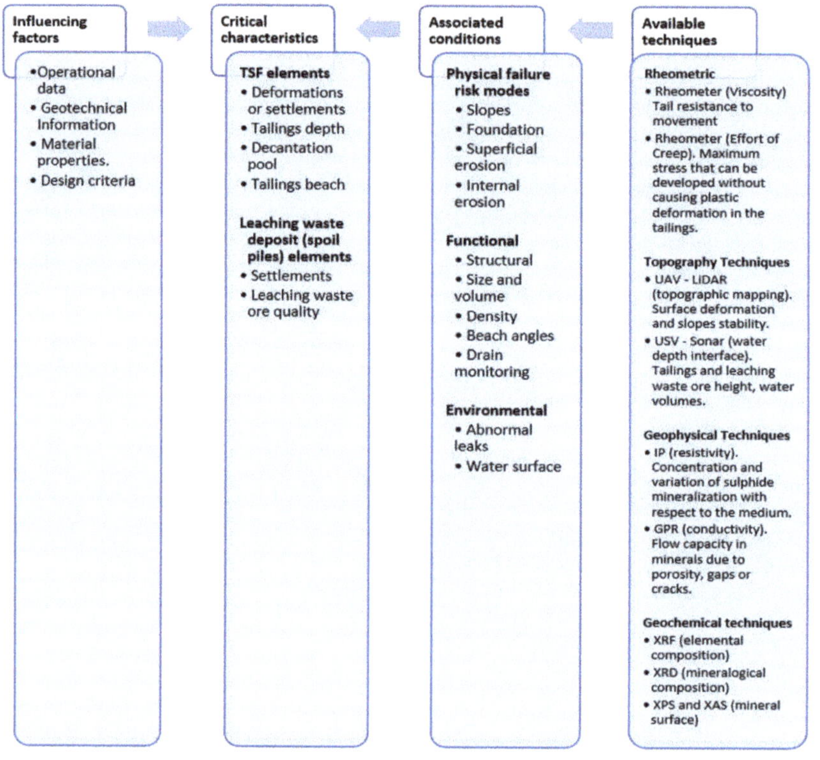

Figure 6. Parameters for a database generation for TSF and LWD.

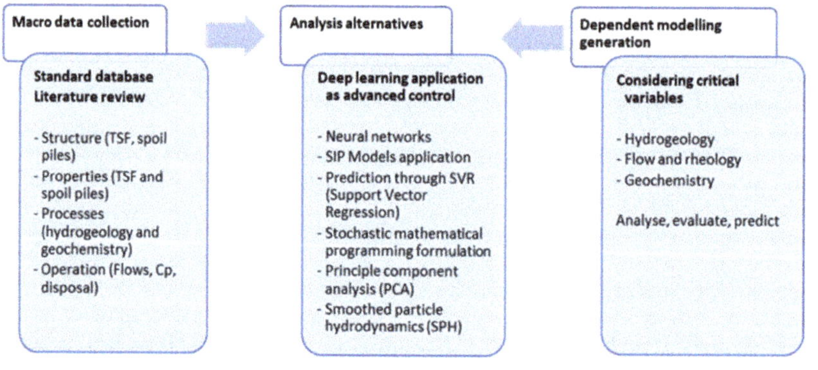

Figure 7. Example of possible correlations between parameters with different soft computing techniques.

5. Conclusions

This research provides compelling evidence for the growing adoption of soft computing as a viable solution in the design, development, and operation of intelligent systems for various phases of mineral processing and metal extraction operations. Soft computing has the ability to adapt, learn, and act independently in the mining industry. By considering the problem at hand, the characteristics of the environment, and the variables involved, it can develop and compute solutions to complex problems. Given the complex and non-linear nature of the processes involved, selecting the appropriate type of soft computing method becomes crucial when considering the various options available, such as artificial neural networks, expert systems, fuzzy algorithms, and more. These alternatives present interesting opportunities for addressing the complex and multivariable behaviours displayed by the processes, opening up new avenues for analysis and optimisation.

This publication presents the considerable potential of sophisticated technologies and tactics in enhancing safety, decision-making processes, and plant performance within the mining and mineral processing sectors. Fuzzy logic, machine vision, artificial neural networks, and advanced measurement technologies are among the key tools discussed. The effectiveness of these approaches is demonstrated in areas such as risk assessment, accident prediction, process modelling, control performance improvement, prediction accuracy enhancement, optimal condition determination, and composition estimation. Additionally, the importance of feature selection, soft computing approach selection, big data utilisation, and database treatment is emphasised. Further research is suggested in areas such as froth image analysis, where, thanks to the fact that image acquisition technologies have been evolving and reducing costs, it is possible to apply soft computing and reliable data linking to help optimise mineral recovery performance and improve profitability, reduce operational costs, and enhance real-time monitoring of froth properties. Overall, the integration of advanced technologies, control strategies, and data-driven approaches holds great promise for optimising processes, enhancing efficiency, and reducing instability in the mining industry. On the other hand, as our investigation shows, there is a real possibility of implementing soft computing in other critical processes for operational continuity in the mining industry, such as the transport and deposit of tailings and spent heap leaching. We have successfully identified key variables that are crucial for the successful implementation of soft computing techniques in various industrial applications. These variables can be effectively monitored and collected at an industrial level. For comminution stages, important variables include the feed rate, screen aperture size, and crusher rotor speed. In deposit studies, variables such as the deposit element grade, tonnage, mining and processing costs, and metal prices play a significant role. In pyrometallurgical processes, variables such as the temperature, gas flow rate, and particle size distribution of the feed material are essential. Additionally, in processes involving froth and leaching, variables such as the bubble size, froth class, viscosity, pH, temperature, leaching time, and concentrations of sulfuric acid and copper in the leach solution are critical factors to consider.

The creation of a comprehensive database derived from diverse mining operations can facilitate the identification of shared characteristics and develop correlations among essential mineralogical parameters. Incremental learning through artificial neural networks can aid in operational decision-making. Analysing variables such as hydrogeology, flow, rheology, and geochemistry can be a good approach to the implementation of this type of technology for the control of mining waste. Finally, the application of this type of technology over time does not depend on the application of the tool itself but on the source, availability, and quality of the data that can be obtained.

Author Contributions: N.H., M.S.G., J.O. and R.M. contributed to the methodology, conceived and designed the experiments; N.H. analysed the data and wrote the paper. All authors have read and agreed to the published version of the manuscript.

Funding: This research received no external funding.

Acknowledgments: The authors sincerely and gratefully acknowledge the support and guidance provided by the Oulu Mining School from the University of Oulu and the Geological Survey of Finland GTK.

Conflicts of Interest: The authors declare no conflict of interest.

References

1. Ramirez, O. *Digitization of Mining: Major Challenges and Motivations*; SPT Mining: Stockholm, Sweden, 2021.
2. European Parliament. *Resource Efficiency: Moving towards a Circular Economy*; European Parliament: Strasbourg, France, 2014.
3. Ge, Z. Review on data-driven modeling and monitoring for plant-wide industrial processes. *Chemom. Intell. Lab. Syst.* **2017**, *171*, 16–25. [CrossRef]
4. DOMO. Data Never Sleeps 10.0. Available online: https://www.domo.com/es/data-never-sleeps (accessed on 24 October 2023).
5. Taylor, P. Amount of Data Created, Consumed, and Stored 2010–2020, with Forecasts to 2025. Statista. Available online: https://www.statista.com/statistics/871513/worldwide-data-created/ (accessed on 24 October 2023).
6. Duarte, F. Amount of Data Created Daily. Exploding Topics. 2023. Available online: https://explodingtopics.com/blog/data-generated-per-day (accessed on 24 October 2023).
7. Usman, M.; Ma, Z.; Zafar, M.W.; Waheed, A.; Li, M. Analyzing the determinants of clean energy consumption in a sustainability strategy: Evidence from EU-28 countries. *Environ. Sci. Pollut. Res.* **2021**, *28*, 54551–54564. [CrossRef] [PubMed]
8. Krzaklewski, M.; Van Laere, H. Opinion of the European Economic and Social Committee on "Digital Mining in Europe: New solutions for the sustainable production of raw materials". European Economic and Social Committee, Consultative Commission on Industrial Change. 2020. Available online: https://eur-lex.europa.eu/legal-content/EN/ALL/?uri=CELEX:52020IE1559 (accessed on 24 October 2023).
9. Harris, S. *Unearthing the Future: How Digital Is Revolutionizing the Mining Industry*; Orange Business Services: Paris, France, 2021.
10. Fundación Chile. *ROADMAP: Digitalización para una Minería 4.0*; Fundación Chile: Santiago, Chile, 2020.
11. Qi, C.-c. Big data management in the mining industry. *Int. J. Miner. Metall. Mater.* **2020**, *27*, 131–139. [CrossRef]
12. Flores, V.; Hadfeg, Y.; Bekios, J.; Quelopana, A.; Meneses, C. A method for automatic generation of explanations from a rule-based expert system and ontology. In *Advances in Intelligent Systems and Computing*; Springer: Berlin/Heidelberg, Germany, 2017; pp. 167–176. [CrossRef]
13. Bergh, L.G.; Yianatos, J.B.; Leiva, C.A. Fuzzy supervisory control of flotation columns. *Miner. Eng.* **1998**, *11*, 739–748. [CrossRef]
14. Sanchez, J.A. *Control Avanzado de Procesos (Teoría y Práctica)*; Diaz de Santos S.A.: Madrid, Spain, 2003.
15. Sbárbaro, D.; del Villar, R. *Advanced Control and Supervision of Mineral Processing Plants*; Springer: London, UK, 2010. [CrossRef]
16. Nad, A.; Jooshaki, M.; Tuominen, E.; Michaux, S.; Kirpala, A.; Newcomb, J. Digitalization Solutions in the Mineral Processing Industry: The Case of GTK Mintec, Finland. *Minerals* **2022**, *12*, 210. [CrossRef]
17. Argyropoulos, S.A. Artificial Intelligence in Materials Processing Operations: A Review and Future Directions. *ISIJ Int.* **1990**, *30*, 83–89. [CrossRef]
18. Shean, B.J.; Cilliers, J.J. A review of froth flotation control. *Int. J. Miner. Process.* **2011**, *100*, 57–71. [CrossRef]
19. Uusi-Hallila, S.; Paavola, M.; Leiviskä, K. *Utilizing Froth Phase Behaviour and Machine Vision to Indicate Flotation Performance*; University of Oulu: Oulu, Finland, 2014.
20. Philip, T.P. Process Control in Metallurgical Plants—From an Xstrata Perspective. *IFAC Proc. Vol.* **2007**, *40*, 377–389. [CrossRef]
21. Brooks, K.; Munalula, W. Flotation Velocity and Grade Control Using Cascaded Model Predictive Controllers. *IFAC-PapersOnLine* **2017**, *50*, 25–30. [CrossRef]
22. Ai, M.; Xie, Y.; Tang, Z.; Zhang, J.; Gui, W. Deep learning feature-based setpoint generation and optimal control for flotation processes. *Inf. Sci.* **2021**, *578*, 644–658. [CrossRef]
23. Aldrich', C.; Moolman, D.W.; Gouws, F.S.; Schmitz, G.F. Machile Learning Strategies for Control of Flotation Plants. *IFAC Proc. Vol.* **1995**, *28*, 99–105. [CrossRef]
24. Hadler, K.; Cilliers, J.J. The relationship between the peak in air recovery and flotation bank performance. *Miner. Eng.* **2009**, *22*, 451–455. [CrossRef]
25. Ai, M.; Xie, Y.; Xie, S.; Zhang, J.; Gui, W. Fuzzy association rule-based set-point adaptive optimization and control for the flotation process. *Neural Comput. Appl.* **2020**, *32*, 14019–14029. [CrossRef]
26. Valera, A.; Vallés, M.; Díez, J.L. Simulación y Control de Procesos Físicos de Forma Remota. In *Revista Iberoamericana de Automatica e Informática Industrial*; Universidad Politécnica de Valencia: Valencia, Spain, 2005; pp. 20–29.
27. Romero, D.; Bernus, P.; Noran, O.; Stahre, J.; Berglund, Å.F. The operator 4.0: Human cyber-physical systems & adaptive automation towards human-automation symbiosis work systems. In *IFIP Advances in Information and Communication Technology*; Springer: New York, NY, USA, 2016; pp. 677–686. [CrossRef]
28. Liu, D.; Yuan, Y.; Liao, S. Artificial neural network vs. nonlinear regression for gold content estimation in pyrometallurgy. *Expert Syst. Appl.* **2009**, *36*, 10397–10400. [CrossRef]
29. Estrada, F.; Cipriano, A. Hybrid Model Predictive Control for Grinding Plants. In Proceedings of the 19th World Congress the International Federation of Automatic Control, Cape Town, South Africa, 24–29 August 2014.

30. Martin, V.; Eng, P.; Fontaine, D.; Cathcart, J. Challenges with conducting tailings dam breach studies. In Proceedings of the Tailings and Mine Waste 2015, Vancouver, BC, Canada, 26–28 October 2015.
31. Fu, Y.; Yang, B.; Ma, Y.; Sun, Q.; Yao, J.; Fu, W.; Yin, W. Effect of particle size on magnesite flotation based on kinetic studies and machine learning simulation. *Powder Technol.* **2020**, *376*, 486–495. [CrossRef]
32. McCoy, J.T.; Auret, L. Machine learning applications in minerals processing: A review. *Miner. Eng.* **2019**, *132*, 95–109. [CrossRef]
33. Hoseinian, F.S.; Abdollahzade, A.; Mohamadi, S.S.; Hashemzadeh, M. Recovery prediction of copper oxide ore column leaching by hybrid neural genetic algorithm. *Trans. Nonferrous Met. Soc. China (Engl. Ed.)* **2017**, *27*, 686–693. [CrossRef]
34. Sun, W.; Xue, Y. An Improved Fuzzy Comprehensive Evaluation System and Application for Risk Assessment of Floor Water Inrush in Deep Mining. *Geotech. Geol. Eng.* **2019**, *37*, 1135–1145. [CrossRef]
35. Danish, E.; Onder, M. Application of Fuzzy Logic for Predicting of Mine Fire in Underground Coal Mine. *Saf. Health Work* **2020**, *11*, 322–334. [CrossRef]
36. Li, M.; Wang, H.; Wang, D.; Shao, Z.; He, S. Risk assessment of gas explosion in coal mines based on fuzzy AHP and bayesian network. *Process Saf. Environ. Prot.* **2020**, *135*, 207–218. [CrossRef]
37. Stange, W. Using Artificial Neural Networks for the Control of Grinding Circuits. *Miner. Eng.* **1993**, *6*, 479–489. [CrossRef]
38. Tessier, J.; Duchesne, C.; Bartolacci, G. A machine vision approach to on-line estimation of run-of-mine ore composition on conveyor belts. *Miner. Eng.* **2007**, *20*, 1129–1144. [CrossRef]
39. Olivier, L.E.; Craig, I.K.; Chen, Y.Q. Fractional order and BICO disturbance observers for a run-of-mine ore milling circuit. *J. Process Control.* **2012**, *22*, 3–10. [CrossRef]
40. Hamzeloo, E.; Massinaei, M.; Mehrshad, N. Estimation of particle size distribution on an industrial conveyor belt using image analysis and neural networks. *Powder Technol.* **2014**, *261*, 185–190. [CrossRef]
41. Umucu, Y.; Deniz, V.; Bozkurt, V.; Fatih Çağlar, M. The evaluation of grinding process using artificial neural network. *Int. J. Miner. Process.* **2016**, *146*, 46–53. [CrossRef]
42. Cai, W.; Dou, L.; Zhang, M.; Cao, W.; Shi, J.Q.; Feng, L. A fuzzy comprehensive evaluation methodology for rock burst forecasting using microseismic monitoring. *Tunn. Undergr. Space Technol.* **2018**, *80*, 232–245. [CrossRef]
43. Olivier, L.E.; Maritz, M.G.; Craig, I.K. Deep Convolutional Neural Network for Mill Feed Size Characterization. *IFAC-PapersOnLine* **2019**, *52*, 105–110. [CrossRef]
44. Moolman, D.W.; Aldrich, C.; Van Deventer, J.S.J.; Stange, W.W. The classification of froth structures in a copper flotation plant by means of a neural net. *Int. J. Miner. Process.* **1995**, *43*, 193–208. [CrossRef]
45. Ramasamy, M.; Narayanan, S.S.; Rao, C.D.P. Control of ball mill grinding circuit using model predictive control scheme. *J Process Control* **2005**, *15*, 273–283. [CrossRef]
46. Chen, X.-s.; Zhai, J.-y.; Li, S.-h.; Li, Q. Application of model predictive control in ball mill grinding circuit. *Miner. Eng.* **2007**, *20*, 1099–1108. [CrossRef]
47. Cortés, G.; Verdugo, M.; Fuenzalida, R.; Cerda, J.; Honeywell, E.C. Rougher Flotation Multivariable Predictive Control; Concentrator A-1 Division CODELCO Norte. In Proceedings of the V International Mineral Processing Seminar, Santiago, Chile, 22–24 October 2008.
48. Aldrich, C.; Marais, C.; Shean, B.J.; Cilliers, J.J. Online monitoring and control of froth flotation systems with machine vision: A review. *Int. J. Miner. Process.* **2010**, *96*, 1–13. [CrossRef]
49. Riquelme, A.; Desbiens, A.; Del Villar, R.; Maldonado, M. Identification of a non-linear dynamic model of the bubble size distribution in a pilot flotation column. *Int. J. Miner. Process.* **2015**, *145*, 7–16. [CrossRef]
50. Ali, D.; Hayat, M.B.; Alagha, L.; Molatlhegi, O.K. An evaluation of machine learning and artificial intelligence models for predicting the flotation behavior of fine high-ash coal. *Adv. Powder Technol.* **2018**, *29*, 3493–3506. [CrossRef]
51. Hoseinian, F.S.; Aliakbar, A.; Bahram, R. Semi-autogenous mill power prediction by a hybrid neural genetic algorithm. *J. Cent. South Univ.* **2018**, *25*, 151–158. [CrossRef]
52. Shean, B.; Hadler, K.; Neethling, S.; Cilliers, J.J. A dynamic model for level prediction in aerated tanks. *Miner. Eng.* **2018**, *125*, 140–149. [CrossRef]
53. Zhang, J.; Tang, Z.; Ai, M.; Gui, W. Nonlinear modeling of the relationship between reagent dosage and flotation froth surface image by Hammerstein-Wiener model. *Miner. Eng.* **2018**, *120*, 19–28. [CrossRef]
54. Zhang, J.; Tang, Z.; Xie, Y.; Ai, M.; Zhang, G.; Gui, W. Data-driven adaptive modeling method for industrial processes and its application in flotation reagent control. *ISA Trans.* **2021**, *108*, 305–316. [CrossRef]
55. Moreno, C.M.; Pérez-Correa, J.R.; Otero, A. Dynamic modelling of copper solvent extraction mixer-settler units. *Miner. Eng.* **2009**, *22*, 1350–1358. [CrossRef]
56. Pang, Q.; Fu, P.; Zhong, X. Fuzzy control of pH value in rare-earth impurity leaching process. In Proceedings of the 2011 International Conference on Electronics, Communications and Control, ICECC 2011—Proceedings, Ningbo, China, 9–11 September 2011; pp. 2065–2067. [CrossRef]
57. Azizi, A.; Ghaedrahmati, R.; Ghahramani, N.; Rooki, R. Modelling and simulation of the cyanidation process of Aghdareh gold ore using artificial neural network and multiple linear regression. *Int. J. Min. Miner. Eng.* **2016**, *7*, 139–154. [CrossRef]
58. Gao, T.; Yin, S.; Qiu, J.; Gao, H.; Kaynak, O. A partial least squares aided intelligent model predictive control approach. *IEEE Trans. Syst. Man Cybern. Syst.* **2018**, *48*, 2013–2021. [CrossRef]

59. Xu, R.; Nan, X.; Meng, F.; Li, Q.; Chen, X.; Yang, Y.; Xu, B.; Jiang, T. Analysis and prediction of the thiourea gold leaching process using grey relational analysis and artificial neural networks. *Minerals* **2020**, *10*, 811. [CrossRef]
60. Gui, W.-H.; Wang, L.-Y.; Yang, C.-H. Transactions of Nonferrous Metals Society of China Intelligent Prediction Model of Matte Grade in Copper Flash Smelting Process. 2007. Available online: www.csu.edu.cn/ysxb/ (accessed on 24 April 2023).
61. Deng, X.; Wang, X. Incremental learning of dynamic fuzzy neural networks for accurate system modeling. *Fuzzy Sets Syst.* **2009**, *160*, 972–987. [CrossRef]
62. Liu, J.-h.; Gui, W.-h.; Xie, Y.-f.; Yang, C.-h. Dynamic modeling of copper flash smelting process at a Smelter in China. *Appl. Math. Model.* **2014**, *38*, 2206–2213. [CrossRef]
63. Savic, M.V.; Djordjevic, P.B.; Mihajlovic, I.N.; Zivkovic, Z.D. Statistical modeling of copper losses in the silicate slag of the sulfide concentrate smelting process. *Pol. J. Chem. Technol.* **2015**, *17*, 62–69. [CrossRef]
64. Puspita, A.N.G.; Surjandari, I.; Zulkarnain; Kawigraha, A.; Permatasari, N.V. Optimization of saprolite ore composites reduction process using artificial neural network (ANN). *Procedia Comput. Sci.* **2019**, *161*, 424–432. [CrossRef]
65. Cardoso, W.; Di Felice, R.; Baptista, R.C. Artificial Neural Networks for Modelling and Controlling the Variables of a Blast Furnace. In Proceedings of the 6th International Forum on Research and Technology for Society and Industry, RTSI 2021—Proceedings, Virtual, 6–9 September 2021; Institute of Electrical and Electronics Engineers Inc.: Piscataway, NJ, USA, 2021; pp. 148–152. [CrossRef]
66. Qian, Q.; Fang, X.; Xu, J.; Li, M. Multichannel profile-based monitoring method and its application in the basic oxygen furnace steelmaking process. *J. Manuf. Syst.* **2021**, *61*, 375–390. [CrossRef]
67. Cardoso, W.; di Felice, R.; Baptista, R.C. Artificial Neural Network for Predicting Silicon Content in the Hot Metal Produced in a Blast Furnace Fueled by Metallurgical Coke. *Mater. Res.* **2022**, *25*, e20210439. [CrossRef]
68. Wang, X.; Hu, T.; Tang, L. A Multiobjective Evolutionary Nonlinear Ensemble Learning with Evolutionary Feature Selection for Silicon Prediction in Blast Furnace. *IEEE Trans. Neural Netw. Learn. Syst.* **2022**, *33*, 2080–2093. [CrossRef]
69. Yang, C.; Jin, F.; Zhao, J.; Wang, W. A Deep-Convolution-Generative-Adversarial-Networks-based Missing Data Filling Method for Blast Furnace Gas System in Steel Industry. In Proceedings of the 2022 IEEE 11th Data Driven Control and Learning Systems Conference, DDCLS 2022, Chengdu, China, 3–5 August 2022; Institute of Electrical and Electronics Engineers Inc.: Piscataway, NJ, USA, 2022; pp. 1183–1189. [CrossRef]
70. Zhao, L.; Zhu, D.; Liu, D.; Wang, H.; Xiong, Z.; Jiang, L. Prediction and Optimization of Matte Grade in ISA Furnace Based on GA-BP Neural Network. *Appl. Sci.* **2023**, *13*, 4246. [CrossRef]
71. Reitsma, F.; Woods, P.; Fairclough, M.; Kim, Y.; Tulsidas, H.; Lopez, L.; Zheng, Y.; Hussein, A.; Brinkmann, G.; Haneklaus, N.; et al. On the sustainability and progress of energy neutral mineral processing. *Sustainability* **2018**, *10*, 235. [CrossRef]
72. Cisternas, L.A.; Ordóñez, J.I.; Jeldres, R.I.; Serna-Guerrero, R. Toward the Implementation of Circular Economy Strategies: An Overview of the Current Situation in Mineral Processing. *Miner. Process. Extr. Metall. Rev.* **2021**, *43*, 775–797. [CrossRef]
73. Wang, L.; Peng, Y.; Runge, K.; Bradshaw, D. A review of entrainment: Mechanisms, contributing factors and modelling in flotation. *Minerals Engineering* **2015**, *70*, 77–91. [CrossRef]
74. COCHILCO. *Sulfuros Primarios: Desafíos y Oportunidades*; COCHILCO: Santiago, Chile, 2017.
75. Laurila, H.; Karesvuori, J.; Tiili, O. Strategies for instrumentation and control of flotation circuits. In *Mineral Processing Plant Design, Practice, and Control Proceedings*; Society for Mining, Metallurgy, and Exploration (SME): Englewood, CO, USA, 2002; pp. 2174–2195.
76. Moilanen, J.; Remes, A. Control of the flotation process. In Proceedings of the V International Mineral Processing Seminar (Procemin 2008), Santiago, Chile, 22–24 October 2008; pp. 318–325.
77. Dunne, R.C.; Lane, G.S.; Richmond, G.D.; Dioses, J. Flotation data for the design of process plants Part 1—Testing and design procedures. *Trans. Inst. Min. Metall. Sect. C Miner. Process. Extr. Metall.* **2010**, *119*, 199–204. [CrossRef]
78. Vallejos, P.; Yianatos, J.; Matamoros, C.; Díaz, F. Mineral solids transport in a two-dimensional flotation froth. *Miner. Eng.* **2019**, *138*, 24–30. [CrossRef]
79. Park, H.; Wang, L. Experimental studies and modeling of surface bubble behaviour in froth flotation. *Chem. Eng. Res. Des.* **2015**, *101*, 98–106. [CrossRef]
80. Åkesson, B.M.; Toivonen, H.T. A neural network model predictive controller. *J. Process Control* **2006**, *16*, 937–946. [CrossRef]
81. Wills, A.; Schön, T.B.; Ljung, L.; Ninness, B. Identification of Hammerstein-Wiener models. *Automatica* **2013**, *49*, 70–81. [CrossRef]
82. Sun, K.; Qiu, J.; Karimi, H.R.; Gao, H. A Novel Finite-Time Control for Nonstrict Feedback Saturated Nonlinear Systems with Tracking Error Constraint. *IEEE Trans. Syst. Man Cybern. Syst.* **2021**, *51*, 3968–3979. [CrossRef]
83. Kallioinen, J.; Heiskanen, K. Effective Flotation of a Difficult Nickel-Ore Based on Intelligent Mineral Technology. *Miner. Eng.* **1993**, *6*, 917–928. [CrossRef]
84. Harbort, G.; de Bono, S.; Carr, D.; Lawson, V. Jameson cell fundamentals—A revised perspective. *Miner. Eng.* **2003**, *16*, 1091–1101. [CrossRef]
85. Carr, D.; Dixon, A.; Tiili, O. Optimizing Large Flotation Cell Performance Through Advanced Instrumentation and Control. In Proceedings of the 10th Mill Operators Conference, Adelaide, Australia, 12–14 October 2009; pp. 299–304.
86. Morari, M.; Lee, J.H. Model predictive control: Past, present and future. *Comput. Chem. Eng.* **1999**, *23*, 667–682. [CrossRef]
87. Adetola, V.; DeHaan, D.; Guay, M. Adaptive model predictive control for constrained nonlinear systems. *Syst. Control Lett.* **2009**, *58*, 320–326. [CrossRef]

88. Liu, L.; Liu, Y.-J.; Tong, S. Neural Networks-Based Adaptive Finite-Time Fault-Tolerant Control for a Class of Strict-Feedback Switched Nonlinear Systems. *IEEE Trans. Cybern.* **2019**, *49*, 2536–2545. [CrossRef] [PubMed]
89. Rosenfeld, A.; Tsotsos, J.K. Incremental Learning through Deep Adaptation. *IEEE Trans. Pattern Anal. Mach. Intell.* **2020**, *42*, 651–663. [CrossRef]
90. Komulainen, T.; Pekkala, P.; Rantala, A.; Jämsä-Jounela, S.L. Dynamic modelling of an industrial copper solvent extraction process. *Hydrometallurgy* **2006**, *81*, 52–61. [CrossRef]
91. Project, T.D. *Sustainable Improvement in Safety of Tailings Facilities a European Research and Technological Development Project Report Tailings Management Facilities—Intervention Actions for Risk Reduction*; University of Leeds: Leeds, UK, 2004.
92. COCHILCO. *Yearbook: Copper and Other Mineral Statistics 2001–2020*; COCHILCO: Santiago, Chile, 2020.
93. Palma, J.H. Operación y control de tranques de relave. In *Seminario Minería Chilena y sus Desafíos: Una Visión Integral*; Pontificia Universidad Católica de Chile: Santiago, Chile, 2016. [CrossRef]
94. Kreft-Burman, K.; Saarala, J.; Anderson, R. Sustainable Improvement in Safety of Tailings Facilities. In *TAILSAFE*; Finnish Environment Institute (SYKE): Helsinki, Finland, 2005.
95. The Mining Association of Canada. *A Guide to the Management of Tailings Facilities VERSION 3.1*; The Mining Association of Canada: Ottawa, ON, Canada, 2019.
96. Parviainen, A.; Kauppila, T.; Loukola-Ruskeeniemi, K. Long-term lake sediment records and factors affecting the evolution of metal(loid) drainage from two mine sites (SW Finland). *J. Geochem. Explor.* **2012**, *114*, 46–56. [CrossRef]
97. Zardari, M.A. Stability of Tailings Dams—Focus on Numerical Modelling. Licenciate Thesis, Luleå Tekniska Universitet, Luleå, Sweden, 2011.
98. Rusu, A.A.; Rabinowitz, N.C.; Desjardins, G.; Soyer, H.; Kirkpatrick, J.; Kavukcuoglu, K.; Pascanu, R.; Hadsell, R. Progressive Neural Networks. June 2016. Available online: http://arxiv.org/abs/1606.04671 (accessed on 6 April 2022).
99. French, R.M. Dynamically constraining connectionist networks to produce distributed, orthogonal representations to reduce catastrophic interference. In Proceedings of the Sixteenth Annual Conference of the Cognitive Science Society, Atlanta, GA, USA, 13–16 August 1994.
100. Herrera, L.J.; Pomares, H.; Rojas, I.; Guillén, A.; Prieto, A.; Valenzuela, O. Recursive prediction for long term time series forecasting using advanced models. *Neurocomputing* **2007**, *70*, 2870–2880. [CrossRef]
101. He, H.; Chen, S.; Li, K.; Xu, X. Incremental Learning from Stream Data. *IEEE Trans. Neural Netw.* **2011**, *22*, 1901–1914. [CrossRef]

minerals

MDPI

Article

A Review on Global Cooperation Network in the Interdisciplinary Research of Geochemistry Combined with Artificial Intelligence

Qianlong Zhang [1,2,3], Yongzhang Zhou [1,2,3,*], Juxiang He [1,2,3], Biaobiao Zhu [1,2,3], Feng Han [1,2,3] and Shiyao Long [1,2,3]

1 School of Earth Sciences & Engineering, Sun Yat-sen University, Zhuhai 519000, China; zhangqlong3@mail2.sysu.edu.cn (Q.Z.); hejx65@mail2.sysu.edu.cn (J.H.); zhubb8@mail2.sysu.edu.cn (B.Z.); hfhanfeng@hotmail.com (F.H.); longshy6@mail2.sysu.edu.cn (S.L.)

2 Guangdong Provincial Key Lab of Geological Process and Mineral Resources Survey, Sun Yat-sen University, Zhuhai 519000, China

3 Center for Earth Environment & Resources, Sun Yat-sen University, Zhuhai 519000, China

* Correspondence: zhouyz@mail.sysu.edu.cn

Abstract: With the rapid development of modern geochemical analysis techniques, massive volumes of data are being generated from various sources and forms, and geochemical data acquisition and analysis have become important tools for studying geochemical processes and environmental changes. However, geochemical data have high-dimensional, nonlinear characteristics, and traditional geochemical data analysis methods have struggled to meet the demands of modern science. Nowadays, the development of big data and artificial intelligence technologies has provided new ideas and methods for geochemical data analysis. However, geochemical research involves numerous fields such as petrology, ore deposit, mineralogy, and others, each with its specific research methods and objectives, making it difficult to strike a balance between depth and breadth of investigation. Additionally, due to limitations in data sources and collection methods, existing studies often focus on a specific discipline or issue, lacking a comprehensive understanding of the bigger picture and foresight for the future. To assist geochemists in identifying research hotspots in the field and exploring solutions to the aforementioned issues, this article comprehensively reviews related studies in recent years, elaborates on the necessity and challenges of combining geochemistry and artificial intelligence, and analyzes the characteristics and research hotspots of the global collaboration network in this field. The study reveals that the investigation into artificial intelligence techniques to address geochemical issues is progressing swiftly. Joint research papers serve as the primary means of contact within a worldwide collaborative network. The primary areas of focus in the ongoing research on the integration of geochemistry and artificial intelligence include methodologies for analyzing geochemical data, environmental modifications, and mineral prospectivity mapping. Geochemical data analysis is currently a significant focus of research, encompassing a range of methods including machine learning and deep learning. Predicting mineral resources for deep space, deep Earth, and deep sea is also a pressing topic in contemporary research. This paper explores the factors driving research interest and future trends, identifies current research challenges, and considers opportunities for future research.

Keywords: geochemistry; big data; artificial intelligence; knowledge graph; analysis of cooperation network; CiteSpace

1. Introduction

Geochemistry is a field that examines the chemical makeup and transformations occurring in the Earth's interior and surface. It has a crucial function in addressing numerous fundamental Earth science challenges. Geochemical data analysis holds immense significance in comprehending the Earth system, identifying mineral reserves, monitoring the environment, and predicting earthquakes and volcanic activity, amongst others, thereby offering a data-supported foundation for Earth's scientific research and associated applications. Since the 1990s, scientists have endeavored to combine geochemical data with computer technology, augmenting the efficiency and precision of geochemical data analysis. Turing Award winner Jim Gary postulates that scientific inquiry has traversed four paradigms: the empirical paradigm, the theoretical paradigm, the computational paradigm, and the data-driven paradigm. Over the past decade, scientific research has shifted from being problem-driven to data-driven, leading to the emergence of the fourth paradigm of scientific discovery: data-intensive scientific discovery [1]. In recent times, the utilization of big data and artificial intelligence algorithms has transformed geology [2–4]. The integration of big data and artificial intelligence techniques with geochemistry has become a prominent research topic in the geochemistry field due to technological advances. This research includes an array of data types, such as geochemical and literature data, and encompasses various fields, including Earth science, environmental science, and information science. Through the application of big data analytics and methods, scientists have achieved significant advances in geochemical data analysis research. One such achievement includes the accurate classification of lithology [5], as well as distinguishing the rock's tectonic environment [6–8], mineral classification [9–11], the genesis of ore deposits [12–14], and mineral prospectivity mapping [15,16].

In previous studies, researchers obtained a large volume of geochemical data on deposits, minerals, rocks, soil, etc., through sampling analysis. Currently, there are various methods for the analysis of geochemical data to identify the frequency and spatial characteristics of these data, including classical statistics such as mean $\pm$ k [17], probability plot [18], exploratory data analysis [19], multivariate statistics [20], geostatistics [21], and fractal and multifractal models [22–24]. However, these data possess high-dimensionality and nonlinear characteristics, and traditional methods have limited ability to handle large volumes of multidimensional data. The information contained in multidimensional space has largely remained unexplored, posing a challenge for geochemical data analysis. With the development of big data and artificial intelligence technologies, machine learning algorithms have increasingly been applied to the field of geochemistry, providing a new perspective for the interpretation of geochemical data. Previous studies have highlighted various methods such as random forests [25,26], support vector machines [27,28], one-class support vector machines [29–31], neural networks [32,33], metric learning [34], and deep learning [35–41]. These studies have demonstrated the tremendous potential of artificial intelligence technologies, accelerating new discoveries and innovative solutions in the field of geochemistry.

The field of geochemistry encompasses a wide range of research areas, ranging from the microscale of atoms and molecules to the macroscale interactions within various spheres of the Earth system. The scale of this research makes geochemistry a challenging field that requires extensive data analysis and computation. The application of big data technology can assist geochemists in collecting, processing, and analyzing data more efficiently. For example, the real-time monitoring and data collection of chemical substances in global soil, water, and atmosphere can facilitate more accurate predictions and control of their impacts on the environment and human health [42–44]. Additionally, big data can be utilized in the development and optimization of geochemical models, enabling researchers to simulate and predict chemical processes within the Earth system more accurately [45–47]. At the same time, the application of artificial intelligence can further enhance the efficiency and precision of geochemical research. Techniques such as machine learning and deep learning can automatically identify and predict patterns and trends in geochemical processes. For instance, artificial intelligence can assist researchers in automatically classifying and identifying outliers in geochemical data, reducing human errors and improving research precision [48–52].

With the emergence of an increasing number of big data methods, the integration of geochemistry with big data and artificial intelligence has become increasingly prominent, demonstrating the vitality of intelligent research in geochemistry. Simultaneously, as globalization continues to advance, international collaboration in the field of research has become more prevalent. By analyzing the structure and dynamic changes of global collaboration networks, insights can be gained into the development trends and frontier hotspots within the field of geochemistry. However, geochemical research encompasses numerous subfields, such as ore deposits, petrology, and mineralogy, each with its specific research methods and objects. This diversity results in challenges in attaining a comprehensive understanding of the depth and breadth of research. Furthermore, existing research often focuses on specific areas or issues due to limitations in data sources and collection methods, lacking a global perspective and future foresight. To gain in-depth insights into the current development of the integration of geochemistry with big data and artificial intelligence, this paper provides a comprehensive review of recent relevant studies. Through in-depth analysis and mining of the related literature and data, it summarizes the research progress of big data and artificial intelligence methods in geochemistry (including mineralogy, petrology, and ore deposits). It also reveals the characteristics of the global collaboration network in the integration of geochemistry with big data and artificial intelligence, as well as the frontier and hotspots in this field. The purpose of this paper is to provide valuable reference information for related research fields, offer new perspectives and ideas to researchers, and promote the deep integration and development of geochemistry with big data and artificial intelligence. The rest of this paper is organized as follows: Section 2 presents the selection of data and research methods. Section 3 analyzes the characteristics of cooperation networks in research. Section 4 discusses the analysis of research hotspots. Section 5 concludes the paper and provides future prospects.

2. Data and Methods

2.1. Data Resources

This study aims to investigate the latest research trends in the field of geochemistry–big data–artificial intelligence (GC-AI) over the past decade. To achieve this, we conducted a literature search using the Web of Science database with the following keywords: "geochemi*," "mineralogy," "petrology," "Ore deposit," "data science," "data-driven," "big data," "artificial intelligence," "machine learning," "deep learning," "neural network," "big data analysis," and "big data method." The search was limited to articles published between 1 January 2013 and 1 August 2023. Initially, we retrieved 2707 articles. After removing duplicates, we ended up with a final dataset of 2421 articles.

2.2. Methods

Bibliometric analysis enables the visualization of knowledge map characteristics in specific fields, providing insights into the development and trends of scientific research. The most commonly used bibliometric visualization software programs include CiteSpace (version 5.4), VOSviewer, and HistCite, which can effectively illustrate the evolving trends in a particular field over time. VOSviewer, while lacking cluster labeling and supporting only one clustering algorithm, has its own strengths. Similarly, HistCite offers fewer visualization analysis methods. Meanwhile, CiteSpace stands out with its clear developmental background, efficient relationship visualization, and comprehensive analysis and display of bibliometric data.

CiteSpace, as an information visualization analysis program, is used for exploring latent knowledge within scientific research [53,54]. It includes cooperative network analysis and keyword analysis. The cooperative network analysis reveals the collaborative relationships between different countries/regions and institutions in the field, while keyword analysis involves co-occurrence, cluster, and burst analysis. Co-occurrence and cluster analysis unveil the key topics, keywords, and frontier knowledge, while burst analysis identifies research hotspots and potential future trends [55,56]. In this study, CiteSpace version 5.4 was utilized for bibliometric analysis in the GC-AI field, and the research workflow is depicted in Figure 1.

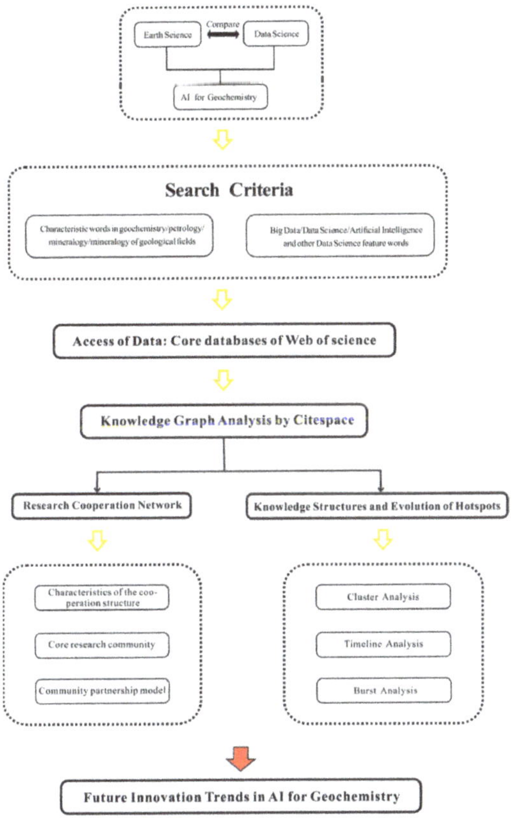

Figure 1. Bibliometric analysis process of GC-AI.

3. Analysis of Research Network Cooperation

3.1. Parameter Settings

CiteSpace has several important parameters and terminology, and adjusting these parameters plays an important role in the analysis of the knowledge graph.

3.1.1. G-Index

The G-index is an index based on citations, which sorts all papers by an author in descending order according to citations, and the largest g value is determined such that the total citations of the top g papers are not less than g^2. Typically, a higher G-index indicates a greater influence of the author in the field. The formula is as follows:

$$g^2 \leq k\sum_{i\leq g} c_i, k \in Z^+ \quad (1)$$

The value of k is positively correlated with the number of nodes.

3.1.2. Top N and Top N%

Top N refers to the selection of the top N citations based on their citation counts, which can be used for analyzing important studies. Top N% refers to the percentage of citations selected.

In this study, the analysis of collaboration networks resulted in a g-value of 15, while the k-value for the literature citation network was 5. The top N and top N% values were set to 50 and 10%, respectively.

3.1.3. Centrality

Centrality is a metric defined for each node within a network, measuring the likelihood of the node being on any shortest path within the network. It is used to identify and quantify the importance of nodes, and nodes with high centrality are often key hubs connecting two different domains.

3.1.4. Pruning

Pruning is a data processing method used to filter out unimportant nodes and edges. This includes using path-finding algorithms such as Pathfinder, minimum spanning tree algorithms such as Minimum Spanning Tree, and network slicing algorithms such as Pruning Sliced Networks to trim nodes for better display of the core structure and main features of the research network. This study is centered around a small-scale network structure. It is worth noting that certain algorithms, such as MST, may not yield a singular solution, resulting in multiple potential pruning scenarios, which could introduce ambiguity in the analysis findings. In contrast, Pathfinder offers a distinct pruning solution that enhances the stability and replicability of the analysis. Moreover, Pathfinder selectively retains nodes of higher significance within the knowledge network, facilitating the improved identification of key nodes. Consequently, Pathfinder was selected for node pruning in this study.

3.2. Analysis of Publications

A statistical analysis of the annual distribution of literature data exported from WOS was conducted to reflect the development trend of research regarding big data and artificial intelligence in the field of geochemistry. As shown in Figure 2, research in this field before 2017 was in a stable and ascending stage, with a small number of average publications ranging between 100 and 160 papers. Since 2017, the number of publications has shown a clear upward trend, with a rapid increase every year. In 2022, the number of publications reached 370, which is 3.7 times that of 2013, indicating that research on this topic has progressed rapidly and has attracted increasing attention and emphasis from researchers. Figure 3 presents the data distribution pattern in the field of GC-AI, with a total of 583 countries/regions, 7318 institutions, and 544 journals with 2421 publications.

Figure 2. Trends in the number of publications of GC-AI.

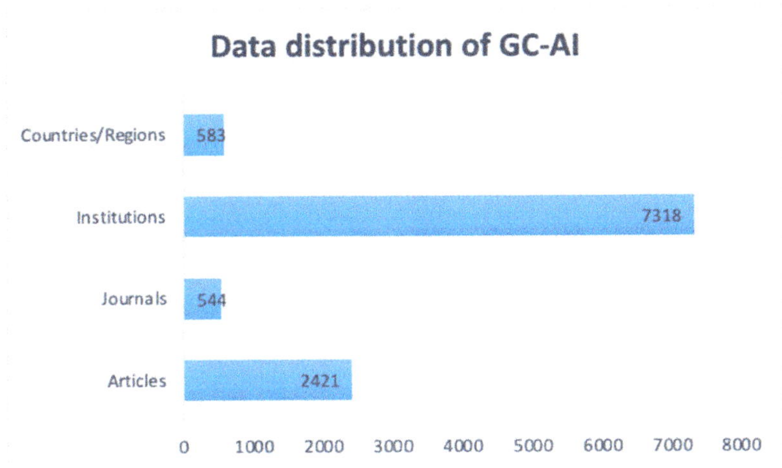

Figure 3. Data distribution in publications of GC-AI.

3.3. *Cooperative Network of Countries/Regions*

Analyzing the cooperative networks among countries/regions using CiteSpace allows us to comprehend the cooperative relationships in a specific research field across different nations. This analysis helps us identify countries with a high level of cooperative awareness and cooperation, thus inferring the international collaborative characteristics of the field. Figure 4 depicts the cooperative network of country/region-based research in the GC-AI field, with Table 1 presenting the top 10 countries/regions based on publication output.

From Figure 4 and Table 1, it is evident that the main research countries are the United States, China, Germany, the United Kingdom, Australia, and Canada. The United States ranks first with 729 papers, followed by China (620 papers), and Germany ranks third with 290 papers. The United Kingdom (270 papers) ranks fourth, with Australia (254 papers) rounding out the top five. The international cooperation among these countries is closely interwoven and exhibits a substantial level of cooperation. Furthermore, Table 1 reveals that Germany (0.23), the United Kingdom (0.16), and Canada (0.15) show the highest centrality, indicating their relative prominence within the cooperative network of countries/regions.

The findings demonstrate that the field of GC-AI exhibits a high level of research proficiency and internationalization, with the cooperative network of countries for research in this field characterized by close and diverse ties.

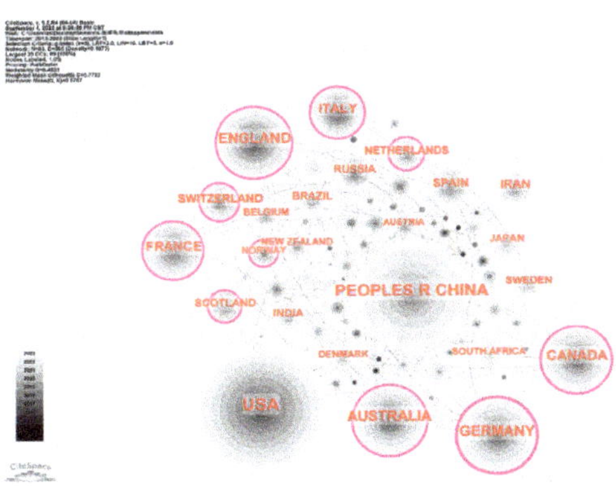

Figure 4. Cooperative network of countries/regions.

Table 1. Top 10 countries/regions in publications.

No.	Counts	Counts/Total	Countries	Centrality
1	729	30.1%	USA	0.1
2	620	25.6%	China	0.02
3	290	12.0%	Germany	0.23
4	270	11.2%	England	0.16
5	254	10.5%	Australia	0.13
6	226	9.3%	Canada	0.15
7	176	7.3%	France	0.12
8	150	6.2%	Italy	0.12
9	92	3.8%	Iran	0.03
10	83	3.4%	Spain	0.07

3.4. *Cooperative Network of Research Institution*

In conducting an analysis of the institutional cooperation network in the GC-AI field, it is observed that there is extensive cooperation between universities, as well as between universities and research institutions (Figure 5). This cooperative network is primarily composed of academic institutions such as universities and research institutes, with three main cooperative groups emerging: China, Europe, and the United States. However, these groups exhibit distinct geographical divisions, with the most cooperation within institutions being limited to the same country, and international cooperation across borders being relatively scarce. The top 10 institutions based on publication output are listed in Table 2, with China University of Geosciences (193), Chinese Academy of Sciences (185), and Centre National de la Recherche Scientifique (122) occupying the top three positions. Among them, the Helmholtz Association demonstrates the highest centrality (0.32), followed by the National Aeronautics and Space Administration and UDICE French Research Universities, with centrality values of 0.27 and 0.24, respectively. These centrality values signify their gross influence, rich research expertise, and available resources in the field.

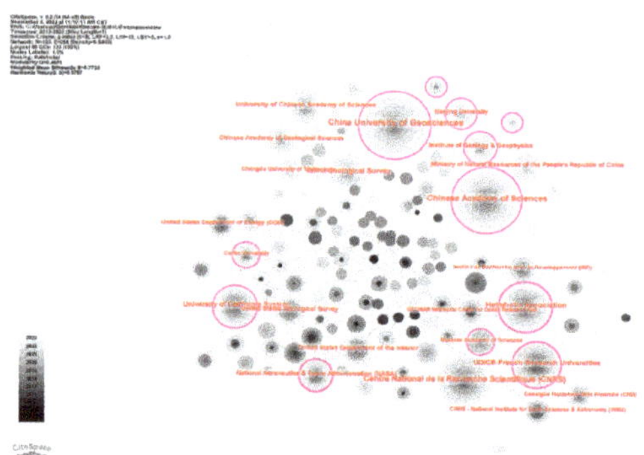

Figure 5. Cooperative network of research institutions.

Table 2. Top 10 institutions in publications.

No.	Count	Institutions	Centrality
1	193	China University of Geosciences	0.11
2	185	Chinese Academy of Sciences	0.13
3	122	Centre National de la Recherche Scientifique (CNRS)	0.07
4	113	Helmholtz Association	0.32
5	89	China Geological Survey	0.04
6	87	UDICE French Research Universities	0.24
7	81	University of California System	0.17
8	61	University of Chinese Academy of Sciences	0.01
9	53	United States Department of the Interior	0.1
10	53	United States Geological Survey	0.1

4. Discussion of Research Hotspots

4.1. Keywords Clustering

High-frequency and highly central keywords illustrate the focal points of most authors within a given time period, indicating the research hotspots and frontiers. The co-occurrence network of keywords is shown in Figure 6, where nodes represent keywords and the size of each node represents the frequency of co-occurrence. The color of the lines between nodes reflects the chronological order of their appearance. It can be observed that the research hotspots in the GC-AI field include climate change, water, geochemical models, geochemical evolution, sediments, machine learning, and mineral prospectivity mapping.

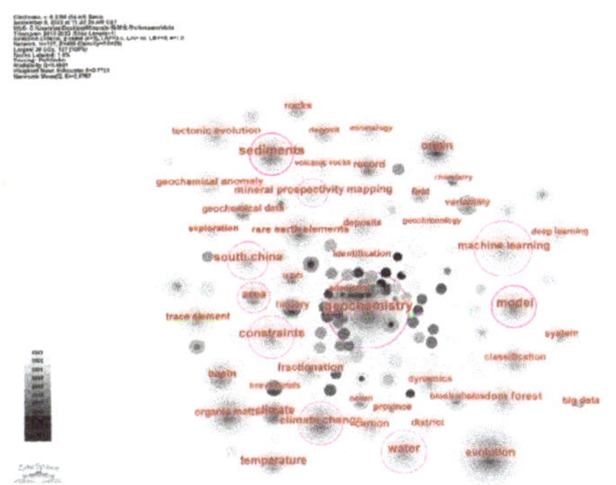

Figure 6. Keywords concurrence networks of GC-AI.

Based on the analysis of the graph data, the top 10 keywords with centrality in GC-AI research literature are presented in Table 3. It can be observed that there are three nodes with centrality values above 0.2 and six nodes with centrality values above 0.15. These nodes can be considered core nodes within the GC-AI field, reflecting the significance and influence of these keywords in the research domain. These keywords govern the development trends in this field.

Table 3. Top 10 keywords for centrality.

No.	Count	Centrality	Keywords
1	165	0.27	Model
2	109	0.24	Geochemistry
3	50	0.22	Sediments
4	69	0.18	Area
5	97	0.16	Water
6	90	0.15	South China
7	42	0.14	Constraints
8	36	0.14	Climate change
9	37	0.1	Machine learning
10	57	0.1	Mineral prospectivity mapping

Meanwhile, in this study, K-means clustering and LSI algorithms were employed to conduct network analysis on the keywords. The silhouette coefficient was utilized to evaluate the similarity between elements within the clusters. Ranging from 0 to 1, the silhouette coefficient measures the compactness of samples within the cluster and the separation between clusters. A value greater than 0.5 indicates a reasonable clustering, while a value exceeding 0.7 suggests a well-defined and convincing cluster. Figure 7 illustrates seven clusters in the order of 0 to 6, namely "machine learning" (0.8) for cluster #0, "climate change" (0.794) for cluster #1, "rare earth elements" (0.705) for cluster #2, "geochemistry" (0.673) for cluster #3, "mineral prospectivity mapping" (0.883) for cluster #4, "uranyl" (0.864) for cluster #5, and "deep sea" (0.909) for cluster #6. The size of clusters and their corresponding silhouette coefficients are summarized in Table 4. It can be observed that clusters 0, 1, and 2 contain the largest number of keywords, resulting in compact samples within the clusters and well-separated samples between clusters. All seven clusters have silhouette coefficients greater than 0.5, with six of them exceeding 0.7, indicating that this method successfully achieved the clustering of the GC-AI field with favorable performance.

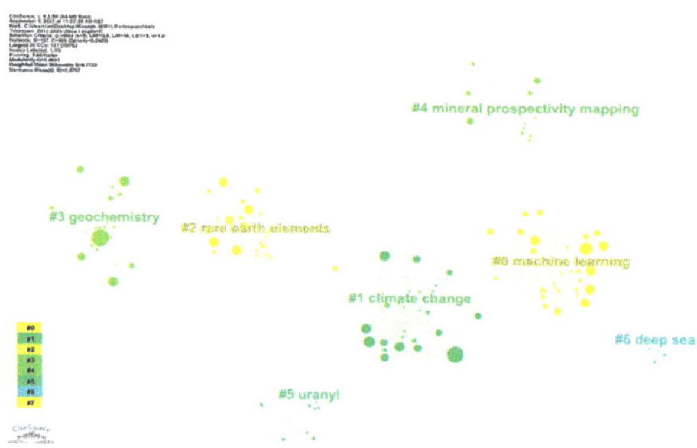

Figure 7. Keyword clustering of GC-AI.

Table 4. Silhouette ranking for keyword clustering.

Clusters	Count	Silhouette
Machine learning	62	0.8
Climate change	55	0.794
Rare earth elements	54	0.705
Geochemistry	30	0.673
Mineral prospectivity mapping	29	0.883
Uranyl	20	0.864
Deep sea	14	0.909

4.2. *Keyword Timeline*

To elucidate the developmental trajectory of this field, we conducted a timeline analysis of keyword progression. Figure 8 illustrates the progression of seven clusters during the past decade. Based on the law of timeline development, it can be inferred that the field of GC-AI has expanded into two crucial directions, from the advancement of research techniques to the breakthrough of exploration areas. This indicates the incessant strides made by the field in terms of theorization and application.

1. In the initial phase of GC-AI, researchers concentrated on utilizing rule-based techniques to recognize and categorize patterns and characteristics in geochemical data applications, including Bayesian networks and decision trees, among others. The field of geochemical artificial intelligence started to shift from rule-based methods to machine learning with the emergence of artificial intelligence (AI) technologies. Machine learning techniques enable automatic learning of data features and patterns without the need for manual feature design and selection, thus saving time. Some examples of these techniques include random forest, artificial neural networks, deep neural networks (i.e., deep learning), and others.
2. In recent years, the study and implementation of geochemical artificial intelligence have broadened to encompass various domains. These involve deploying machine learning algorithms for ore formation prophecy, the automated detection and prediction of groundwater contamination, the quantitative examination of rare earth elements' distribution patterns, climate change forecasting and prevention, and more.

Currently, the application domain is expanding continuously, and investigations on the deep Earth, deep space, and deep sea are the primary focus of upcoming research.

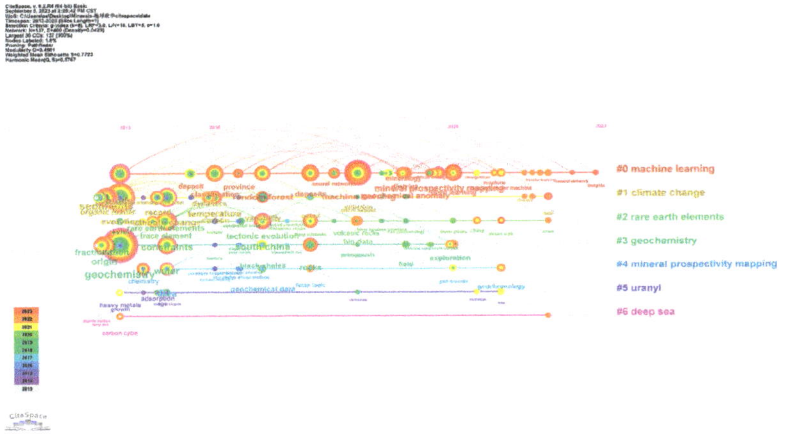

Figure 8. Keywords clustering timeline of GC-AI.

4.3. Keyword Brust Analysis

The field of GC-AI has undergone research priorities and changes at different time periods. To ascertain research hotspots in this field, a keyword burst analysis was undertaken. The burst word detection function was implemented to scrutinize keywords that had significantly changed in frequency in short-term research articles, demonstrating research topic hotspots at various time periods and exploring future research development trends. This study identified 37 keywords that had experienced bursts (Figure 9). "Strength" denotes the burst strength of the burst term, while "Begin" and "End" signify the start and end years of the burst term, respectively. The blue section denotes the timeline, while the red section reflects the significance and focus of keywords in this field of research. A longer burst length implies that the popularity of the keyword is sustained for a longer duration, thus indicating a stronger research front. From the diagram, it can be observed that the greatest surge in GC-AI research was in the areas of climate, fluid flow, and sediments. However, their research popularity has waned, suggesting that these were the most popular research subjects in the early stages of this field.

From the perspective of evolutionary trends, burst words emerged during three intensive periods, namely 2013–2014, 2016–2018, and 2019–2021, indicating the development of research areas and methods. In the field of GC-AI, specific keywords such as "model" and "evolution" burst relatively early. The appearance of these keywords suggests that researchers in the field have been interested in using data analysis techniques to address geochemical problems since the early stages. Over time, novel keywords, including "artificial neural networks" and "support vector machine", have gradually surfaced, highlighting the field's emphasis on artificial intelligence methods. Initially, research efforts were directed towards studying climate, crust, rocks, and other related areas. Nevertheless, with the advent of breakthroughs in research technology and the demands of social development, there is now an emerging focus on mineral resource research, which has become a hot topic in the field.

Figure 9. Keywords burst of GC-AI.

4.4. Discussion

Scientific research is currently experiencing its fourth paradigm [1], and data science has brought about significant advancements in various industries. Big data mining and artificial intelligence are necessary research topics for advances in geosciences [2,3,45]. Geochemical analysis based on big data has been gradually acknowledged by geochemical researchers in recent years due to the intersection of geochemistry and data science [57–60]. The use of a global cooperative network has emerged as the primary model for cooperation in the field of geochemistry and artificial intelligence. An examination of pertinent research and data reveals that this model displays characteristics of multidisciplinary intersection, spanning diverse data types such as geochemical and bibliographic data, and covering varied fields, including Earth science, environmental science, and information science. Simultaneously, the worldwide cooperative network is increasingly adopting co-authored papers as the primary means of contact, leading to the creation of a network based on regions, institutions, and themes. Consequently, these cooperative networks serve as a platform for scientific researchers to connect and cooperate, endorsing the sharing and refinement of scientific research resources and advancing innovation and the development of scientific research.

Although the field of geochemical artificial intelligence has made many advances, there are still many challenges and problems that need further research and resolution. For example, how to improve the generalization and robustness of models, how to handle imbalanced data distribution, and how to apply advanced algorithms to practical scenarios are issues that need further research and resolution in this field [4,45,61–64]. In addition, compared with other fields, research achievements and applications in geochemical artificial intelligence are relatively limited, requiring more innovation and breakthroughs.

In the future, new geochemical artificial intelligence technologies and applications can be further explored and developed. For example, generative models such as generative adversarial networks can be used for geochemical data generation and processing, and transfer learning technology can be applied to transfer models from other fields into geochemistry. Strengthening international cooperation and communication in the field of geochemical artificial intelligence to promote the development and application of this field will make greater contributions to solving global environmental issues.

5. Conclusions and Prospect

This article presents an all-encompassing evaluation of the status and research hotpots concerning the global cooperative network that merges geochemistry with artificial intelligence. Through knowledge graph analysis of geochemistry and artificial intelligence publications published within the past 10 years, we gained the following insights:

1. Research trend: Based on the analysis of publications in the literature, it was observed that the research in this field witnessed a gradual increase before 2017, followed by a notable surge in the number of publications thereafter, with a steep annual rise. By 2022, the volume of publications had surged to 3.7 times that of 2013, highlighting the increasing significance of artificial intelligence techniques in the research toolbox of geochemists. The study of artificial intelligence techniques for geochemistry is quickly progressing within the GC-AI field.
2. Characteristics of cooperative network: A network analysis was conducted on the national/regional and institutional cooperative network to uncover research collaboration patterns in this area. The results indicate a high research standard and internationalization level. The cooperative network has the characteristics of being close-knit and diverse. Research institutions from China, the United States, and Europe are prominent in this field, with many cooperative relationships and research outcomes. Notably, the National Aeronautics and Space Administration and Université libre de Bruxelles exhibit a high level of research and impact.
3. Research hotspots: Through analysis of keyword co-occurrence networks and clustering, timeline, and burst word detection, it was found that some clusters have high relevance and similarity in the field of geochemical artificial intelligence. The application of artificial intelligence technology for mineral resource prediction is currently a research hotspot, and deep space, deep Earth, and deep sea mineral resource exploration will become future research trends.

Author Contributions: Conceptualization, Q.Z. and Y.Z.; methodology, Q.Z.; software, Q.Z.; validation, Q.Z., J.H. and F.H.; formal analysis, Q.Z.; investigation, Q.Z.; resources, Y.Z.; data curation, B.Z.; writing—original draft preparation, Q.Z.; writing—review and editing, Q.Z.; visualization, Q.Z. and S.L.; supervision, Y.Z.; project administration, Y.Z.; funding acquisition, Y.Z. All authors have read and agreed to the published version of the manuscript.

Funding: This research was financially supported by the National Key Research and Development Program of China (Grant No. 2022YFF0801201), the National Natural Science Foundation of China (Grant No. U1911202), and the Science and Technology Project of Guangdong Provincial Department of Science and Technology (Grant No. GDKTP2020053500).

Data Availability Statement: Data are available upon request from the authors.

Acknowledgments: We express our utmost gratitude to Fan Xiao and Pengpeng Yu from Sun Yat-sen University. Furthermore, we extend our sincere appreciation to Renguang Zuo, the esteemed Guest Editor of this Special Issue, for his exceptional organization and guidance.

Conflicts of Interest: The authors declare no conflict of interest. We declare that we do not have any commercial or associative interest that represents a conflict of interest in connection with the work submitted.

References

1. Tolle, K.M.; Tansley, D.S.W.; Hey, A.J.G. The fourth paradigm: Data-intensive scientific discovery. *Proc. IEEE* **2011**, *99*, 1334–1337. [CrossRef]
2. Zhou, Y.; Zuo, R.; Liu, G.; Yuan, F.; Mao, X.; Guo, Y.; Xiao, F.; Liao, J.; Liu, Y. The great-leap-forward development of mathematical geoscience during 2010–2019: Big data and artificial intelligence algorithm are changing mathematical geoscience. *Bull. Miner. Pet. Geochem.* **2021**, *40*, 556–573. [CrossRef]
3. Zhang, W.G.; Ching, J.Y.; Goh, A.T.C.; Leung, A.Y.F. Big data and machine learning in geoscience and geoengineering: Introduction. *Geosci. Front.* **2020**, *12*, 327–329. [CrossRef]
4. Zhang, W.; Gu, X.; Tang, L.; Yin, Y.; Liu, D.; Zhang, Y. Application of machine learning, deep learning and optimization algorithms in geoengineering and geoscience: Comprehensive review and future challenge. *Gondwana Res.* **2022**, *109*, 1–17. [CrossRef]

5. Petrelli, M.; Bizzarri, R.; Morgavi, D.; Baldanza, A.; Perugini, D. Combining machine learning techniques, microanalyses and large geochemical datasets for tephrochronological studies in complex volcanic areas: New age constraints for the Pleistocene magmatism of central Italy. *Quat. Geochronol.* **2017**, *40*, 33–44. [CrossRef]
6. Jiao, S.; Zhang, Q.; Zhou, Y.; Chen, W.; Liu, X. Gopalakrishnan. Progress and challenges of big data research on petrology and geochemistry. *Solid Earth Sci.* **2018**, *3*, 105–114. [CrossRef]
7. Han, S.; Li, M.; Zhang, Q.; Song, L. An Automated Method to Generate and Evaluate Geochemical Tectonic Discrimination Diagrams Based on Topological Theory. *Minerals* **2020**, *10*, 62. [CrossRef]
8. Liu, X.; Zhang, Q.; Zhang, C. Identification of the Original Tectonic Setting for Oceanic Andesite Using Discrimination Diagrams: An Approach Based on Global Geochemical Data Synthesis. *J. Earth Sci.* **2022**, *33*, 696–705. [CrossRef]
9. O'Sullivan, G.; Chew, D.; Kenny, G.; Henrichs, I.; Mulligan, D. The trace element composition of apatite and its application to detrital provenance studies. *Earth Sci. Rev.* **2020**, *201*, 103044. [CrossRef]
10. Saha, R.; Upadhyay, D.; Mishra, B. Discriminating tectonic setting of igneous rocks using biotite major element chemistry—A machine learning approach. *Geochem. Geophys. Geosyst.* **2021**, *22*, e2021GC010053. [CrossRef]
11. Zheng, D.; Wu, S.; Ma, C.; Xiang, L.; Hou, L.; Chen, A.; Hou, M. Zircon classification from cathodoluminescence images using deep learning. *Geosci. Front.* **2022**, *13*, 101436. [CrossRef]
12. Wang, C.-L.; Zhu, W.-N.; Wang, Y.-T. Genesis of the Kiruna-type Nixintage iron deposit, Chinese Western Tianshan, NW China: Constrains of ore geology, geochemistry and geochronology. *J. Geochem. Explor.* **2022**, *243*, 107094. [CrossRef]
13. Chen, N.; Mao, J.; Zhang, Z.; Duan, Z.; Santos, A.; Li, H. Arc magmatic evolution and porphyry copper deposit formation under compressional regime: A geochemical perspective from the Toquepala arc in Southern Peru. *Earth-Sci. Rev.* **2023**, *240*, 104383. [CrossRef]
14. Zhang, P.; Zhang, Z.; Yang, J.; Cheng, Q. Machine Learning Prediction of Ore Deposit Genetic Type Using Magnetite Geochemistry. *Nat. Resour. Res.* **2023**, *32*, 99–116. [CrossRef]
15. Zuo, R.; Xiong, Y. Big data analytics of identifying geochemical anomalies supported by machine learning methods. *Nat. Resour. Res.* **2018**, *27*, 5–13. [CrossRef]
16. Chen, Y.; Zhao, Q. Mineral exploration targeting by combination of recursive indicator elimination with the $\ell2$-regularization logistic regression based on geochemical data. *Ore Geol. Rev.* **2021**, *135*, 104213. [CrossRef]
17. Hawkes, H.E.; Webb, J.S. *Geochemistry in Mineral Exploration*; Harper and Row: New York, NY, USA, 1962.
18. Sinclair, A.J. Selection of threshold values in geochemical data using probability graphs. *J. Geochem. Explor.* **1974**, *3*, 129–149. [CrossRef]
19. Tukey, J.W. *Exploratory Data Analysis*; Addison-Wesley: Reading, PA, USA, 1977.
20. Reimann, C.; Filzmoser, P.; Garrett, R.; Dutter, R. *Statistical Data Analysis Explained: Applied Environmental Statistics with R*; John Wiley & Sons: Hoboken, NJ, USA, 2008.
21. Matheron, G. *Traité de Géostatistique Appliquée*; Editions Technip: Paris, France, 1962.
22. Cheng, Q.; Agterberg, F.P.; Ballantyne, S.B. The separation of geochemical anomalies from background by fractal methods. *J. Geochem. Explor.* **1994**, *51*, 109–130. [CrossRef]
23. Cheng, Q.; Xu, Y.; Grunsky, E. Integrated spatial and spectrum method for geochemical anomaly separation. *Nat. Resour. Res.* **2000**, *9*, 43–52. [CrossRef]
24. Cheng, Q. Mapping singularities with stream sediment geochemical data for prediction of undiscovered mineral deposits in Gejiu, Yunnan Province, China. *Ore Geol. Rev.* **2007**, *32*, 314–324. [CrossRef]
25. Gonbadi, A.M.; Tabatabaei, S.H.; Carranza, E.J.M. Supervised geochemical anomaly detection by pattern recognition. *J. Geochem. Explor.* **2015**, *157*, 81–91. [CrossRef]
26. Parsa, M.; Maghsoudi, A. Assessing the effects of mineral systems-derived exploration targeting criteria for random Forests-based predictive mapping of mineral prospectivity in Ahar-Arasbaran area, Iran. *Ore Geol. Rev.* **2021**, *138*, 104399. [CrossRef]
27. Zuo, R.; Carranza, E.J.M. Support vector machine: A tool for mapping mineral prospectivity. *Comput. Geosci.* **2011**, *37*, 1967–1975. [CrossRef]
28. Maepa, F.; Smith, R.S.; Tessema, A. Support vector machine and artificial neural network modelling of orogenic gold prospectivity mapping in the Swayze greenstone belt, Ontario, Canada. *Ore Geol. Rev.* **2020**, *130*, 103968. [CrossRef]
29. Chen, Y.; Wu, W. Mapping mineral prospectivity by using one-class support vector machine to identify multivariate geological anomalies from digital geological survey data. *Aust. J. Earth Sci.* **2017**, *64*, 639–651. [CrossRef]
30. Chen, Y.; Wu, W.; Zhao, Q. A bat-optimized one-class support vector machine for mineral prospectivity mapping. *Minerals* **2019**, *9*, 317. [CrossRef]
31. Xiong, Y.; Zuo, R. A positive and unlabeled learning algorithm for mineral prospectivity mapping. *Comput. Geosci.* **2020**, *147*, 104667. [CrossRef]
32. Ziaii, M.; Pouyan, A.A.; Ziaei, M. Neuro-fuzzy modelling in mining geochemistry: Identification of geochemical anomalies. *J. Geochem. Explor.* **2009**, *100*, 25–36. [CrossRef]
33. Yu, X.; Xiao, F.; Zhou, Y.; Wang, Y.; Wang, K. Application of hierarchical clustering, singularity mapping, and Kohonen neural network to identify Ag-Au-Pb-Zn polymetallic mineralization associated geochemical anomaly in Pangxidong district. *J. Geochem. Explor.* **2019**, *203*, 87–95. [CrossRef]

34. Wang, Z.; Dong, Y.; Zuo, R. Mapping geochemical anomalies related to Fe–polymetallic mineralization using the maximum margin metric learning method. *Ore Geol. Rev.* **2019**, *107*, 258–265. [CrossRef]
35. Xiong, Y.; Zuo, R. Recognition of geochemical anomalies using a deep autoencoder network. *Comput. Geosci.* **2016**, *86*, 75–82. [CrossRef]
36. Luo, Z.; Xiong, Y.; Zuo, R. Recognition of geochemical anomalies using a deep variational autoencoder network. *Appl. Geochem.* **2020**, *122*, 104710. [CrossRef]
37. Luo, Z.; Zuo, R.; Xiong, Y.; Wang, X. Detection of geochemical anomalies related to mineralization using the GANomaly network. *Appl. Geochem.* **2021**, *131*, 105043. [CrossRef]
38. Yang, N.; Zhang, Z.; Yang, J.; Hong, Z.; Shi, J. A convolutional neural network of GoogLeNet applied in mineral prospectivity prediction based on multi-source geoinformation. *Nat. Resour. Res.* **2021**, *30*, 3905–3923. [CrossRef]
39. Yang, N.; Zhang, Z.; Yang, J.; Hong, Z. Mineral Prospectivity Prediction by Integration of Convolutional Autoencoder Network and Random Forest. *Nat. Resour. Res.* **2022**, *31*, 1103–1119. [CrossRef]
40. Wang, Z.; Zuo, R. Mineral prospectivity mapping using a joint singularity-based weighting method and long short-term memory network. *Comput. Geosci.* **2022**, *158*, 104974. [CrossRef]
41. Li, T.; Zuo, R.; Zhao, X.; Zhao, K. Mapping prospectivity for regolith-hosted REE deposits via convolutional neural network with generative adversarial network augmented data. *Ore Geol. Rev.* **2022**, *142*, 104693. [CrossRef]
42. Lyons, K.J.; Ikonen, J.; Hokajärvi, A.-M.; Räsänen, T.; Pitkänen, T.; Kauppinen, A.; Kujala, K.; Rossi, P.M.; Miettinen, I.T. Monitoring Groundwater Quality with Real-Time Data, Stable Water Isotopes, and Microbial Community Analysis: A Comparison with Conventional Methods. *Sci. Total. Environ.* **2023**, *864*, 161199. [CrossRef]
43. Huang, J.; Wang, D.; Zhu, Y.; Yang, Z.; Yao, M.; Shi, X.; An, T.; Zhang, Q.; Huang, C.; Bi, X.; et al. An overview for monitoring and prediction of pathogenic microorganisms in the atmosphere. *Fundam. Res.* **2023**. [CrossRef]
44. Siddiqi, S.A.; Al-Mulla, Y. Wireless Sensor Network System for Precision Irrigation using Soil and Plant Based Near-Real Time Monitoring Sensors. *Procedia Comput. Sci.* **2022**, *203*, 407–412. [CrossRef]
45. Zuo, R.; Wang, J.; Xiong, Y.; Wang, Z. The processing methods of geochemical exploration data: Past, present, and future. *Appl. Geochem.* **2021**, *132*, 105072. [CrossRef]
46. Parsa, M.; Sadeghi, M.; Grunsky, E. Innovative methods applied to processing and interpreting geochemical data. *J. Geochem. Explor.* **2022**, *237*, 106983. [CrossRef]
47. Gelencsér, O.; Árvai, C.; Mika, L.; Breitner, D.; LeClair, D.; Szabó, C.; Falus, G.; Szabó-Krausz, Z. Effect of hydrogen on calcite reactivity in sandstone reservoirs: Experimental results compared to geochemical modeling predictions. *J. Energy Storage* **2023**, *61*, 106737. [CrossRef]
48. Wang, J.; Zhou, Y.; Xiao, F. Identification of multi-element geochemical anomalies using unsupervised machine learning algorithms: A case study from Ag–Pb–Zn deposits in north-western Zhejiang, China. *Appl. Geochem.* **2020**, *120*, 104679. [CrossRef]
49. Zhang, S.; Bourdeau, J.; Nwaila, G.; Ghorbani, Y. Advanced geochemical exploration knowledge using machine learning: Prediction of unknown elemental concentrations and operational prioritization of re-analysis campaigns. *Artif. Intell. Geosci.* **2022**, *3*, 86–100. [CrossRef]
50. Zuo, R.; Wang, J.; Yin, B. Visualization and interpretation of geochemical exploration data using GIS and machine learning methods. *Appl. Geochem.* **2021**, *134*, 105111. [CrossRef]
51. Zuo, R.; Xiong, Y.; Wang, J.; Carranza, E.J.M. Deep learning and its application in geochemical mapping. *Earth-Science Rev.* **2019**, *192*, 1–14. [CrossRef]
52. Mkono, C.N.; Chuanbo, S.; Mulashani, A.K.; Mwakipunda, G.C. Deep learning integrated approach for hydrocarbon source rock evaluation and geochemical indicators prediction in the Jurassic–Paleogene of the Mandawa basin, SE Tanzania. *Energy* **2023**, *2023*, 129232. [CrossRef]
53. Chen, C. CiteSpace II: Detecting and visualizing emerging trends and transient patterns in scientific literature. *J. Am. Soc. Inf. Sci. Technol.* **2006**, *57*, 359–377. [CrossRef]
54. Chen, C.; Ibekwe-SanJuan, F.; Hou, J. The structure and dynamics of co-citation clusters: A multiple-perspective co-citation analysis. *J. Am. Soc. Inf. Sci. Technol.* **2010**, *61*, 1386–1409. [CrossRef]
55. Dan, Z.; Jiapeng, X.; Yizhu, Z. Study on sustainable urbanization literature based on Web of Science, scopus, and China national knowledge infrastructure: A scientometric analysis in CiteSpace. *J. Clean. Prod.* **2020**, *264*, 121537.
56. Zuo, Z.; Cheng, J.; Guo, H.; Li, Y. Knowledge mapping of research on strategic mineral resource security: A visual analysis using CiteSpace. *Resour. Policy* **2021**, *74*, 102372. [CrossRef]
57. Yu, S.; Deng, H.; Liu, Z.; Chen, J.; Gu, X.; Li, J.; Xiao, K.; Mao, X. Identifying multivariate geochemical anomalies via tensor dictionary learning over spatial-elemental dimensionalities. *Comput. Geosci.* **2022**, *165*, 105153. [CrossRef]
58. Doucet, L.S.; Tetley, M.G.; Li, Z.-X.; Liu, Y.; Gamaleldien, H. Geochemical fingerprinting of continental and oceanic basalts: A machine learning approach. *Earth-Science Rev.* **2022**, *233*, 104192. [CrossRef]
59. Salgado, L.; López-Sánchez, C.; Colina, A.; Baragaño, D.; Forján, R.; Gallego, J. Hg and As pollution in the soil-plant system evaluated by combining multispectral UAV-RS, geochemical survey and machine learning. *Environ. Pollut.* **2023**, *333*, 122066. [CrossRef]

60. Lindsay, J.J.; Hughes, H.S.; Yeomans, C.M.; Andersen, J.C.; McDonald, I. A machine learning approach for regional geochemical data: Platinum-group element geochemistry vs geodynamic settings of the North Atlantic Igneous Province. *Geosci. Front.* **2021**, *12*, 101098. [CrossRef]
61. Sadeghi, B.; Cohen, D.R. Decision-making within geochemical exploration data based on spatial uncertainty–A new insight and a futuristic review. *Ore Geol. Rev.* **2023**, *161*, 105660. [CrossRef]
62. Engle, M.A.; Chaput, J. Visualizing high dimensional structures in geochemical datasets using a combined compositional data analysis and Databionic swarm approach. *Int. J. Coal Geol.* **2023**, *275*, 104303. [CrossRef]
63. Parsa, M.; Shirazy, A.; Shirazi, A.; Pour, A.B. 4–Processing and interpretation of geochemical data for mineral exploration. In *Geospatial Analysis Applied to Mineral Exploration*; Elsevier: Amsterdam, The Netherlands, 2023; pp. 171–188, ISBN 9780323956086. [CrossRef]
64. Wang, J.; Zuo, R. Model averaging for identification of geochemical anomalies linked to mineralization. *Ore Geol. Rev.* **2022**, *146*, 104955. [CrossRef]

MDPI AG
Grosspeteranlage 5
4052 Basel
Switzerland
Tel.: +41 61 683 77 34

Minerals Editorial Office
E-mail: minerals@mdpi.com
www.mdpi.com/journal/minerals

www.ingramcontent.com/pod-product-compliance
Lightning Source LLC
Chambersburg PA
CBHW061436210926
51850CB00008B/1529
* 9 7 8 3 7 2 5 8 5 4 6 3 9 *